INTO THE UNKNOWN

INTO THE UNKNOWN

The X-I Story

Louis Rotundo

Smithsonian Institution Press

Washington and London

To

DICK FROST

my project engineer in writing this book

© 1994 by Louis Rotundo
All rights reserved

This book was edited by Initial Cap Editorial Services
Production Editor: Jenelle Walthour
Designer: Alan Carter

Library of Congress Cataloging-in-Publication Data
Rotundo, Louis C.
Into the unknown: the X-1 story / by Louis Rotundo
p. cm.
Includes bibliographical references and index.
ISBN 1-56098-305-1 (acid-free paper)
1. High-speed aeronautics—History. 2. Bell X-1 (Supersonic
planes) 3. Supersonic planes—Research—United States—History.
I. Title.
TL551.5.R68   1994
629.132′305′0973—dc20   93-15989

British Library Cataloguing-in-Publication Data available

A paperback reissue (ISBN 1-56098-915-7) of the original cloth edition

Manufactured in the United States of America
08 07 06 05 04 03 02 01 5 4 3 2 1

⊗ The paper used in this publication meets the minimum requirements
of the American National Standard for Permanence of Paper for
Printed Library Materials Z39.48-1984.

For permission to reproduce illustrations appearing in this book,
please correspond directly with the owners of the works, as listed
in the individual captions. The Smithsonian Institution Press does
not retain reproduction rights for these illustrations individually, or
maintain a file of addresses for photo sources.

Publisher's note: The descriptions of otherwise undocumented
personal incidents and the recollections of episodes and persons
are entirely the author's. Every effort has been made to verify
details and ensure corrrectness; inaccuracy if it occurs is regretted.

Contents

Foreword by Charles E. "Chuck" Yeager   vii

Preface and Acknowledgments   ix

Abbreviations   xii

Introduction   1

1   In the Beginning   5

2   The Pinecastle Tests   53

3   The Program Takes a Pause   90

4   The Goodlin Era   127

5   The Changing of the Guard   179

6   The Air Force Takes Over the Program   234

Epilogue   285

Appendix   289

Notes   291

Index   319

Foreword

I am pleased to write the foreword to this new book on the history of the X-1. Although the story of the X-1's record-breaking flights has been told on many occasions, this is the first time that a complete flight history has been integrated with the behind-the-scenes agency meetings and the public events that shaped so much of the programmatic decisions.

It hardly seems possible that over forty-six years have passed since that October day when the mystery of the sound barrier was finally put to rest. I was privileged to participate in the last thirteen flights chronicled in this book. The X-1 program afforded me the professional opportunity to take part in two historic milestones: first, as the pilot of the airplane that pierced the supposedly impenetrable sound barrier, I experienced first-hand the enormous challenges that the X-1 team faced in exploring the unknown transonic region. As the pilot, I got a large share of the glory, but I can truly say it was a team effort to construct, instrument, and fly the fifty X-1 flights that led to the conquest of the sound barrier. Second, my flights represented the return of the Air Force Flight Test Division to research flying. Colonel Al Boyd, my boss at Wright Field, made sure that I was especially aware that my failure in such a high-visibility program would have had serious consequence for future Air Force flight testing of research aircraft. The successful conclusion of the X-1 program demonstrated the capabilities of military test pilots. It also opened the door to increased military participation in the research testing of new airplanes that led to the future creation of the Air Force Flight Test Center at Edwards Air Force Base.

The limited aerodynamic and flight information available in the mid-1940s regarding the sound barrier provided the rationale behind the decision to build a transonic research aircraft. The author's detailed reconstruction of the X-1 story sheds significant new light on many of the previously discussed facts of the supersonic program. Even as an X-1 participant, I was surprised to learn of the continuing twists and turns that the program experienced before we achieved Mach 1. With the passage of time, this new information provides additional perspectives on the importance of the first transonic research airplane. The flights of the X-1 opened up a new era in aviation. They provided America with a wealth of aeronautical experience and technical capabilities. This information provided the solid foundation for many of our later achievements, here on earth and in outer space. I am proud to have played a part in that story.

Charles E. "Chuck" Yeager

Preface and Acknowledgments

The story of the X-1, Capt. Charles E. "Chuck" Yeager, and the breaking of the sound barrier have etched a vivid picture in the minds of the American public. Thanks to continuing coverage of the U.S. space program, and books such as *Yeager* and *The Right Stuff* (as well as the movie of the same name), Americans have a grand vision of our past aviation achievements. The X-1 story forms a substantial part of America's heroic aviation history, which began at Kitty Hawk, dominated the skies over Europe and Asia, and eventually took U.S. astronauts to the moon and back. It is a complex story of harrowing challenge, of determined men whose decisions forecast further advances of significance for technological progress, a story that cannot be told in brief. Much of the documentation for the X-1 remains available, a great deal of which has rarely been seen. This primary source material forms the foundation for this book. As such, the reader may note the lengthy list of notes that accompany this work. At the risk of becoming too detailed with some facts that are common knowledge, and some events that seem quite technical in content, I wanted to provide a complete basis for the statements and conclusions contained within the book, especially since many conclusions are quite different than the accepted story of the X-1. Readers desiring additional information are directed to the detailed listing of sources.

Two clarifications on the terminology in this book may assist the reader in understanding the material. The terms *indicated* and *recorded* are used in discussing X-1 flight data. *Indicated* refers to information seen on the cockpit gauges. *Recorded* is the data collected by the NACA instruments.

The airplane that currently hangs in the Smithsonian Air and Space Museum is known to history as the X-1. But in the initial period of its existence (specifically, the period covered by this book), it was designated as the XS-1. To be consistent with the various documents of that time, I refer in the manuscript to the plane as it was known at that time—the XS-1.

In preparing this book, I would like to acknowledge my deep thanks to three individuals whose earlier works provide the foundation for any serious study of the X-1. This volume in no way replaces Richard Hallion's superb *Supersonic Flight*, Jay Miller and Ben Guenther's detailed *Bell X-1 Variants*, and Jay Miller's *The X-Planes: X-1 to X-31*. Rather, this book builds upon their works, amplifying and clarifying details unknown at the time of their writing and finally providing what I hope are the concluding chapters of the X-1 supersonic story. Additionally, Ben Guenther was kind enough to assist me by providing a copy of his unpublished day-by-day X-1 events log. This invaluable manuscript and the archival documents he supplied from the NASA holdings made sense of the vast amount of NACA information available for this work.

I am especially indebted to several organizations and individuals. Without their cooperation and kind assistance, this book could not have been written. Specifically, I would like to thank Dr. James Hansen and the staff of the NASA Langley Research Center/Historical Program, and Dr. James Young and his staff at the Air Force Flight Test Center History Office. Most of the primary source documents come from these two repositories. (Unless otherwise noted, all NACA, most Bell, and some AMC documents are from Langley's archives.) Dr. Young kindly read the entire manuscript and made many helpful suggestions as to form and content. He also made available a copy of his *Supersonic Symposium: The Men of Mach 1*, which provides an excellent overview of the events, and a taped transcript of a discussion by the participants. A further source of material is found in the records of the Air Materiel Command. These documents are cataloged as the "Sara Clark Collection" at the National Archives Records Center in Suitland, Maryland. Additional records were provided by Susan Dewberry and Bruce Parham of the National Archives, Lee Saegasser of the NASA Historical Office in Washington, Alphonse Salandra and Stanley Smolen of Bell Aerospace, and Frank Mendola of the Orange County Historical Museum.

As to individuals, a very special thanks is due to Richard Frost for allowing numerous telephone interviews, reading and correcting several portions of the manuscript, and providing his personal flight log and many Bell Aircraft memorandum. Chalmers "Slick" Goodlin, Brig. Gen. Chuck Yeager (Ret.), Jack Russell, and Charles Hall all consented to several interviews or read and commented on specific points of the manuscript. Goodlin also contributed valuable documents from his personal files. Mrs. Irene Dow provided copies of her late husband Harold Dow's flight log and valuable photographs. Joel Baker took time to be interviewed several times and also provided his personal flight log. In addition, Walter Williams, Douglas Rumsey, O. Norman Hayes, Charles Taylor, Gerald Truszynski, Alvin "Tex" Johnston, Joseph Cannon, Mary Woolams, Doris Stanley Proebstel, Jay Demming, Don Thomson, Lt. Gen. Laurence Craigie (Ret.), Maj. Donald Eastman (Ret.), Col. Ezra Kotcher (Ret.), Col. David Pearsall (Ret.), Maj. Gen. Osmund Ritland (Ret.), and Maj. Gen. George Smith (Ret.) were interviewed and gave freely of their time, expertise, and photographs. The Greater Orlando Aviation Authority generously provided funding to assist in the preparation of the Pinecastle information. Material for this chapter is from my unpublished manuscript, "Forgotten Dawn: The X-1 at Pinecastle." Finally, I would like to thank Felix Lowe, director of the Smithsonian Institution Press, for his counsel and encouragement, and my editor, Therese Boyd, for her patience and skill in repairing my prose and lapses of clarity. Without the cooperation of all those mentioned, this project could not have been completed.

Abbreviations

AAF	U.S. Army Air Forces
AERL	Aircraft Engine Research Laboratory, NACA, Cleveland
AFFTC-HO	Air Force Flight Test Center History Office
AMC	Air Materiel Command at Wright Field
ATSC	Air Technical Services Command
GPO	Government Printing Office
LMAL	Langley Memorial Aeronautical Laboratory, NACA, Virginia
MC	Materiel Command
NACA	National Advisory Committee for Aeronautics
RMI	Reaction Motors Incorporated
USAF	U.S. Air Force

Introduction

The story of the quest to first break the sound barrier has been told many times. Unfortunately myth, recollections, and hearsay have become accepted parts of the historical legacy. This book seeks to correct those myths and accurately record the events and decisions that shaped the X-1 program. It is an exciting story, with equal parts courage, initiative, and ingenuity, which reflects how civilization continues to achieve scientific progress. The interagency and industry cooperation that helped break the sound barrier and later put people into space did not develop overnight. Rather, it was the result of over 20 years of work begun in the 1940s and brought together in the 1950s to propel our fledgling space efforts to new heights in the 1960s. These successes were then rechanneled into new aviation technology, including the space shuttle, the B-2 stealth bomber, and the proposed Transatmospheric vehicle.

To fully understand the quest for supersonic flight and the origins of the X-1 program, it is necessary to go back to the early days of World War II. Prior to the war, American aviation lagged behind European efforts, due to a lack of adequate funding and the failure of U.S. military leaders to reach agreement on the proper air mission for our forces. When American rearmament began in 1939, the U.S. Army Air Corps possessed no modern aircraft with the exception of the B-17. However, all the famous American aircraft that won World War II were already under development or in production by the time of Pearl Harbor. The Japanese aerial attack on Hawaii in 1941 accelerated the demands for intensive cooperation among industry, government, and science to develop and provide the necessary aircraft to establish American aviation supremacy.

Before the X-1, efforts among government, aircraft contractors, and industry dealt with specific fixes for unique aviation problems.

Wartime needs transformed the U.S. government's National Advisory Committee for Aeronautics (NACA) from an agency devoted to basic aeronautical science research to one that focused on testing airplanes, cleaning up existing problems, and refining new aircraft models. The problems of here and now—such as redesigning the P-42 cowling to fix cooling problems, or using wind tunnel tests to correct the P-38's inability to recover from steep dives—took precedence over the longer-term research that the agency formerly had claimed as its mandate. The spirit of cooperation fostered in winning World War II added to the development of an interlocking agenda between government and research. The experiences of these efforts formed the foundation for the X-1 program. But while wartime cooperation came from wartime experiences, the partnership did not develop in a manner foreseen before the war.

Prior to World War II, government funding for aeronautical research and development only covered a small portion of what actually was accomplished. Since much of the aviation equipment developed for the Army Air Corps had direct or, sometimes, indirect commercial application, U.S. industry was willing to provide a major share of the development costs. During the war this situation was altered and expanded, both through the significant levels of funding provided by the National Defense Research Council (NDRC), and through amortizing research costs in the huge production and supply contracts that were available as part of the war effort. The speedy development of new products became more important than the actual costs involved in the project. The end of the war, however, drastically changed this picture. Cost-cutting and a return to peacetime conditions became the new congressional policy. The NDRC was abolished and large aircraft production contracts simply ceased to exist. But the rapid and continuing evolution of aviation technology had substantially changed the ground rules for aeronautical research. The U.S. aeronautical industry could foresee little commercial application for many of the newest technologies and the resulting products. As a consequence, the U.S. government became the lead agency in fostering a continuing research and development partnership between the military services, industry, and science. That partnership was recast even before the end of World War II and received its new foundation in the supersonic aircraft program.

Research problems in aviation did not begin with the X-1 program. Significant risks had always existed in flying airplanes. Some were instructional problems, such as those caused by pilot error; some represented mechanical issues, such as powerplant requirements or engine failure; many involved production quality problems, such as defective equipment. But the issues relative to the speed of sound seemingly implied something new. The sonic "wall" inferred that a physical barrier existed and that humans might not possess the technical capabilities to overcome this physical restriction. The pre-1945 stories about the "wall" that supposedly represented the sound barrier were based on fearsome predictions of aircraft failure and death. Even some prominent scientists held out little technical hope that a solid object could pierce the sound barrier and survive. Today computers, flight simulators, and precise mathematical calculations provide a substantial basis for all advanced aeronautical research. But in 1945, the technical capabilities for fully understanding the uncertainties of the transonic zone and the sound barrier simply did not exist. The only alternative to wind tunnel theory was actual flight testing. In the end, engineers built a plane that scientists *later* verified through wind tunnel data as feasible.

The X-1 was the first airplane constructed solely for high-speed research purposes. The progress of the project defined, and then solidified, the coalition among America's military services, aircraft manufacturers, and the NACA. The evolution of the program through its developmental stages became the foundation for many later research aircraft projects. The design, construction, and utilization of high-speed research planes provided the necessary information for many later models of U.S. production aircraft and became the hallmark of America's test flight program. With the success of the X-1 project, this pattern of development resulted in a literal explosion of new research aircraft programs and information that culminated in the pioneering flights of the X-15. The X-plane experience, coupled to the development of trained personnel and program managers, new equipment, and fundamental flight research techniques, paved the way for the success of America's manned space flight efforts in the 1960s and beyond.

While the X-1 was recognized and acclaimed as an important airplane in its time for being the first to break the speed of sound, the true significance of its achievements and the politics of the decisionmaking that shaped our aviation future have received less attention. The X-1 high-speed research project provided the necessary focus for two larger, yet interconnected, issues. First, the divergent NACA/Army

Air Forces (AAF) views on the conduct of postwar aviation research provided an intellectual framework for the desire to develop a greatly expanded U.S. research capability. Second, the creation of the independent U.S. Air Force (USAF) spurred the opportunity to define and achieve the service's own niche in aeronautical inquiry. The flights of the X-1 overlay these issues. After 45 years, it is possible to place in perspective the broader effects the project's policy decisions had on the development of American aviation research. In the final analysis, the X-1 program acted as a defining moment in changing the execution of aviation research in the United States. It propelled research and development funding to higher levels than previously witnessed. It provided the aura of solid achievement necessary to cap the requests for such funds. It laid the foundations of our space program and confirmed American aviation supremacy in the latter half of the twentieth century. This is the story of that beginning.

1

In the Beginning

The story of the first effort to fly faster than the speed of sound began long before an Air Force pilot flew the X-1 on that fateful October day in 1947. The Austrian physicist and mathematician Ernst Mach (1838–1916) had discussed the theory of the speed of a solid object moving through a gas and its relationship to the speed of sound as early as the nineteenth century. Subsequently, this principle was identified as a Mach number, with Mach 1 representing an object moving at the speed of sound.

Problems arise when researching the speed of sound because it varies based upon altitude and temperature conditions. Generally, the higher in altitude an airplane goes, the lower the speed of sound would be. Thus, at sea level a plane must exceed 760 MPH to break the speed of sound, while at an altitude of 36,000 feet the same aircraft must only fly faster than 660 MPH. Regardless of altitude, when an object moves through the air, the changes in speed and pressure create disruptions in the smooth flow of air around the object. At slower speeds, the air has more warning that an object is approaching by the forward displacement of the air molecules as the object pushes through the air mass. As the object approaches transonic speeds, the disturbances in the air pressure cause significant disruptions of the air flow over the surface of the object. As higher speeds are achieved, this pressure shock wave precedes the moving object. The resulting compression of air into a space in front of the object creates a loss of lifting power and thus control and stability problems for that airfoil. This phenomenon is called compressibility and represents a serious problem for any aircraft attempting to fly faster than transonic speed. Research on the phenomenon associated with an object moving through the resistance of air had intrigued aviation researchers since the 1920s, long before aircraft had the capability to approach speeds relevant to the problem.

In 1935, reporters asked the British aerodynamicist W. F. Hilton what research problems he was currently working on in the United Kingdom's National Physical Laboratory's high-speed wind tunnel. Hilton replied that his work was on problems relative to airflow. Pointing to an airfoil drag plot, he said, "See how the resistance of a wing shoots up like a barrier against higher speed as we approach the speed of sound." This straightforward analysis of a research problem had unintended consequences. By the next morning, many British newspapers had misinterpreted Hilton's remarks and coined the phrase "sound barrier" to describe the phenomenon. Although Hilton was pessimistic regarding the possibilities of supersonic flight based upon the horsepower/Mach number requirements, he did not intend the misrepresentation of his remarks alluding to a physical barrier in flight.[1]

During the late 1930s, aviation technology was based on propeller-driven technology. The struggle to increase the power and efficiency of conventional prop aircraft dominated the search for technological supremacy in the interwar years. With the continuing political tensions in Europe prior to World War II, aviation witnessed giant strides in technological knowledge and subsequent aircraft performance. Speeds and capabilities once thought astonishing became routine. As the developmental requirements for potential combat aircraft increased, the aviation industry engineers strained to keep pace with the heightened demands for useful data on aircraft performance. The absence of data restricted the margin of design error. Designers had to learn how to merge the new powerplant capabilities into the existing technology of airframes and wings. While safety remained inherent in all aircraft research and design tests, it became increasingly difficult to obtain the necessary research information in a timely manner. The demands for the new aircraft, and the approach of war, made some research a "learn as you go" proposition.

With the outbreak of World War II in September 1939, advances in aviation research literally exploded off the drawing boards as production aircraft performance increased substantially. New planes with greater capabilities poured out of the factories of all the warring countries. During the war, American fighter aircraft routinely exceeded 400 MPH in flight. The performance of the Lockheed P-38 Lightning, the Republic P-47 Thunderbolt, and, later, the North American P-51 Mustang provided U.S. military forces with a decided advantage in air-to-air combat. When those pilots executed steep dives in combat, the concurrent increases in speed brought them close to the predicted area of transonic flight. At that point, pilots returned (sometimes) with harrowing tales for industry engineers of the alarming loss of aircraft control. The need to investigate high-speed flight issues appeared concurrently with another, far more serious technological research problem. As early as 1941, U.S. Army Air Corps leaders had become aware of the significant jet and rocket research efforts of Nazi Germany. By midwar, Allied military intelligence began to hear stories of advanced German aircraft utilizing the new rocket and jet-engine propulsion systems. These new German aircraft models promised significant performance advantages over conventional propeller-driven aircraft. Both the United States and Great Britain immediately expanded their previously minimal jet aircraft development programs to test the new technology, although neither Allied program reached a useful combat deployment prior to the end of the war. They did, however, provide substantial information on higher performance aircraft. Based upon these continuing wartime advances in powerplant design and subsequent aircraft performance, the issues related to high-speed flight could no longer be postponed. As a result, the accelerating pace of aviation demands finally brought U.S. researchers back to an oft-discussed subject: a transonic research program to explore the issues of high-speed performance and safety.

Although the subjects of compressibility and transonic speeds would not demand full attention until after World War II, they first received significant international recognition at the 1935 Fifth Volta Congress on High Speeds in Aviation, held in Rome.[2] The congress, sponsored by the Accademia d'Italia, consisted of representatives from all the major industrial nations. Against the backdrop of an increasingly military presence in aviation, this meeting marked the beginnings of the possibilities of supersonic flight. Two concepts received prominent discussion: supersonic wind tunnel research and swept-wing research. High-speed wing design and air motion in a wind tunnel had been explored prior to the conference by researchers in Germany, Italy, and the United States. The German researcher Dr. Adolf Busemann suggested sweptback wings as a way to overcome the problem of shock stall. The swept design of the wings would delay the sudden loss of lift due to compressibility by preventing the airflow from striking the wing at right angles. Theodore von Kármán, a representative for the United States at the conference, returned to Cal Tech determined to stimulate wind tunnel research on transonic aircraft problems. Little interest was generated immediately, since the von Kármán approach required a wind tunnel capability that greatly exceeded any known requirement for propeller-driven aircraft. But the ongoing discussions within the aviation community provided an important foundation for later research efforts.

Some of those discussions on high-speed aircraft were already occurring within the offices of the National Advisory Committee for Aeronautics (NACA) and the Army Air Corps Engineering School at Wright Field, Ohio. During the mid-1930s, conversations on transonic and supersonic problems led two engineers, NACA's John Stack (head of the Langley Memorial Aeronautical Laboratory [LMAL] Compressibility Division) and Wright Field's Ezra Kotcher (then a civilian, later to retire as a lieutenant colonel), to conclude that breaching the sound barrier was not a problem of penetrating a "wall," but rather a problem of developing a set of solutions to a series of aerodynamic and powerplant questions. Data, not ability, was the factor missing from the efforts to understand the so-called sound barrier. With better understanding, it would be possible to develop technology to allow an airplane to fly faster than sound.

Stack had been investigating the effects of compressibility on high-speed flight since 1927. In 1933–34, he produced a conceptual model of an experimental airplane for compressibility research. Although the aircraft was never built, the proposal did generate significant discussion on the possibility of flying at speeds greater than 500 MPH. It stressed the need for further understanding of the aerodynamic effects on aircraft operating in the region just below supersonic speeds. In 1942, Stack recommended to NACA headquarters (HQ) in Washington, D.C., that the agency pursue development of a transonic research airplane. NACA HQ did not enthusiastically embrace the new idea, but the initial discussions led to the formation of a research team headed by Stack to develop the design requirements for such a new aircraft.

At the same time that Stack was developing his ideas, Kotcher was working at Wright Field. In 1939, Kotcher submitted a report to the prestigious Kilner-Lindbergh Board on the future of aeronautical research. In the document, Kotcher advocated a "comprehensive flight research program" to allow comparisons between wind tunnel data and actual flight performance data. Kotcher also advocated the use of gas-turbine or rocket propulsion systems to allow greater performance in the new aircraft than that achievable with propeller technology. At that time, neither technology was seriously being considered for high-speed flight. The board provided the report to the Army Air Corps (Army Air Forces as of June 20, 1941) chief of staff Maj. Gen. Henry H. "Hap" Arnold. Armed with this information, Kotcher began to advocate a transonic research effort reinforced by a comprehensive wind-tunnel research program. While some questioned the type of powerplant for such experiments, Kotcher was consistent in his advocacy of rocket propulsion for the program to escape the inherent limitations of propeller-driven aircraft. Throughout the war, Kotcher continued to press for a sound research program on flight in the transonic range. In this, he was not alone. In 1943, Brig. Gen. Franklin O. Carroll, the chief of the Engineering Division at Wright Field asked Theodore von Kármán to investigate the potential for building an airplane capable of Mach 1.50 (1,000 MPH). Several days later, von Kármán reported that it was feasible to design such a high-speed airplane. The Army Air Forces (AAF) would make good use of that information.[3]

While Ezra Kotcher pursued his concept of a high-speed airplane, LMAL's John Stack developed his earlier proposals into a small turbojet aircraft theoretically capable of Mach 0.80–1.0. By the summer of 1943 he was ready to unveil his work. His efforts coincided with continuing flight difficulties with the new production models of the P-39, P-47, and the Navy's SB2C divebomber. All three models were experiencing fatal failures of the tail assemblies in high-speed dives. Stack's work, the problems with the production aircraft, and recent Allied intelligence information on new German aircraft changed the AAF's thinking regarding the need to pursue research on high-speed flight characteristics.

On December 18, 1943, a conference was held at the NACA HQ to discuss jet propulsion advancements in England. Representatives of the AAF, Navy, NACA, and eight American industrial firms attended the meeting. Many of the attendees had traveled to England during the summer/fall of 1943 and the meeting opened with their review of the British experiences. Although most confined their remarks to the agreed agenda, researchers Robert E. Wolf of Bell Aircraft and Eastman N. Jacobs of the NACA expressed reservations as to the current direction of U.S. research, specifically, the continuing efforts to fit new powerplants to existing conventional aircraft. Jacobs called for a unified, cooperative program among the military services, aircraft manufacturers, and the NACA to develop extreme performance airplanes suitable to exploit the new powerplants. The test results would increase the acquisition of data from level flight, which was safer than the hazardous dives currently executed to obtain the same data. Wolf stressed the necessity for such data; he pinpointed the industry's need for additional information on the flight characteristics of aircraft at high speeds if manufacturers were to respond to the pace of aviation change. He recommended that the NACA take the lead in this effort, a suggestion that met receptive ears at the NACA HQ. On February 8, 1944, the NACA's Dr. George W. Lewis, director of aeronautical research, informed Henry J. E. Reid, Langley Lab's engineer-in-charge, that he wished to establish a high-speed panel to coordinate research at the three NACA laboratories.[4]

The first meeting of the High-Speed Panel (as it became known) occurred on March 2–3, 1944, at LMAL in Hampton, Virginia.[5] Russell G. Robinson (NACA HQ's chief of research coordination) was designated as chairman, with H. J. Allen, John Stack, Eastman Jacobs, and R. E. Littell as panel members. Others at the meeting included Dr. Lewis, Henry Reid, and John W. Crowley (Langley's chief of research). Robinson opened the meeting by reiterating the purpose of the panel: to review the compressibility research program of the NACA toward the goal of supersonic flight. Lewis added that he wanted the group to be the most forward-thinking group at LMAL. After a review of the existing equipment and methods at Langley— conventional plane, special jet, nonairplane objects, projectile tests, supersonic wind-tunnel tests, subsonic wind-tunnel tests, water channel for hydraulic analogy, and analytical methods—those present agreed that research would focus on the conventional plane and special jet categories. Jacobs indicated a need for three types of airplanes: subsonic propeller, jet-propelled, and supersonic. Three candidates were selected for the tests: for the propeller-driven aircraft, rockets were suggested as a means to increase speed; for the high subsonic speeds, the NACA wanted to procure a new AAF P-80 jet or modify an existing P-59.

With the two easy choices out of the way, the group addressed the issue of a supersonic aircraft. All agreed there was little or no information available to even design such an airplane. To date, no wind tunnel tests on an airfoil had been conducted. Stack recommended that tests on wings, bodies, and stabilizing surfaces proceed in the NACA supersonic wind tunnel. But the issues involved with this subject caused conversation to return to the jet-propelled research airplane. Several individuals spoke to the need to procure a small airplane powered by four Westinghouse jet engines that would be capable of re-search flights of one-half hour duration. To conclude the meeting, Lewis suggested that this research recommendation be submitted to the Army for review.

While the NACA continued to pursue its interest in high-speed aircraft, the Material Command at Wright Field conducted its own studies of the issue of high-speed aircraft. In late 1943, Kotcher was selected as project officer on the Northrop XP-79B rocket-propelled flying wing project. Although the program never reached the produc-tion stage (the only one built, as project MX-324, was quickly de-stroyed in a crash), it did lead the way toward additional research on rocket aircraft. In early 1944, Kotcher undertook a further study of the issue of rocket versus turbojet power. The report, somewhat humor-ously labeled the "Mach 0.999 study" (supposedly referring to the impenetrable sound barrier), defined the potential of both powerplants and indicated the potential superiority of the newer rocket designs. The rocket motor's principal advantage over the turbojet was thrust. The greater power of the rocket engine provided superior performance at higher altitudes than the turbojet engine. The ability to reach higher altitudes would allow for increased performance and less potential strain on the aircraft as it attempted to break the sound barrier. It also allowed the attempt to pierce the barrier to be conducted in level flight rather than in a dive. Many engineers believed this point to be ex-tremely important, however difficult it might be to achieve through contemporary aircraft performance. Should trouble develop, a diving airplane would be significantly harder to control from a safety stand-point than one in level flight or climbing. By April, the Wright Field Design Branch had assembled its information into a theoretical air-plane. The proposed model included a bullet-shaped fuselage, mid-wing design, and conventional tail surfaces (on the fuselage and not on the vertical fin as on the later XS-1), a smoothly faired canopy, and a liquid-fueled rocket engine of 6,000-pound thrust.[6] While the Materiel Command worked to conclude its study, other meetings continued the progress toward a detailed design program.

On March 16, 1944, a conference on the issue of compressibility was held at Langley Laboratory.[7] Present were representatives of the Army, Navy, and NACA. During the meeting, John Stack stressed the need for research at the level near Mach 1.0. Unfortunately, Stack reported, the availability of the needed data was restricted due to the continuing problems with wind-tunnel performance tests. Specifically, the data that could be obtained was considered reliable only for theoretical research purposes because of the "choking" effects within the tunnel. During World War II, the state of contemporary technology for wind tunnels was still considered primitive. While wind tunnel testing was fine for lower speed (below Mach 0.70) and very high speed (up to Mach 2.0) research, the zone near transonic flight presented engineers with a major technical problem. Models placed in a wind tunnel developed air shock waves during simulated tests above Mach 0.70 and below Mach 1.30. These shock waves reflected off the tunnel walls and disrupted ("choked off") the air flow over the model. Thus, the data developed in the tests did not represent reality, but rather an anomaly of the testing procedure. This problem was seen as quite serious for the technical complexities of prediction. As a result, some in the military and scientific communities viewed the major NACA commitment to wind tunnels as somewhat misplaced. They believed the NACA was fearful of the negative consequences to their political budgetary base if they could not resolve this technical problem. Opponents of vast expenditures on wind tunnel testing stated that if pure field testing of aircraft could produce results faster than that obtained by building an additional wind tunnel—one that might be incapable of producing results—then, so the argument went, why bother? Although later engineering work would produce a special slotted tunnel wall to negate the effects of choking, the development of the modified tunnel actually occurred *simultaneously* with the XS-1 supersonic tests at Muroc Dry Lake, California, in 1947. But in 1944, no one could foresee a rapid solution to the problem. Thus, although theoretical corrections could be made for Mach number variations, seemingly there was no substitute for actual flight testing.

In recognition of the technical problem with wind tunnel research, Stack urged the pursuit of transonic data through actual flight testing. Regardless of the Mach number, he indicated, "it was unreasonable to assume no lift would be obtained if an airfoil were placed at an angle to the airstream." Stack concluded his remarks by stating that previous aircraft inability to execute a pullout from a dive near high Mach numbers had been due to the plane's own stability potential. In summation of the NACA position, John Crowley proposed that several airplanes be produced, and procured by the NACA, so that an actual test program could proceed under rigorous research conditions. The aircraft should be constructed overstrength to meet the anticipated higher loads and stresses at high Mach number conditions. The absence of armament or tactical equipment would allow the extra weight to be devoted to strengthening and instrumenting the research aircraft. The additional load factor strength would allow such high-speed phenomena as buffeting, increased instability, and high-speed stall to be investigated with reduced fear of structural failure. The acquisition of flight data near Mach 1.0 would allow the cancellation of flight restrictions that currently handicapped existing aircraft even though higher performance was available to some of these planes.

Since the NACA could not (by virtue of its charter) build airplanes and the AAF/Navy seemed open to commissioning the work through private companies, a tentative program was outlined. However, each service made it clear to the NACA that it viewed the program in a different light. For the AAF, the proposed plane would simply be "a major developmental step toward higher operating speeds extending upward through Mach 1." For the Navy, the plane would be a vehicle that could counteract "the myth of an impenetrable barrier and [provide] needed high-speed data." The final recommendation of the meeting was that the Army and Navy would each furnish one airplane. Five days after the March 16 meeting, Langley proposed in writing a general test program to the Navy's Bureau of Aeronautics (BuAer) and the AAF's Materiel Command at Wright Field. On March 31, a study request by two Wright Field engineers, captains F. D. Oranzio and G. W. Bailey, was sent to the chief of the Aircraft Laboratory at Wright Field, outlining the AAF's interest in pursuing a transonic research aircraft.[8]

Developing a research airplane was new territory for Wright Field.[9] The AAF's desire for additional expertise in flight research was a direct outgrowth of the postwar perception by several high-ranking military figures of the perceived failure of the NACA to keep up with the immense technological changes in the immediate prewar and war years. Particularly disturbing to the AAF was the discovery of the sweeping advances made by Germany in aviation science during the war. The German application of turbojet and rocket powerplants to specific production aircraft and the utilization of sweptback wings provided an unwelcome development for Allied propeller-driven aircraft. The wartime disclosures of such airplanes raised serious concerns within the American military services about past reluctance to allow the Army/Navy to develop and apply the latest aviation technology.

Objectively, the NACA was not alone in the blame for America's technological backwardness in certain areas of propulsion and aerodynamic research. The AAF contributed to its own lack of awareness through its narrow research base and the absence of sufficient numbers of skilled scientific personnel. Prior to the war, the Army Air Corps had maintained small research facilities at Wright Field, but these facilities and the personnel that worked there were focused more on the development issues of production aircraft than on future research problems. Total personnel in July 1939 was only 2,000 (90 percent civilian), although this figure would grow during wartime to over 17,000. During the late prewar and war years, very few officers at Wright Field possessed a scientific background. Nor were the necessary high-speed research facilities available for the Air Corps's research programs. Yet the Air Corps personnel were supposed to work with outside agencies and industry to evaluate and develop new technology. This situation was a contradiction that many senior AAF officers wanted to change.

More was at stake, however, than a mere lack of properly educated personnel or adequate facilities. The politics of organizational control also played a role in the emerging struggle over aviation's future. In 1939, Gen. Hap Arnold proposed a 20-foot 40,000-horsepower wind tunnel to be built at Wright Field (although advanced at the moment of conception, it would be quickly rendered obsolete by the advances in technology). The NACA HQ immediately objected to this proposed expenditure as wind tunnels had traditionally been the domain of the NACA. On behalf of the agency, Dr. Lewis stated any such tunnel should be built and operated by the NACA. A political and budgetary battle developed over the funding necessary to provide the facility. (Although the proposed facility was eventually built at Wright Field, the ongoing political struggle, of which more will be said later, opened the door for funding for additional needs at the NACA laboratories.) But, as a result of the increased budgetary emphasis, America's prewar focus and structure for aviation research changed substantially in the late war and early postwar years. The NACA program that entered World War II focused on basic science research did not have the same agenda after the struggle.

The same could be said for the AAF and its facilities at Wright Field. During the late war period, the AAF leadership, convinced that the service should be more involved in research, increased the numbers of promising young officers it sent to colleges and universities to continue their advanced training. The final culmination of this effort to increase AAF expertise was the creation of the Air Force Development Command (AFDC) in 1950, an event that marked an important milestone in the technological maturity of military aviation research. But all of this was in the future. In 1943–44, the AAF was still struggling to formulate a policy on the question of research and development.

In May 1944, the AAF, in preparing for its postwar future,[10] proposed Project B-7 covering its plans for fiscal year 1946/50. This plan provided an orderly transition from the vast wartime expenditures on aviation to a stable research base for the postwar period and called for the largest single allocation for experimental and developmental aircraft. But General Arnold recognized that the hardware plans for a postwar AAF lacked the close cooperation and technological input from the scientific community to assist in spending that research and development funding. Their invaluable input during the war had provided many key innovations for the AAF and Arnold wanted to ensure that this close interaction would remain a foundation of AAF thinking. As a result, in November 1944, Arnold asked von Kármán to form a distinguished panel of scientists to chart long-range research and development needs for the AAF. The Scientific Advisory Group (SAG) was charged with producing a program that would provide security for America over the next 10 to 20 years. It was a big order, but the group von Kármán assembled represented many of the very best in American science. The resulting SAG study in August 1945, entitled *Where We Stand*, incorporated many of the relevant discussions of the AAF leadership and the best of the German technological discoveries. It recommended a comprehensive wind-tunnel program and recognized the importance of high-speed aircraft research. The report bluntly stated that America "cannot hope to secure air superiority in any future conflict without entering the supersonic speed range." Long-range research, an effectively trained scientific staff, and a high-speed flight program were the keys to American supremacy. But the immediate problem in 1944 was how best to enter the supersonic race.

While Major Kotcher continued his search for industry support, he received a helping hand from a familiar partner. A well-known aviation company was moving ahead on its proposed high-speed design activities through its own conversations with the NACA. On April 14, 1944, Lawrence D. "Larry" Bell, the president of Bell Aircraft Corporation, sent the NACA a copy of a three-view drawing of a proposed research airplane. At the same time Paul Emmons, the chief aerodynamicist for Bell Aircraft, requested that the NACA provide assistance with wind-tunnel model tests. Emmons believed this data would enable Bell Aircraft to incorporate the NACA's findings into Bell's latest recommendations to Wright Field. Although Bell Aircraft, the NACA, and the Materiel Command at Wright Field seemed to be pursuing the same objective, it remains somewhat unclear to this day as to the specific daily interaction among the groups.[11]

On April 20, a meeting was held between representatives from the NACA and Wright Field, including an AAF team from the Materiel Command's Engineering Division, representatives from the NACA's Aircraft Engine Research Laboratory (AERL), members of the NACA's LMAL research group, and the High-Speed Panel. The Materiel Command's group indicated their interest in obtaining a plane to pursue high-speed research work. Captain Kotcher stressed the power requirements desired for the airplane and how a rocket-powered model could meet that demand. However, questions were immediately raised by the NACA group as to the reliability and flight duration capabilities of a rocket-powered airplane. As an alternative, the NACA team discussed their work on jet-propulsion aircraft. A visit to Langley was arranged for Kotcher on May 8 to view the NACA proposals for a jet-powered high-speed aircraft.[12]

A series of meetings during May 1944 indicated just how fast the progress on the program for a high-speed research aircraft was advancing. From May 15 to 20, 1944, AAF personnel held meetings in Washington to discuss the fiscal year 1945 research agenda of its Materiel Command's Engineering Division. During the gathering, compressibility research by the NACA was assigned a number 1 priority. The first phase would focus on diving aircraft, the second phase on a P-80A aircraft, and the final phase on an AAF-supplied supersonic airplane. On May 15–16, a small group of AAF and NACA personnel met at Langley Laboratory to review the NACA high-speed research airplane design. It called for an airplane of 6,440 pounds gross weight powered by four Westinghouse 19XB units (6,400 pounds of thrust). Kotcher continued to indicate the Materiel Command's interest in such a proposal.

Finally, during May as well, the NACA's Committee on Aerodynamics reviewed the agency's options as to the priorities for high-speed aircraft research. The NACA's John Crowley informed the committee that a three-stage investigation process would be utilized: research on a conventional airplane; constructing and instrumenting a high-speed research plane capable of 10 to 15 minutes flying time; and finally, flying the airplane at supersonic speeds to obtain vital high-speed data. At the conclusion of the committee's discussions, Dr. Lewis indicated his support and stated the final design for such a plane would be turned over to the AAF or the Navy since the NACA had no intentions of finalizing a design or constructing such an aircraft. The AAF representative present at the meeting indicated the Materiel Command's continued interest in procuring a high-speed aircraft.[13]

On August 9, the NACA representatives of the High-Speed Panel met at Wright Field with AAF personnel to review proposals for various designs of a high-speed research airplane.[14] The following conclusions were recommended for any final design:

1. A final contract should include provisions for a wing of 8 to 9 percent thickness chord (although some discussion was given to a 12 percent thickness chord wing).
2. The airplane should retain a conventional tail assembly.
3. Initial contacts should be made with North American Aircraft and Republic Aircraft (both believed to be interested in the project).
4. The wing should be stressed for at least 12g or perhaps even 15g conditions.

A further meeting was scheduled within the next several weeks to review the proposals.

By the end of the summer of 1944, Major Kotcher was ready to solicit cooperation from U.S. industry. Although initially both North American Aviation and Republic Aviation were thought to be interested in the project, the U.S. aviation industry was already fully committed to the massive construction programs necessary for the war effort. Building a one-of-a-kind research plane might sound attractive, but such work remained very time-consuming in terms of both personnel and space. Throughout the fall, Kotcher could find no aviation company interested in pursuing this special project. However, on November 30, Robert Woods of Bell Aircraft visited Wright Field to discuss existing corporation contract work. One of the founders of Bell Aircraft as well as chief design engineer for the company, he had a long and successful track record designing aircraft for the U.S. government. During the visit, Woods dropped in to see Major Kotcher. The conversation quickly turned to Kotcher's frustration at efforts to get the transonic aircraft program moving forward. Kotcher found not only an avid listener in Woods, but also a fellow enthusiast on the subject of transonic research. Upon learning that the Air Technical

Services Command (ATSC, as the Materiel Command and the Air Service Command were redesignated on August 31, 1944) would only require the transonic airplane to be safe and controllable up to Mach 0.80, Woods committed Bell Aircraft to construct such an airplane.[15]

The Design Takes Shape

By late 1944 the McDonnell Aircraft Company had joined Bell Aircraft in the continuing ATSC discussions to develop a transonic research program. These conversations concluded on December 13–14, 1944, with a conference held at Langley Field.[16] Present for the meeting were Major Kotcher and captains G. W. Bailey, and R. B. Pearce of Wright Field; Russell Robinson of NACA HQ; Jean Roche and Col. Carl F. Greene of the Langley ATSC liaison office; and Floyd Thompson, Robert Gilruth, John Stack, Milton Davidson, and Eugene Draley of Langley Laboratory. The focus of the conference included a discussion of the Langley analysis of the proposed transonic aircraft and the existing design studies, following up on the meetings of March 16 and May 15. During the meeting, Kotcher proposed moving beyond the review stage and indicated that ATSC wanted Langley to establish design requirements for an experimental airplane to assist the aircraft companies in their construction efforts.

Responding to the request, Stack outlined the Langley problems in identifying the technical coefficients for the aircraft. He indicated the NACA's belief that no consideration should be given to high-speed flight research in a diving configuration since that posed other flight safety problems. Stack also stressed that flights utilizing a "mother" aircraft to assist in drop operations would further complicate the research mission and should be rejected. The Wright Field ATSC contingent agreed on both of these points.

Kotcher continued to insist that a rocket engine must be the powerplant for the proposed aircraft. The NACA disagreed and attempted to demonstrate the utility of using jet engines as opposed to the heavier and less fuel efficient rocket engines.[17] The NACA also expressed concerns over the extreme altitude necessary to get the maximum advantage from rockets as well as the difficulty in handling the fuel. Kotcher disagreed, reminding the audience that Wright Field was developing a new pressure suit that should handle the altitude problem and that the ATSC would be responsible for handling the fuel requirements of the airplane. No final decision was reached on either point. Kotcher agreed, however, that flight research in a level flight configuration was the best safety regime.

The proposal submitted by McDonnell (MCD-520) was quickly dismissed since it incorporated both a vertical dive technique and the use of a mother ship for air-launch. When the group turned to the Bell Aircraft proposal (MCD-524), Stack raised several concerns regarding the amount of instrument space on the airplane. Captain Bailey indicated that space had been set aside for the NACA instrumentation for the flights. Stack also questioned the location of the wheels and rocket engines under the wings; both would complicate any necessary repair and replacement work. Kotcher reminded Stack that the current Bell proposal was not considered final and thanked him for his comments. Both Stack and Kotcher agreed the necessary design requirements were established well enough to serve as guidelines for a final design. As a result, the group settled upon the Bell Aircraft concept for further research studies.

The next day, December 14, Kotcher opened the meeting by outlining ATSC requirements for the new project: the research aircraft must be a transonic—not supersonic—plane, of conventional configuration; and it must be a civilian research (rather than a military) airplane. The conventional configuration requirement addressed the issue of whether the new airplane should have straight or swept wings. The ATSC was already aware in 1944 of the German development of swept-wing aircraft. In fact, Kotcher, von Kármán, and Robert T. Jones of the NACA had at various times discussed swept-wing aircraft and their potential. But while the concept seemed promising, to the Americans it remained unproven technology at that time. The new U.S. P-80 jet fighter possessed straight wings, as did a number of other potential aircraft designs. Since the new research aircraft was to assist in the theoretical knowledge for future aircraft designs, it appeared logical to continue with a proven useful design. As a result, the ATSC mandated straight wings for the new airplane. (Swept wings were included in the design for the later XS-2.)

Additional points of agreement among the conferees included: that based on the previous day's disagreement over rocket versus jet power, an alternative propulsion systems analysis would be conducted;

the AAF would provide the crew and the facilities for the rocket fuel operations; maximum power should be available for two minutes; and the AAF would supply the pilots. In discussing the characteristics of the airplane, the conferees disagreed on the question of wing thickness, which was important relative to possible shock stall in the transonic zone (conventional planes were then built with 15 to 18 percent wings). The NACA recommended a thin 12 percent thickness chord airfoil for the wing, but the ATSC objected. The groups finally agreed that an attempt would be made to utilize a new thin wing/tail combination on the airplane. As a result, the conferees recommended that the plane would possess a 10 percent wing and 8 percent tail thickness chord airfoil.

A great deal of discussion centered on the special problems the tail presented. It was decided that the tail would be of conventional configuration of three surfaces, with conventional controls. However, it was stipulated that the tail would include one special feature, an arrangement for adjustment of the horizontal stabilizer. (The final success of the program would eventually rest on this decision.) Last, the conferees agreed the airplane's major components should have significant interchangeability for replacement purposes.

As soon as the December 13–14 meeting ended, the NACA and Bell Aircraft began work on the new requirements. On December 18, the NACA High-Speed Panel gathered to review the results of the recent Langley meeting. The NACA team agreed to provide Bell Aircraft with a new set of specifications for a transonic airplane so that the company could prepare a detailed design. It would be a government-controlled aircraft, with a flight duration of two minutes at 35,000 feet; ground takeoff would be by dolly and landing would be by skid. Already, the NACA had completed (as best they could) a supersonic wind-tunnel model test. Further tests in the subsonic eight-foot high-speed tunnel were planned to assist in determining, as nearly as possible, the practical flight characteristics of the design. The panel also discussed the ATSC offer to loan a specially prepared P-80 for high-speed data collection. The AAF had planned to utilize the P-80 for speed runs at its Muroc Dry Lake research facility.

On December 20 and 21, Major Kotcher and Bell Aircraft representatives Bob Woods, and engineers Benson Hamlin, Paul Emmons, and Vladimir Morkovin met at Wright Field. Agreement was reached on the following characteristics for the new plane:

1. The airplane should be practical.
2. The plane should be capable of carrying 500 to 1,000 pounds of instruments.
3. The plane would be for transonic research but, if possible, supersonic flight would be considered.
4. The airplane should be capable of using a 7,000-foot runway. It would take off in 5,000 feet at 150 MPH and land at 90 MPH.
5. The pilot would be seated, not prone, in the plane.
6. The plane should be capable of 10 minutes of flight to allow it to reach 35,000 feet.
7. The airplane should be constructed within one year.

Woods informed the ATSC that Bell Aircraft would build such a plane. The question was, Woods asked, "who would fly it?" Bell Aircraft would shortly reveal that it had another idea beyond the scope of the December 14 agreement on AAF pilots.[18]

The conclusion of the December 20–21 meeting did not find everyone in agreement with the stated plan. John Stack continued to express dissatisfaction with the choice of a rocket powerplant and other program details. Since the AAF was paying the bills for construction of the supersonic aircraft, it was logical that they would have the final voice regarding the plane. But Stack believed reliance on rocket propulsion would provide ample possibilities for failure. Although defeated at Wright Field, Stack had a fall-back plan. In March 1944, the NACA had detailed Milton Davidson to work with the Navy's BuAer on their supersonic flight program. Over the summer the NACA and the Navy continued to view opportunities for a new research airplane. On September 22 these plans came together when 1st Lt. Abraham Hyatt, a Marine Corps officer, proposed by memorandum that the Navy pay for a research airplane to provide information on the issues of transonic flight to include drag, flight loads, and stability and control problems. On December 19, BuAer issued an intent-to-procure letter that provided Stack with the airplane he really believed was needed for transonic research. The Navy outlined the basic requirements as provided in close consultation with the NACA: jet propulsion; minimum high speed of 650 MPH at sea level; landing and takeoff ability; thin 10 percent thickness chord wings; complete instrumentation for flight research; and excellent low-speed handling to provide data for future production aircraft. The Navy selected Douglas Aircraft as its prime contractor for the new airplane. The proposed NACA/Navy research plane presented a significant departure from the proposed Army aircraft. Aside from jet power, the new plane seemed to be directed more toward research in the transonic zone up to supersonic speeds rather than a true supersonic aircraft.[19]

Although no formal Wright Field/Bell agreement had been signed by the beginning of 1945, it was quite apparent that Bell Aircraft was moving rapidly to close a deal for the Army's new transonic aircraft program. Bell Aircraft focused on three simple objectives in their review of a potential aircraft design.[20] First, the plane should have more control power than any conventional airplane of that time. Second, the plane should be more rugged than any model currently in existence. Third, the plane should possess sufficient thrust to allow it to pass through the speed of sound as rapidly as possible. To achieve those goals, Bell engineers spent the remainder of December and the beginning of January meeting with the ATSC and NACA personnel to discuss the design criteria for the new aircraft. The NACA urged that the design be kept as simple as possible with total flexibility to research differing powerplant and aerodynamic characteristics. However, the NACA again cautioned that someone must decide if this was a subsonic or supersonic flight program since the two types most probably would not possess the same characteristics. Second, the Bell Aircraft/Army meetings at Aberdeen Ordnance Proving Grounds resulted in a close look at the behavior of a .50-caliber bullet in flight. As the projectile was known to be stable in supersonic flight, it was believed its shape might be a suitable candidate for possible aircraft design application. But by the middle of January the Bell Aircraft team still knew very little about what the design should incorporate. Most believed their own ideas were as solid as any from the various meetings.

By the end of the winter, Bell Aircraft had completed its initial design for the new aircraft. Between March 15 and 17, 1945, a conference attended by the NACA, Bell Aircraft, and the ATSC was held at Wright Field.[21] The purpose of the meeting was twofold: to review the Bell Aircraft proposal to build a transonic research airplane; and to discuss the outlines of the formal test program. Present at the meeting were Larry Bell, Bob Woods, Robert Stanley (Bell's new chief of engineering), and the other Bell engineers involved with the program. Bob Stanley was already well known to Wright Field personnel as the former chief test pilot and director of research for Bell Aircraft, now serving as chief engineer for the corporation. Stanley possessed a distinguished resumé of accomplishments in the aviation field, including helping to design planes for Douglas Aircraft while still at California Institute of Technology. He later became a naval aviator, America's national sailplane champion, and in October 1942 at Muroc Dry Lake, the United States' first jet pilot. Acknowledged as a brilliant and talented aeronautical engineer, Stanley was a driven, complex person. Few people ever saw him relaxed. The ATSC was represented by Lt. Col. H. Sibert, Capt. David Pearsall (soon to be XS-1 project officer), and other officer and civilian advisors. Because of John Stack's absence overseas, the NACA was represented by John Becker, who possessed extensive knowledge of wind-tunnel testing procedures. A large part of the discussions were to focus on the latest NACA wind-tunnel test data and a proposed wind-tunnel test program to accompany the MX-653 project (as the MX-524 project was now to be designated).

During the initial conversations, the Bell Aircraft team unveiled a sketch of its proposed airplane. The experimental craft was to have a 28-foot wingspan and weigh 13,550 pounds loaded (65% of takeoff weight would be fuel). The proposed rocket engine would have a combined thrust of 6,000 pounds. (It was unclear at this time which Reaction Motors Inc. [RMI] engine combination currently under development for the Navy would be ready in time for the tests.) The fuel would consist of liquid oxygen–alcohol–water, with a turbine-driven pump to provide the 300 pounds per square inch pressure for the fuel system. If such a pump was not ready by the time the airplane was ready to fly (within six months), Bell Aircraft would then substitute a pressurized fuel system and accept a performance loss due to the increased weight of the system. The fuel capacity of the plane would provide a 4½-minute flight duration of which 1½ minutes would be available for level-flight high-speed test runs. Wing loading at takeoff would be 93 pounds per square foot and at landing about 35 pounds per square foot. The Bell Aircraft design incorporated significant NACA design information.

During the follow-up discussion, considerable conversation focused on the fall-back plan to use pressurized fuel tanks if the turbine pump was unavailable at flight time. Several participants expressed concern that in the event of a crash the pressurized tanks would ensure an explosion. Speaking for Bell Aircraft, Bob Stanley responded that this situation would also hold for a plane using the turbine pump. In fact, the Bell representatives were more concerned with the safety and reliability of the rocket motor than with the dangers of the pressurized fuel system. They indicated their desire to thoroughly develop and test the engine at Bell Aircraft before any flights. The Bell team recommended that the AAF contact the Navy and RMI to obtain all possible information on the current development work. After review, Bell Aircraft suggested the AAF might also wish to undertake an engine development program.

When the conferees focused on the actual performance of the airplane, they quickly discovered a problem. The Bell representatives advised the conference participants that the proposed design would carry a total fuel load of 6,740 pounds of propellant. As fuel was consumed by the aircraft in its climb to altitude, the decrease in overall weight of the plane would increase its rate of climb and allow additional level-flight research time. Thus, straight calculations of fuel/flight time were somewhat misleading. But the bottom line was there would not be any possibility of 10 minutes of flight capability as stipulated at the December 20–21 meeting between Bell and ATSC. Bell engineers indicated their calculations suggested a theoretical top speed of 800 MPH (Mach 1.20). However, the NACA's Becker indicated that his calculations, based on the NACA test data, revealed that a top speed of Mach 1.0 was probably the best the aircraft was capable of achieving in level flight. The ATSC requested that Bell Aircraft rework its calculations using the NACA data. The debate was not entirely academic; in effect, the NACA data indicated the plane was only marginally capable of supersonic speed. This NACA conclusion would reappear on numerous occasions during the flight program and would have continuing negative consequences for NACA's attitude about the Bell airplane's potential.

In view of the reduced performance estimates of the plane, the ATSC now asked whether the proposed design was acceptable. The Bell Aircraft team stated the plane had the maximum possible performance capabilities with the means of propulsion that were likely to be available in the foreseeable future. The only other alternative was to test the plane in a diving configuration, which was a situation the Bell plane design had initially been selected to avoid. Perhaps sensing the program was in jeopardy, Becker then stated the existing Bell proposal represented a significant leap forward in aviation technology over anything currently available. Since the "primary objective" of the new plane was supposed to be to provide data in the range of Mach 0.80 to 1.0, Becker added, it appeared the Bell design would meet the project specifications. With that seemingly supportive statement, the seeds of future disagreement were planted. After additional discussion, the ATSC accepted the Bell proposal based upon the potential design advances incorporated in the plane. Bell Aircraft was not required to guarantee a high-speed performance figure. However, it certainly appeared that Major Kotcher's supersonic plane had just become, at best, a transonic one.

In turning to the flight test program, Bob Stanley indicated that Bell's chief test pilot, Jack Woolams, had expressed a strong desire to fly the plane.[22] Woolams was already well known to the ATSC personnel as an experienced and very capable test pilot. Born in 1917, he grew up leading a very active life, playing football and baseball and running track in high school, as well as punching cattle and breaking horses on his father's ranch. After high school Woolams attended the University of Chicago for two years before joining the Army Air Corps. Obtaining his wings in 1938, he served one and one-half years active duty before returning to the University of Chicago where he received his degree in economics in June 1941. Shortly thereafter, on June 19, Woolams joined Bell Aircraft Corporation. Within three months he moved from production test flying to the experimental research department under then chief test pilot Bob Stanley.

In his short aviation career, Woolams managed to achieve a number of notable firsts and to demonstrate excellent flying skills on numerous occasions. On September 28, 1942, Woolams became the first man to fly a fighter aircraft nonstop across the United States when he piloted a specially prepared Bell P-39 from Los Angeles to Washington, D.C., in 11 hours. In 1943, Woolams spent eight months at Muroc Army Air Field in California testing the XP-59, America's first jet aircraft. During the summer, Woolams set a new American altitude record (unofficial because of the war) of 47,600 feet, beating the old record of 43,000 feet. Previously, he had established a world's record in going to 46,000 feet in an XP-59 without a pressurized cabin or heater. In 1944, he became chief test pilot for Bell Aircraft Corporation. Known equally for his outstanding skills as a pilot and as a fun-loving prankster, Woolams was also seen as the epitome of the breed of postwar test pilots made famous by Tom Wolfe in *The Right Stuff*. Business was business, but business was such a pleasure to Jack Woolams.

In moving to a discussion of where to test, Stanley stated that Bell Aircraft would like to fly the test flights at or near its Niagara Falls plant. The ATSC expressed agreement with the Bell proposals and indicated that an additional contract would be issued to cover the initial acceptance flight tests. (This agreement represented a substantial departure from the conversations of December 1944.) Both participants, however, expressed a strong desire for the NACA to participate in the entire program, from design through to a proposed flight test program and data collection on the flights. Becker stated that the Langley Laboratory would prepare a detailed flight research program to be ready for Bell Aircraft's March 26 visit to the Virginia research facility.

The rest of the meeting concerned the NACA wind-tunnel test program. After "considerable discussion," a three-phase program was agreed to and specific project milestones were set. Bell Aircraft agreed to build the models to be tested in the wind tunnels.

On March 16, 1945, an official contract was signed between the ATSC and Bell Aircraft. The contract (W-33-038-ac-9183) specified the construction of three experimental aircraft designed to explore transonic research issues. Total initial estimated cost plus change orders was $4,278,537. The Army assigned three serial numbers (46-062, 46-063, 46-064) to the aircraft. The aircraft program, initially designated MX-524 and now changed to MX-653, retained an internal Bell Aircraft identification as Model 44. It was to remain exclusively identified as the MX-653 project throughout the summer and fall of 1945, until the airplanes themselves were designated as XS-1 (Experimental Supersonic Contract #1). On April 12, ATSC notified all involved parties that the MX-653 program was to be classified confidential and references to it were to be handled accordingly. The ATSC further noted that all performance data in regard to the program was to be classified as secret.[23] Both classifications were to experience some unique twists during the next two years.

As Bell and the ATSC finalized their arrangements for the new airplane, another actor in the process was moving to keep open its preferred option. In February and March, 1945, the Navy and Douglas Aircraft were also finalizing their research project. Over two months, Douglas gave the BuAer its proposals and drawings for the newly designated D-558 high-speed aircraft. In an April 13 submission to the BuAer, Douglas engineers outlined its entire proposed program for the Navy's transonic research. The first phase included six airplanes for research up to Mach 0.89; in phase 2 Douglas would modify two planes for increased turbojet and rocket power for research up to Mach 1.0; and finally, in phase 3, Douglas would provide a combat aircraft mockup. Phase 1 delivery would commence within one year of contract at a total cost for the six aircraft of $6,888,444.80. On May 9, the Navy approved the Douglas program and issued a letter of intent on June 22, 1945.[24] The Navy program had started and NACA now had an alternative if the Army program continued to disappoint them.

Building a Program

During late March and early April 1945, a continuous series of meetings was held among the NACA, Bell Aircraft, and the ATSC to ensure complete agreement on the technical requirements for the experimental transonic aircraft. A proposed MX-653 flight program was prepared on March 30 by the NACA, and formed the basis for subsequent discussions throughout the summer and fall. During the summer, Bell Aircraft's Bob Woods and Jack Woolams traveled to Europe to look at the information recently captured on the German aviation industry. The Bell duo found that German scientists had gathered significant information on the flight regime only up to the area of Mach 0.90; above that remained an area of unknown. Germany, however, had done more than simply analyze theoretical data. The Nazi aviation industry had produced and deployed the Me-163, a swept-wing rocket-powered aircraft, for combat in Europe. This revolutionary airplane provided valuable insights, until then unavailable, on high-speed aerodynamic characteristics. German scientists informed Woods that swept-wing aircraft would allow passage through the transonic range without the dangerous trim changes probable with conventional aircraft. Woods was impressed with the German information and believed it would provide useful data for the Bell Aircraft development efforts.[25]

By April 16 Bell Aircraft was able to inform the NACA and the ATSC that the mockup of the MX-653 airplane was ready for an informal inspection. Roy Sandstrom, Bell's project (chief of preliminary design) engineer for the MX-653, requested that the ATSC at Wright Field inspect the mockup before Bell Aircraft proceeded with the actual construction. Because of the unusual nature of the project, Sandstrom cautioned, this inspection would not be like any previous formal mockup inspection. The new MX-653 model consisted of the fuselage from the wing trailing edge forward, including the cockpit controls, landing gear, and instrumentation. Additionally, Bell Aircraft advised the ATSC of its conversations with RMI, designated as the builder for both the rocket motor and the turbine pump. RMI continued to target July 1 as the completion date for the necessary powerplant equipment.[26] However, that date was to slip continually as the completion deadline approached.

The XLR-11 powerplant work performed by RMI under separate contract with the Army (originally for the Navy) called for the construction of four 6,000-pound-thrust rocket motors. Concurrently, RMI was attempting to design a promising new highly efficient turbine-driven dual-propellant pump to work with the powerful rocket engines. Unfortunately, the new pump was not scheduled to be ready until September 1, or three months after delivery of the engines. Since Bell engineers anticipated completion of the MX-653 airplane by September, they feared that any additional delay in the RMI schedule would complicate the flight test program. Thus, they decided that the MX-653 be designed so that the pumping unit plus fuel tanks could be replaced by pressurized fuel tanks. In recognition that this was probably going to occur, the Bell Aircraft engineers decided that the first MX-653 would rely on pressurized tanks as the risk was simply too great that setbacks would occur in the RMI project.[27] Unfortunately, the proposed use of pressure tanks would seriously erode the already badly degraded MX-653 flight time. There were also other consequences from this engineering decision, as will be seen later. History has, however, revealed how wise Bell Aircraft was to make this painful decision so early in the program.

From April 26 through 28, the NACA team visited Bell Aircraft's Niagara Falls facility to view the mockup.[28] The group from Langley included a large contingent of engineers to discuss operational questions with their Bell counterparts. The first day's meeting opened with a briefing by Bob Stanley. After viewing the mockup and holding detailed technical conversations between engineers, the two groups gathered to discuss the key issues of the program. Specifically, Stanley informed the LMAL team that Bell Aircraft believed the RMI fuel pumps would be unavailable in time for flight tests. Thus it was necessary for Bell to continue construction of the MX-653 airplane to incorporate high-pressure nitrogen gas tanks to force-feed the liquid oxygen and alcohol into the engine. This pressurized system would deliver approximately the same volume of thrust, but it would only provide about one-half the powered flight duration time (about 2½ minutes) of the proposed RMI pumps. The problem with the pressurized fuel system was the heavier weight and volume of the twelve nitrogen tanks and the high-strength steel fuel tanks necessary to handle the internal pressures. Not only was vehicle landing weight increased by 2,000 pounds, and wing loading significantly increased, but the fuel capacity was reduced from 8,160 pounds to 4,680 pounds. As a result, Bell Aircraft indicated, the first flights would probably occur at either Muroc Dry Lake or Wendover Field, Utah, because both airfields exhibited higher elevations, which would assist in flight testing the now limited experimental plane's endurance. The NACA group confined most of their comments at this time to relocation of some of the test instrumentation. On April 27, a sizable team of people from the ATSC arrived to be briefed on the project. The review process and a summation of the agreements between the NACA and Bell Aircraft were reviewed for the ATSC personnel.

Everyone in the MX-653 program was concerned that the smaller fuel loads meant correspondingly shorter flight duration times. If ground-launch of the aircraft was utilized, it would require substantial fuel use simply to climb to a test altitude. The technical inability to reach 35,000 feet would reduce the possibility of piercing the sonic barrier. While a flight could be conducted at lower altitudes, it was only above 35,000 feet that the MX-653 could fulfill its basic performance requirements. Currently, a P-80 aircraft in a shallow dive could already provide some useful information near the transonic range and Langley Laboratory had recently demonstrated wind tunnel capabilities up to Mach 0.93. If the MX-653 could attain only similar speeds, then the project was merely an expensive novelty. Further, research time at level flight speeds would be adversely affected since the total testing time would remain very limited even if the experimental aircraft was carried aloft for launch. On May 1, John Stack advised LMAL's chief of research that the new plane "may prove quite unsuitable."[29] The airplane was in danger of failing to meet the minimum NACA requirements. Although the lower fuel volume would allow low-altitude research flights, it was only above 35,000 feet that the new airplane was expected to obtain "speed greatly in excess of the critical." Thus the continuing problems encountered by RMI with the turbine pump threatened to disrupt the entire flight test program. Either a suitable pump had to be found or a different method of flight research for this project had to be adopted. Stack attempted to settle the issue by asking Crowley to urge Wright Field to seek other engineering organizations in addition to RMI to develop the turbine pump.

The difficulty over the turbine pump was not the only bad news for the NACA during the month of May. On May 31, Col. George F. Smith, chief of the service engineering subdivision of the Engineering Division at Wright Field, wrote to Dr. Lewis at NACA HQ to advise him that the AAF welcomed the NACA assistance on the MX-653 program.[30] The NACA cooperation had been conveyed to ATSC on May 10, 1945, in a letter entitled "Flight and Wind Tunnel Investigation of Aerodynamic Characteristics of Bell MX-653 Transonic Airplane." But, in further remarks, Smith provided potentially troubling news for the possibility of flight testing at LMAL. Smith informed Lewis that while it was desirable to "conduct tests in the East, it appears that the extremely high take-off speed and the probability of power-off landings *will dictate* that the initial tests be made either at Muroc, California or Wendover, Utah, where large expanses of flat terrain exist" (emphasis added). This was not the only bad news for the NACA. Smith concluded his letter by stating that "it has been the intention of the Air Technical Service Command to place a separate contract with Bell Aircraft Corporation for the MX-653 research flights." This contract would be executed after the acceptance flights as ATSC believed they would have a better understanding of the costs and the location for the tests after they had seen the performance of the airplane in the initial contractor tests. This very important letter would have significant repercussions for the program. If this was the outline of the project as of June 1, 1945, the plan was soon to be subject to major alterations within a matter of months.

While the potential technical delays did not seem decisive to Bell Aircraft, events in progress were working to confine the MX-653 program flexibility within very narrow parameters. To a degree, this could be anticipated, as schedules for new aircraft projects seldom went as planned. Many times these delays occurred for reasons not likely to be foreseen. On June 19, Bell Aircraft's Stanley Smith, the new MX-653 project engineer, reported to the ATSC Wright Field that while the Bell Corporation was impressed with the recent progress at RMI on the rocket engines and turbine pump, Bell still was not sure that RMI's completion date would meet the MX-653 needs.[31] This statement came as no surprise at Wright Field and adjustments by Bell had already been considered—specifically, the pressurized tanks. However, no one at Bell Aircraft had anticipated the AAF's cancellation of the corporation's P-63 fighter plane program, which required Bell to lay off valuable experimental shop employees working on the MX-653 project so that more senior employees from the P-63 program could retain their jobs. Smith estimated the retraining time alone would mean a slippage of one month (to October 1) in the scheduled completion of the MX-653.

While these delays continued, Bell Aircraft notified the ATSC that it would follow up Wright Field's suggestions regarding remote base locations. Bob Stanley and Bell test pilot Alvin "Tex" Johnston would tour Muroc Dry Lake and Wendover Field as potential test sites. This tour was in keeping with the discussions held at the MX-653 mockup inspection on April 26.[32] Both airfields possessed a higher elevation (approximately 2,300 feet for Muroc and 4,500 feet for Wendover above sea level) than many other airfields. This fact would assist in addressing the XS-1's short flight duration time and the desirability of obtaining higher altitudes for increased speed numbers. Bell Aircraft notified ATSC it was anticipated that Stanley and Johnston would leave around July 1, and that they would make a test-site recommendation to Wright Field by July 15. Unmentioned in the conversations was the original Langley Lab proposal to test at its airfield.

Although official ATSC notification by Bell Aircraft that it would avoid testing the XS-1 at Langley Field only occurred on June 19, knowledge of Bell's intentions had a way of preceding formal channels. The decision was certainly not one the NACA desired to hear. The idea of testing the MX-653 at either Muroc Dry Lake or Wendover Field met with unfavorable reactions from the NACA HQ and, as might be expected, the project officials at Langley were particularly upset. On June 14, John Stack vented these feelings in a interoffice memo to John Crowley, the chief of research at Langley Lab.[33] In summary Stack rebutted the suggestion that the aircraft be flight tested at a western location.

> This airplane originated here as did the P-80 program. If we are to do research of this kind we must have the airplane here. I do not believe we should again be treated as a service organization as was the case with the P-80. If shifting of this aircraft to a western station materializes I propose that we transfer *all work* beginning right now so that we can free our people to do research in our present equipment [emphasis in original].

On June 19, Crowley replied to Stack in an interoffice memo outlining his views on the question of location for the test flights.[34] In summary, he indicated that in spirit he agreed with Stack's comments. Testing at Muroc Dry Lake had not eliminated accidents in past test programs. Although Langley Field might not be suitable, there still existed nearby Cherry Point Marine Corps Field, North Carolina, which possessed a sufficiently long runway to handle proposed MX-653 requirements. Bell Aircraft had originally indicated a desire to test the new aircraft at Niagara Falls, but in any event, Crowley reminded Stack, the NACA could always make additional arrangements after the initial acceptance tests. In the short term, Crowley stated, no decisions needed to be made simply because Bell Aircraft was only to perform the agreed contractor tests before the operational transonic tests would occur. But while Stack was being rebuffed in one area, a key NACA concept for the new experimental plane was receiving acceptance by Bell Aircraft.

The NACA engineers had long accepted the concept that slender wings exhibited low compressibility effects.[35] Studies done as far back as the mid-1930s had revealed important information on flow characteristics of low aspect ratio wings. In early 1945, LMAL scientist Robert T. Jones refined the NACA data to demonstrate the ability of sweptback wings to lower the effective Mach number (that is, the airplane's actual speed would be greater than the Mach effects on the wing. Seemingly, no one working at Langley remembered Busemann's concept unveiled at the Volta High-Speed Conference). Although sweptback wings promised significant advantages in efforts to achieve supersonic flight, no mention of the concept occurred at the ATSC meeting in March that resulted in the ATSC/Bell contract for the MX-653. However, by June, Bell had decided to accept the NACA recommendation regarding the use of very thin wings. Thin wings would minimize the buffeting, loss of lift, and control problems anticipated near Mach 1.0. In mid-1945, NACA had developed two possible concepts for the type of thin wing to be utilized on the new airplane. John Stack recommended a 12 percent thickness chord wing as a good all-around compromise among landing efficiency, penetration into the supercritical region, and strength. Another LMAL engineer, Robert Gilruth, recommended a greater margin of safety for the first airplane. As a result, Gilruth and his team proposed a very thin wing of 5 percent thickness chord. The disagreement was resolved by LMAL's assistant chief of research, Floyd Thompson, who reviewed the data and decided to err on the side of safety by recommending Gilruth's thinner wing. However, when Stack proposed that two sets of wings with differing thicknesses be procured for the experimental airplane a compromise was reached. The thickness of the wings would be increased from the very thin recommendation of Gilruth, but not to as great as that proposed by Stack.

A similar debate raged over the thickness for the tail airfoil. If the horizontal tail airfoil was the same thickness as the wing, then compressibility effects would be reached on both simultaneously, with the possible result total loss of airplane control. To avoid that circumstance, the NACA proposed another compromise. The tail airfoil would be of a different thickness chord than the wing and would feature a totally adjustable stabilizer to prevent the anticipated loss of elevator effectiveness due to the formation of high-speed shock waves.

On June 20, the NACA formally replied to the ATSC with recommendations for the wing/tail for the MX-653. Bell had inquired of LMAL on June 6 as to what should be the thickness for the alternative wing/tail sections. A 10 percent thickness chord had already been selected for the initial MX-653 airplane. Although Bell suggested either a 8 percent or 6 percent thickness chord, the NACA countered that same day with a recommendation for a 12 percent and an 8 percent thickness chord wing. On June 11, Major Kotcher telephoned the LMAL ATSC liaison office to inquire why NACA still recommended a 12 percent thickness chord wing in spite of the severe performance penalties with such a conservative approach. The June 20 Reid letter to ATSC attempted to outline the case for the more conservative approach. Ironically, the data relayed to Kotcher made a strong case for the higher performance potential of the thinner wing. These conversations continued into the fall. At that time, Bell authorized the 8 percent airfoil and canceled the 6 percent thickness chord straight wing in favor of a sweptback wing (for the proposed XS-2).[36]

While the technical details of the program continued to be developed, the broader aspects of high-speed flight also received attention. On September 6, Brig. Gen. Laurence C. Craigie, chief of the Engineering Division at Wright Field, convened a special conference on continuing the development of a high-speed research program. This was a follow-up to the initial work agreed on in November 1944 regarding the potential of jet-powered bombers. While acknowledging the excellent work done during the war, Craigie stated that the ATSC "had long felt the need for additional research on high speed aerodynamic phenomena." In the past, wind tunnel testing had focused on then-current AAF aircraft. Certainly such a policy had won the war. But now a larger effort was needed. The NACA was authorized to undertake a six-month lengthy series of tests. Although the conference certainly led to increased cooperation among the NACA, industry, and the ATSC, the wording of Craigie's opening remarks could be interpreted by some as an acknowledgment that NACA was not capable of handling all aspects of a supersonic research program. Indeed, the XS-1 wind tunnel model did not even arrive at LMAL until March 1946 and results of the NACA testing appeared simultaneously with the ongoing Muroc flight test program. In October, the AAF HQ would pointedly ask NACA why no mention had been made of swept-wing potential for the XS-1.[37]

On September 14, Henry Pearson and William Gracey, both of the NACA Langley Lab, visited Bell Aircraft Corporation to discuss the status of the MX-653 project.[38] By this time, the aerodynamic issues had been settled and the main activities on the aircraft now focused in the construction shop. During that visit, Pearson and Gracey met with Stan Smith. The Bell project engineer indicated the first MX-653 flights were anticipated around November 1 and would consist of low-speed glide tests to check stability and control characteristics. At this time, Smith stated, it was difficult to visualize the RMI rocket motor being ready anywhere near the expected completion time for the aircraft. The urgency of proceeding without engines for the MX-653 was in line with the ATSC's interest in putting the project in motion as rapidly as possible. Smith further advised the NACA personnel that Bell Aircraft did not intend to proof-load the first aircraft. The second airplane, calculated to be ready by January 1, would be proof-loaded to about 50 percent of design data for the NACA tests. Thus, the strain gauge installation could not be calibrated for the first tests. Pearson disagreed with this procedure since it rendered NACA participation for this type of data meaningless. He informed Smith that the NACA intended to recommend such a procedure to the ATSC. Finally, Smith outlined what he hoped would be the ultimate program for the three experimental aircraft. Smith stated that Bell Aircraft desired to operate one airplane, let NACA have another, and allow the ATSC to fly the third. For that reason, three sets of NACA instruments should be readied. All of this was news to the NACA team.

Smith further indicated that Bell Aircraft was considering operating the MX-653 from a Boeing B-29 mother ship to increase flight duration time. The issue of air- versus ground-launch for the MX-653 airplane was simple: if it was ground-launched with a full load of fuel, the resulting wing loading stress on the aircraft would present serious hazards for the first few seconds after takeoff. By utilizing an air-launch technique from a B-29, the danger was reduced since the pilot could, in the event of problems, use the higher launch altitude as a safety cushion to either jettison fuel or bail out. Smith also reiterated that an air-launch procedure would tremendously increase MX-653 performance by lengthening flight time by a factor of 4. The issue of flight duration time was not a new problem, having been discussed as far back as the April 26 mockup review. This proposed drop-launch technique avoided the problem of MX-653 fuel capacity by simply taking the experimental plane to a higher altitude and letting it use its fuel primarily for powered level flight runs at the transonic and supersonic regions. During World War II, Germany had successfully utilized a piggyback technique to allow two aircraft to go aloft to attack Allied bomber formations. Recent static tests and research drops using weighted models had seemed to indicate the feasibility of this procedure. Nor was this the first time air launch had been recommended for the MX-653. Prior to the end of World War II, Bell had considered the prospect of towing aloft the proposed experimental craft. However, the concept had been discarded due to the difficulty in obtaining a B-29. Now, with the war in the Pacific over, this was no longer the case. With B-29 aircraft being brought home simply to be junked, Bell Aircraft believed one should be readily available for this purpose, although acquisition could take some time. The acquisition process in itself, however, was cause for some alarm, given the turnaround times being discussed for the MX-653 testing. But what seemed logical to Bell Aircraft was not what the NACA expected to hear. The air-launch technique added additional complications to the project: complications rejected as far back as the original McDonnell Aircraft proposal. For that reason, Paul Emmons advised the NACA personnel that he thought the air-launch concept was still only a distant possibility. But it was not past rejections that the Langley team now wanted to consider.

The LMAL engineers realized that any discussion of an air-launch concept again opened the possibility that Langley Field could be utilized for XS-1 flight tests. Such a circumstance would certainly simplify the NACA work. However, Smith quickly dashed those hopes by indicating that Bell's chief test pilot, Jack Woolams, was already scouring the country looking for a potential airfield. Those sites now under consideration had expanded to include Muroc Dry Lake, Wendover Field, Salina Field (Kansas), Daytona Beach (Florida), Marietta Field (Ohio), and several airfields in Texas. Smith indicated that in his opinion Muroc was simply too far away and that he was inclined to go with Daytona Beach.

Although perhaps not as visible at the time, the interplay over the site for flight tests had more to it than project management or distance. Without the rocket engines, the MX-653 could utilize many airfields, but with them the margin of safety would require longer runways and unpopulated areas. If Muroc Dry Lake was chosen, the flight tests had to be conducted before the rainy season; these rains, which generally started in December, could leave the lakebeds covered with water for several weeks or even two months at a time. Thus the time to completion of the MX-653 weighed as much as the location of the tests. Finally, there was the issue of distance. Muroc may well have seemed too far away to Stan Smith. Certainly the NACA could readily agree with this sentiment. But there was something also troubling to chief engineer Bob Stanley about utilizing Muroc—something other than distance and primitive conditions at the base—something Stanley did not share with most of the participants. California and the western states had become home to a booming aviation industry and competition had already become fierce for flight personnel. Stanley was afraid a trip to Muroc would result in his team leaving for competing aviation companies.[39] Thus a site convenient to the eastern seaboard, with a remote locale for security reasons, would also be ideal for keeping the flight personnel involved in these tests concentrated on this program. Such a site was not yet agreed upon.

On October 10, 1945, the ATSC Wright Field and NACA personnel traveled to Bell Aircraft for a two-day formal program review. It did not take long for the policy disagreements to surface. Based upon their inspection, the NACA engineers estimated the first MX-653 aircraft was about 70 percent complete. During the following conference, Bob Stanley informed the visitors that the first flight was now expected to occur between December 15 and January 1. No powered flights would be attempted under that schedule even though the experimental Bell rocket engine had just been completed. Further, Stanley expected to have the MX-653 make between ten and twenty flights to familiarize the pilot with the basic handling characteristics of the airplane. When questioned about the NACA instrumentation for the airplane, Stanley advised the visiting group that if the NACA desired, the plane could be instrumented. However, "the Bell company did not care if no instruments were aboard for the glide flights." This statement can only have surprised the NACA since the primary purpose of their participation in the project was the data the instrumentation was to record. But even more disappointing news surfaced in response to an NACA question over the future disposition of the number 1 airplane. Stanley informed the group it was his impression that Bell would receive an ATSC follow-on contract for the supersonic flights. The NACA quickly referred Captain Pearsall's attention to the LMAL letter of May 5, 1945, to NACA HQ on the research authorization for NACA participation. Specifically, NACA indicated their acceptance that the initial test flights would be made at the contractor's plant or at a remote location. NACA would participate and instrument the airplane. But, that was with regard to ground-launch of the MX-653; now circumstances had changed. Thus, in quoting from the NACA letter:

> It is also the understanding of Langley that the airplane will be brought to Langley Field after satisfactory completion of acceptance tests and that the research phase of the flight program will be conducted at Langley Field under supervision of the staff of the Langley Laboratory.

Captain Pearsall agreed to review the documents and notify the participants of the role NACA would play in regard to the first airplane. But it was quite clear from the discussion that a major philosophical difference was beginning to crack open the cooperation everyone supposedly desired this program to exhibit. In fact, the real question was why no one acknowledged the ATSC letter from Col. George Smith of May 31, 1945. Perhaps Pearsall was unaware of it, but the letter clearly stated the ATSC intentions regarding the follow-on contract. Either Dr. Lewis chose not to inform LMAL or someone had forgotten the earlier information.[40]

On October 22, 1945, Bob Stanley paid a follow-up visit to Langley Laboratory to discuss current conditions regarding the MX-653 project.[41] Although no documents have been found relative to the purpose of the visit, it is clear from the issues covered that the reason for the meeting was to clear up the procedures to be followed for the MX-653's later flights. The large NACA contingent was led by Floyd L. Thompson. To open the meeting, Stanley indicated it was Bell Aircraft's understanding that the NACA did not care to operate the MX-653 and that Bell would receive a contract *under NACA supervision* to conduct the research flying. It was only at the October 10–11 meeting that Bell Aircraft received the message that the NACA wished to operate the plane. Now, Stanley said, Bell Aircraft needed a definite decision on the NACA plans for the transonic research flights since the corporation was about to activate an MX-653 test station. The NACA plans would directly affect the size and content of the Bell draft of personnel. Answering for the NACA, Melvin N. Gough (chief of the Flight Research Division and a respected NACA test pilot) said it was still very difficult to give any definitive answer as to the NACA plans. The NACA still expected Bell Aircraft to fly the plane to Mach 0.80. At that time, the NACA would take over the flight testing and operate the plane above Mach 0.80 "until such time as the aircraft evidenced characteristics that made it uncontrollable and extremely hazardous to fly. At that time a contract may be negotiated to fly the airplane at higher speeds." Gough specifically asked Stanley if Bell Aircraft intended to exceed Mach 0.80 during its contractor tests. Stanley replied in the negative and stated that the risks were simply too great to take such a chance.

With the biggest issue still not resolved, the meeting turned to the other issues, including the status of the rocket engine and the fuel system to be used on the experimental airplane. To the former issue, Stanley stated that the RMI engine demonstration seen on October 12 provided promise that the MX-653 had finally found its powerplant. However, the Bell rocket engine was still available on standby. To the fuel system issue, Stanley reiterated the Bell decision to provide a pressurized system in the absence of the turbine pump.

Stanley mentioned that the B-29 mother ship had arrived at Niagara Falls and Bell modifications on it would commence shortly. Initially, it had been anticipated that a Boeing B-29 constructed under license at Bell's Marietta (Georgia) plant would be the plane used as the mother ship for the XS-1 tests. Instead, a B-29 from Boeing Aircraft's Renton (Washington) facility was now found to be more suitable.[42] Thus it was a Renton B-29 (#45-21800) that was ferried to Niagara Falls for modification by the Bell engineers. The planned modifications at Bell Aircraft came as a surprise to Langley staff as it was a reversal of the procedure discussed in the October 10–11 meetings; at that time, it had been indicated that the bomber modifications were to be made at Wright Field by the ATSC. Stanley again indicated that no final decision on a test site had been made. However, Bell Aircraft had narrowed the serious choices to two locations: Orlando and Daytona Beach, Florida. Each potential site had advantages. For Daytona Beach, the hard sand beach was already known for auto racing, but security posed somewhat of a problem. For Orlando, the single 10,000-foot runway and the remote location fulfilled two prime criteria. In any event, no final decision had been made when the NACA sprang its surprise.

During the long search by Bell Aircraft for a suitable test site, the NACA Langley had never given up hope that the flights could be nearer their location. Now, they once again brought up the subject of proximity to the LMAL facilities and personnel for ease of operations. With that in mind, Mel Gough suggested the Marine Corps airfield at Cherry Point, North Carolina, with its unusual radial-configured runway 16,000 feet long and 300 feet wide.[43] Taken off balance, Stanley indicated that Cherry Point had not been considered but that he would examine the location as an option. However, one possible objection might be the intensity of training flights known to occur at that location. Further, the cold, unpredictable weather in December/January in North Carolina could present a serious problem. Such a disruption of the MX-653 schedule would certainly cause ATSC resistance since Wright Field was still pressing for utmost speed in the program. Although the decision on a site was seemingly left pending, in reality Stanley had already closed the door on any testing near Langley. Worse, the NACA had other problems at this moment that certainly provided distractions from the immediate needs of the supersonic program.

All the complications of the MX-653 program would have been manageable if taken in isolation, but in light of postwar military appropriations, nothing was taken in isolation. The necessity to sharply cut military spending and the requirement to hold the line on all government expenditures made Congress both a welcome friend and a wary opponent to the research requirements of supersonic flight. Despite the overwhelming victory achieved by U.S. military forces in World War II, Congress proved ready to significantly overhaul the research and development procedures utilized by the NACA. A special congressional committee investigation (chaired by Sen. James M. Mead) highlighted the growing criticism of NACA in view of the enormous information becoming available regarding German aviation advances. Although the Mead committee recognized the NACA's substantial contributions to the final victory, its use of such phrases as "timidity" and "lack of forcefulness" in pursuing new research programs stood as important indictments of past agency efforts.[44] At the same time, several senior AAF officers became much more open in their remarks regarding their views on NACA failures to keep pace with advances (notably jet-related) in aviation technology. As a result, by early 1945 the AAF and NACA postwar research plans began to diverge.

In April 1945, the NACA first considered a plan to construct an

"Altitude and Supersonic Research Laboratory." In November, with captured German information now readily available, the NACA began to investigate all potential sites for the construction of new supersonic wind tunnels. By December, the NACA program had expanded to include a proposed wind tunnel of enormous (four times greater than the latest German models) power requirements. The costs for the new facility would be twice as large as for *all* the other NACA facilities combined. However, while NACA was pursuing its dreams of new facilities, the AAF had also been investigating the concept of new dedicated military research facilities, intending to prevent the possibility of future failures by NACA to keep pace with the state of aviation technology. Such a program was in keeping with the recommendations of von Kármán's *Where We Stand* and supported the growing AAF belief that they and not the Navy were now the first line of America's defense. By the end of 1945, the AAF had completed its plans for a proposed $100-million air engineering development center. The new center would consist of five wind tunnels, one capable of speeds between Mach 8.0 and 10, located in the Rocky Mountains. The realization by both parties that separate programs were now in danger of becoming involved in a budgetary battle did not halt the process. Rather, the controversy in some respects served as an underlying and largely hidden motive for many of the disputes in the supersonic project. The evolving wind tunnel program proposals could not disguise the immediacy of the budgetary problems the NACA and the AAF faced in the postwar cash-strapped environment of Congress.

The war years had seen an almost fourfold increase in NACA funding.[45] (In 1941, the NACA received $11.2 million versus $40.9 million in 1945.) With the needs for continued research on the promising new technologies, many believed the NACA would be spared the deep cuts experienced by other agencies. Thus, it was with some dismay that NACA learned that its fiscal year 1946 budget would only be $24 million. Already by the fall of 1945, the NACA had seen overtime cuts and personnel reductions at all its facilities. But NACA did not intend to take the cuts without a fight. To underscore the concern, the NACA HQ participated in hearings before the House Appropriations Committee on October 29 to indicate the need for the NACA to press on with its high-speed wind tunnel work. Without such research, Russell Robinson, chief of research coordination for the NACA, indicated that the industry would be severely restricted in their design work on supersonic aircraft. The absence of transonic wind tunnel information,

Robinson testified, meant that the NACA could provide little factual research data to assist in answering those hard questions that arise in every design procedure, but that were especially critical when the work represented a leap into the unknown such as supersonic flight. The bottom line was that the NACA had the German World War II research results; it had some preliminary design data, and it had the mandate to go forward with producing a supersonic aircraft. What it did not have, and what was obviously the purpose of the appearance, was enough personnel and equipment to fulfill the existing need for research assistance. In short, the NACA needed more money.

On November 8, Stan Smith filed his latest progress report with the ATSC (renamed the Air Materiel Command [AMC] in November).[46] In the report he recommended to Wright Field that Pinecastle Field in Orlando, Florida, be accepted as the site for the initial MX-653 glide tests. Smith also stated that work on the B-29 conversion to mothership status was progressing satisfactorily and the MX-653 landing gear drop tests by the Bendix Aviation Corporation were now complete. Smith also submitted to the AMC the cost and plans for the final piece of the loading package for the MX-653: the experimental craft's loading pit expected to be dug at Niagara Falls. This simple concrete hole in the ground with wheel slats for use in lowering the airplane into the pit would allow the B-29 to move over the experimental craft, and then permit Bell technicians to simply raise the MX-653 up into the belly of the mother ship. This type of operation required an increase of approximately 60 pounds in the weight of the MX-653 to permit the installation of the necessary supports and release mechanisms for B-29 separation. If a decision was made by the AMC to test without the mother ship, then 33 pounds of the new weight could still be easily removed from the experimental craft.

Pinecastle Field had been selected for a variety of reasons. Not all were explicitly stated in any memorandum, but recorded conversations and the timing of the choice imply the rationale for the Bell decision. First, the necessity to rapidly advance the MX-653 program indicated that the tests should begin as soon as possible. The rainy season at Muroc Dry Lake rendered testing at that location very improbable during the anticipated test time. Second, Pinecastle Field possessed a single 10,000-foot runway sufficient for the glide test. Third, the remote location of the field and the continuous presence of AAF security personnel made Pinecastle an ideal location for a secret project. Fourth, a test location on the East Coast of the United States allowed convenient travel back and forth among Niagara Falls, Langley, and Wright Field. Fifth, the weather in Florida could be anticipated to be much better than most of the rest of the United States during the middle of winter. Finally, the use of an underutilized airfield and the close proximity of an already existing AMC facility (at Orlando), as well as its personnel, meant that flight time and additional help were available when needed. By the completion of the first tests, experience would alter many of these perceptions. But initially the choice seemed a suitable one for the needs of the program. However, just because a site had been selected did not mean everyone could find it.

As a result of a follow-up LMAL phone call on November 13 to Bob Stanley at Bell Aircraft the confusion over the final glide flight testing location was still apparent.[47] Mel Gough advised Langley's chief of research on November 15 that the initial glide flights would be made at the "Army Air Base, Orlando," thus seemingly confusing the downtown airfield east of the city with the more remote Pinecastle Field location some distance southeast of the metropolitan area. The tests were now tentatively scheduled between December 25 and January 25. As time was growing short, Gough believed it was best to send the Langley C-45 airplane direct to Niagara Falls to allow installation of the NACA flight recording instruments at the Bell Aircraft plant. These instruments, Gough indicated, would be ready in two weeks. This was cutting the preparation time very thin, but Gough had another objective in mind—to have the NACA team check on the progress at Bell Aircraft.

With the conclusion of the Bell phase of his memorandum, Gough turned to the issue of the upcoming tests. He recommended that Langley plan to provide its C-45 to transport a team of the NACA specialists to Orlando for the flight tests. In addition to the several technicians from the Instrument Research Division needed for the operation of the radar and telemetry equipment, this team would include an instrument mechanic to service the instruments and load and process the film. Gough further recommended that this individual carry along a significant inventory of spare parts. In addition, Gough felt that a flight engineer would be helpful in working with the Bell Aircraft pilot on the operational aspects of the flight and analyzing and processing the data. Finally, Gough proposed that the NACA have its own pilot on the scene to study and discuss the progress of the flights. This would expedite the learning process at such time when the NACA took over the operation of the MX-653. Due to the distance to Orlando (about 650 miles) and the number of NACA personnel involved, Gough concluded by recommending that the C-45 be assigned to the Pinecastle project to facilitate rapid transportation to and from Langley for technicians and any equipment, spare parts, or repairs that might become necessary.

On the same day, November 15, the AMC made official what Gough had previously outlined as the potential NACA procedure for the MX-653 project.[48] In a lengthy document, Wright Field asked that the NACA Langley take over responsibility for providing the tracking, data acquisition, and analysis personnel for the Pinecastle tests. The detailed letter from Col. George Price to the AMC liaison at Langley provided the first noted reference that the MX-653 project would have a new name. Price's letter identified the MX-653 project's airplane by its somewhat more historically familiar designation, XS-1.

The confusion over the appropriate roles for each program participant was not resolved by the conversations between Langley and Wright Field. To close the loop, on November 23 AMC Wright Field wrote to the Washington NACA director, Dr. George Lewis, to advise him of the understandings in regard to the MX-653 project. While conceding the previous desire "of all persons concerned" that the test flights be made at Langley so as to facilitate data acquisition and handling, the AMC reiterated the need for a runway of at least 10,000 feet. Since Bell Aircraft had made a "thorough search" and selected Pinecastle Field, the AMC was ready to accept this decision as the most practical location for the tests. The AMC still expected the tests to go forward in December and January. While Bell Aircraft would be responsible for the preliminary flight test program and would furnish the pilot, the AMC requested that the NACA cooperate in planning the general flight test program and supervise the data collection through instruments supplied by that agency. An NACA pilot was cleared to fly the airplane "as fast as it appears safe." Having given the bitter medicine, the AMC provided the prize by holding out the prospect that the purpose of these glide tests was to "prove it practical to make the acceptance tests and flight research tests *at Langley Field*" (emphasis added). Finally, the AMC recognized that the continuing difficulties with the turbine pump would necessitate the program moving forward without waiting for its final development. As a result, the NACA would receive two aircraft "with pressure tank fuel systems for the duration of the flight research program."[49] In summary, the AMC had run out of time and patience and expected the NACA to live with the results. The NACA would not see its role in that light, as would soon become obvious.

In the December 6 Bell Aircraft Corporation progress report, Stan Smith notified AMC that conversion work on the B-29 mother airplane was now complete.[50] The modified bomber was ready to serve as a bus to carry the XS-1 to its necessary drop altitude. To achieve this objective, the standard B-29 supplied by the AAF had its bomb-bay doors removed. Cuts were made into the metal of the fuselage to accommodate the XS-1's wings and nose. As a result, the structural integrity of the B-29 was weakened. To avoid any rigidity problems, a steel rib was inserted running the full length of the bomb bay and beyond so as to stiffen the B-29 platform. Within the aircraft, a D-4 shackle was retained in the bomb bay from which the XS-1 could be suspended. An adjustable ladder was attached to the bomb-bay wall to assist the XS-1 pilot in aircraft entry during the B-29's climb to drop altitude. Further changes included removing the main landing-gear doors to provide additional ground clearance and replacement of the gun turrets.

But Smith's good news was again balanced with problems in the delivery date for the NACA to install their instruments (now December 12) and the difficulties regarding the acceptance tests for the RMI engine. The tests were started during the week of November 17, but immediately suspended because of an explosion in the number 2 chamber. The following week, additional problems with surge in the lox and fuel lines caused further delays. Unfortunately, the fix required additional work in the cockpit to ensure the pilot could handle the problem in flight if necessary. Although progress had been made, the rocket engine was still not ready.

With the completion of the XS-1 near at hand, the Bell Aircraft contingent for the series of flight tests was also made ready. Bob Stanley would be in overall charge of the initial XS-1 test effort. It was assumed at Bell Aircraft that Jack Woolams would fly the XS-1 up to the delivery specifications and later, if Bell got the follow-on contract, the flights through the sound barrier. Woolams would be in charge of the test station personnel at Pinecastle. Stan Smith would continue as project engineer during the Pinecastle tests. To man the B-29, Stanley selected Joseph Cannon and Harold Dow to alternate as pilots. Both Cannon and Dow were experienced test pilots in their own right and had been with Bell Aircraft for several years at the B-29 production facility in Georgia. Since the early fall they had been back in Buffalo in anticipation of the XS-1 test flights. Accompanying them on the B-29 would be Bill Miller as flight engineer, Ivan Hauptmann as crew chief, and Frank Nicholas as assistant crew chief. Forming the XS-1 ground team would be Charles MacLean "Mac" Hamilton as crew chief, and assistants William Means and Herman Schneider.[51] Together they represented an experienced team of skilled aviation technicians. If the XS-1 program had a problem, they would find and fix it.

The Bell XS-1

On December 12, with a blanket of snow on the ground, the finished XS-1 received its unofficial rollout from Bell Aircraft's Niagara facility.[52] Photos taken at the time reveal that neither the striking saffron-colored paint nor the tail identification number, 6062, had yet been applied. The experimental craft measured 30 feet, 11 inches in length. In height it was just over 10 feet (10′ 10″). Wing surface area was only 130 square feet. The extremely aerodynamic shape of the airplane featured a conventional semimonocoque aluminum alloy stressed skin construction.

Structural design criteria for the XS-1 was a matter of calculated guesswork and hence arbitrary at best. There simply was not enough research data in 1945 on the anticipated stress loads and possible buffeting to reach supersonic flight to give certainty to the numbers used in design. No airplane had ever done what this plane was being designed to do. As a result, it was decided to err on the side of safety in construction. The NACA provided data to Bell Aircraft that led to the decision to design a maneuver load factor for the airplane of 12g limit and 18g ultimate. These figures were based on weight at the beginning of the high-speed run and at the top of the climb (8,400 pounds). In simple terms, the XS-1 had to be capable of supporting 18 times the weight of the fuselage and its contents without breaking apart under dynamic flight stress. This design load was more than twice the 7g stress factor anticipated as a maximum load in a conventional fighter. The Bell engineers realized that it would be very difficult to generate these types of load in the thinner air where the XS-1 was to be tested. However, the extra margin of safety proposed by NACA was designed to meet the stresses of the unknown turbulence anticipated as the plane passed through the sonic barrier. It proved to be a major example of overdesign.

To meet the high-strength requirements of the 18g loads, the wings for the XS-1 were machined out of solid aluminum plate. As discussed earlier, the XS-1 series of aircraft was designed to utilize thinner thickness than heretofore normal wing/tail airfoils. Two sets of differing thickness wing/tail assemblies were to be constructed for the first two airplanes. For the initial aircraft, an NACA 65–110 airfoil was constructed with a wing chord thickness of 10 percent and an elevator chord thickness of 8 percent. The thinner 8 percent wing and 6 percent tail were still under construction at the time of the Pinecastle tests. The unique movable horizontal stabilizer was capable of being trimmed (5 degrees up and 10 degrees down) by a nitrogen gas-actuated electric motor in the tail of the airplane. The airplane was equipped with conventional control surfaces and a vertical tail and rudder. The left wing was specially prepared at NACA request to contain 240 pressure orifices (with a tube in each orifice to a recording manometer) and house 12 strain gauges. The former were designed to collect pressure distribution data and the latter to obtain air loads data.

Entry into the XS-1 was through a rectangular door on the right side of the fuselage. During takeoff and climb to altitude the XS-1 hatch door was stored on the B-29. No ejection seat was utilized due to weight and difficulty of high-speed exit. The pilot was provided with a conventional backpack-type parachute. If an emergency occurred in low-speed flight, the configuration of the entry hatchway provided some chance of bailing out beneath the leading edge of the wing. However, several pilots believed "some" was entirely too generous a choice of words and felt "slim or no chance" was a better description since the wing was so close to the exiting pilot. The XS-1 cockpit, designed to meet Jack Woolams's height specifications (six feet), was a pressurized cabin of ogive shape. However, the width of the cockpit left Woolams (and every one of its later pilots) cramped for space. In fact, if the dangers from the wing were not already prohibitive for escape, an additional problem was that the pilot was supposed to collapse the control wheel to assist in emergency exit. (One obvious question that immediately arose was how to fly the plane with no control wheel?) The aircraft's windshield did not protrude but rather conformed to the ogive shape of the nose. As might be expected from this configuration, downward visibility, especially on landing, was rated marginal. Care was taken to address the strength of the windshield by installing a double thickness plexiglass unit (plexiglass was selected over plate glass due to the difficulties in producing double curvature safety glass). To reduce the danger of fogging, the space between the two panels was dehydrated. The pilot's breath was similarly dehydrated through a special canister in his oxygen mask. Cabin pressure was at 3 psi (pounds per square inch) above atmosphere and was set through a tube from the B-29 just prior to launch. The cockpit cabin would be pressurized through the utilization of nitrogen gas. The rate of pressure loss was set at 1 psi per hour.

A control wheel, rather than a conventional stick, was selected for the XS-1's flight so that Woolams could use both hands effectively in case of an emergency. The H-shaped yoke contained the controls for thrust selector, instrumentation switches, stabilizer control, and emergency power shutoff. Cockpit instrumentation consisted of (among others) a Machmeter, an airspeed indicator, oxygen system regulator, four propellant system gauges, a g meter, and an altimeter. An attitude gyro was added after the Pinecastle tests.

The XLR-11 rocket engine provided for the two XS-1 aircraft was a 345-pound four-cylinder unit providing 6,000 pounds of maximum thrust. Each chamber provided 1,500 pounds of thrust at a chamber pressure of approximately 230 psi. The engine was not throttleable, although each chamber could be fired individually. Fuel for the engine was provided by 293 gallons of ethyl alcohol and 311 gallons of liquid oxygen as oxidizer stored in two stainless-steel internal tanks. The motor was regeneratively cooled by circulating the fuel through a cooling jacket before it was injected into the combustion chamber. In the absence of the turbine pump, Bell Aircraft installed nitrogen pressure tanks to utilize high-pressure (330 psi) gas to force the liquid oxygen and alcohol into the rocket engine. The nitrogen was contained in onboard spherical tanks. Although the pressure system would not be needed at Pinecastle since the rocket engine was not yet ready, the technique for locating center of gravity within the airplane under simulated loads would provide an important part of the learning process. Due to the size and weight of the nitrogen spheres, the total fuel volume for the airplane was only 4,680 pounds rather than the initially anticipated 8,160 pounds. This provided an engine running time of 2.5 minutes. The estimated top speed the XS-1 would be able to obtain at 50,000 feet was just over 900 MPH. In short, this was not the airplane proposed in 1944 and contracted for in 1945. But it appeared to be one that could fulfill the requirements of breaking the sonic barrier.

Final Preparations

Once the B-29 modification work was complete, Bell engineers arranged for a demonstration flight to examine the bomber's airworthiness. On December 14, the B-29 was tested to a speed of 250 MPH at an observed 30,000 feet. No objectionable shaking was noted. However, it was found necessary to use a nose-down trim setting for potential release of the XS-1. As a result, the Bell Aircraft engineers discussed with Stan Smith the advisability of using an ejector installation to assist in generating a positive separation pressure on the XS-1.[53]

After its unofficial rollout on December 12, the XS-1 was moved to the experimental shop hangar for installation of its instrumentation. Since work on the XS-1 instrumentation was still continuing as of December 14, the decision was made to put off the AMC engineering and acceptance inspection of the finished aircraft. The recently proposed dates of December 19–20 were now moved to January 3–5. In the final analysis, the NACA instrument requirements caused the real delay to last until the end of December. After testing the instruments, the NACA removed the valuable recording devices for shipment back to LMAL. It was Langley's decision that the instruments would be safer with the NACA personnel rather than making the trip to Pinecastle on the XS-1.

While the XS-1 was in the hangar, final adjustments continued. After ground tests, Woolams recommended that the aileron irreversible mechanism be removed due to objectionable play in the system (backlash in the worm gears was targeted as the culprit). Woolams recommended that a conventional aileron system be installed for the first glide flights. Problems also occurred with the landing gear. The system problems were so numerous that rework and testing required extensive additional time. At the same time, RMI continued to press ahead with its engine work. In mid-December the rocket engine was ground-tested at RMI. While the initial tests appeared satisfactory, later inspection of the cylinders found further problems, causing RMI to schedule additional engine tests before sending it to Bell Aircraft.

On December 27, the XS-1 was rolled out of the experimental hangar for its first official viewing, less than 10 months since the Army/Bell Aircraft contract had been signed. The brilliant saffron-colored paint (applied to assist aerial observation) provided a striking first impression. On December 28, the loading pit at Bell Aircraft received its first test. The XS-1 was moved to the pit and lowered into the ground, and the B-29 positioned over it. Minor modifications of the pit were necessary, but in general the handling and loading characteristics of the procedure were deemed acceptable and the XS-1 was hoisted into the belly of the B-29. The XS-1 had already received the initial checkout of its brakes and landing gear, accomplished by simply towing the aircraft with a truck.[54]

As the XS-1 neared completion, the NACA requested that their personnel be allowed to begin the process of instrumentation. To that end, on December 13 William Gracey, Norm Hayes, and John Lekas of LMAL's Flight Instrument Section arrived at Bell Aircraft to install and calibrate the recording and telemetering instrumentation that formed the heart of the NACA's interest in the first flights.[55] Installation delays caused the work to stretch out past the initial planned number of days. When the NACA requested to be allowed to work on Saturday and Sunday, Stan Smith expressed a willingness to cooperate, but not without overtime pay being authorized by the AMC. Wright Field resisted, so the NACA group extended their stay into the next week. They finally departed as per LMAL instructions with the realization that more work would be needed in Florida.

To assist the NACA with the XS-1 data collection, LMAL requested that AMC provide a radar unit for its use at Pinecastle Field. Previous NACA efforts during late November to obtain a standard SCR-584 radar unit had proven futile. Thus, on December 6, Maj. Donald Eastman, acting AMC liaison officer at LMAL, contacted Major Kotcher at Wright Field regarding the possibility of NACA obtaining an SCR-584 radar unit to assist with XS-1 tracking.[56]

While the NACA awaited the AMC's response, on December 18 Walt Williams, LMAL's designated team manager for the Pinecastle tests, telephoned Stan Smith at Bell Aircraft to advise him that the NACA expected *Bell* to be responsible for acquiring the radar unit and its personnel. Smith expressed reservations since Bell Aircraft did not have access to a radar unit either, but said that Bell would ask the AMC about obtaining a radar unit for the tests. While Williams and Smith were on the phone, Major Eastman entered the LMAL office and advised Williams that arrangements had been made with the Orlando Army Air Base to provide the necessary equipment. This information was so new that Eastman had not yet even advised AMC at Wright Field, which he did the next day in a memo to Captain Pearsall (acting as Wright Field XS-1 project officer).[57] Eastman advised Pearsall that the AMC liaison officer at Orlando AAF Base had agreed to provide an SCR-584 radar unit and an M-2 optical tracker for the tests. These units were drawn from the 526th Anti-Aircraft Artillery Battalion stationed at Orlando. The local unit would also provide the equipment operators. Eastman requested Captain Pearsall to ensure that the equipment and operators had priority clearance to work on the XS-1 project since they were also committed to several other potential projects. Also, Eastman asked Pearsall to obtain clearances for NACA personnel to work with the AAF radar personnel and that the LMAL group be allowed to modify the radar unit to fit the unique requirements of the XS-1 tests. These modifications included installing a parabola camera and automatic recording observer to ensure an accurate time history record of the flight path.

On December 28, the LMAL's John Stack prepared a special memo for Langley's chief of research. Once again it demonstrated the fragile nature of the understandings governing the XS-1 test program. Two items in particular on the three-page submission referenced the NACA's desire to establish control over the actual flight test program. Specifically, Stack indicated the purpose of the unpowered flights at Pinecastle Field as "to determine the feasibility of operation from Langley Field." Further, Stack reiterated the NACA goal that after acceptance tests were finished LMAL would take over the pilot responsibilities on the tests. However, Stack acknowledged that use of the B-29 raised difficulties in the possible testing at LMAL that had not been foreseen by the NACA in its original deliberations. Thus, Stack informed the chief of research the NACA would have to lean more heavily on the contractor than originally anticipated. Although NACA pilots were expected to fly the plane as "desirable or necessary" during the contractor tests, it did not imply a change in the NACA position on the eventual acceptability of the XS-1. The NACA would not take possession of the XS-1 project until the XS-1 could be flown with engines, turbine pump, and reserve fuel, and be ground-launched. The NACA, Stack told Crowley, must remain firm on the latter stipulations. On January 8, 1946, the NACA HQ repeated the LMAL conditions in a memorandum to Wright Field.[58]

On January 3, the AMC inspection of the XS-1 occurred at Bell Aircraft's Niagara Falls facility.[59] The three-day conference focused not only on the XS-1 but also on issues relating to high-altitude flight and to the development of the proposed Bell XS-2. The first day of the inspection dealt specifically with the XS-1 and the upcoming tests in Orlando. In a brief discussion of the flight test operations, it was reiterated that Bell Aircraft would be in charge of the Orlando flights. The purpose of the flights would be for Bell's test pilot, Jack Woolams, to become familiar with the aircraft's handling and flight characteristics. Captain Pearsall of Wright Field was placed in charge of securing a P-51 chase plane to accompany the drops and a photographic plane that would film all the tests. If, after the conclusion of the Bell Aircraft familiarization flights, the NACA wished to have its pilot fly the XS-1, it would be made available to them at Pinecastle. The AMC also requested that the pilot's supply of oxygen be increased both during the entry phase of the XS-1 (by means of a station near the ladder) and during the climb to higher altitude. Stanley agreed that Bell engineers would complete both items before the B-29 departed for Orlando, an event anticipated "on or about Jan. 15, 1946."

Even without engines, the XS-1 test flights at Pinecastle Field retained an important mission. As with any new aircraft, testing of the plane to determine flight characteristics was a common and very necessary procedure. With an aircraft built to fly supersonic, with the anticipation for the unknowns built into the structure, and with the unique launching configuration required by the limited, fuel-enforced flight time, the first XS-1 missions took on added significance. Even seemingly minor details took on an importance not normally accorded in flight preparations. As a result, not all the joint conversations went smoothly. A discussion of the project's drop operations received considerable attention. The only conclusion reached at the meeting was to add two guide rails to the front sway pads, one on each side of the bomber-mounted XS-1. These guide rails were provided with a rearward sweep for the dual purpose of preventing lateral movement of the XS-1 into the B-29 after launch and for stopping any rearward drift of the experimental plane after separation from the bomber. At the end of the conference, the AMC gave acceptance approval for the XS-1. All was now clear to proceed with the contractor flights.

The first week of January was very busy for the XS-1 participants. On January 7, Dow and Jack Woolams took the B-29 up on a short 30-minute test flight. On January 8, they repeated the procedure for a 1-hour, 25-minute flight designed to check out the onboard cameras. On January 9, Major Eastman notified Stan Smith at Bell Aircraft that Norm Hayes would be arriving the next day to finish the final adjustments to the installed XS-1 instruments.[60]

On January 10, Stanley authorized a captive flight using the B-29/XS-1 to analyze the mated airplanes flight characteristics.[61] The B-29 flew for 2 hours and 40 minutes. During the flight test, the Bell engineers gathered data on the pressure distributions over the wing and tail surfaces of the XS-1 by using the pressure orifices in the XS-1 and a manometer mounted in the B-29, which was set in a position where it could be photographed in flight by a camera mounted in the B-29. The Bell engineers also checked the B-29 cameras positioned to record the release characteristics of the XS-1. The engineering concerns over the air turbulence under the B-29 at the moment of drop was a reason for continuing caution at Bell Aircraft. However, with the successful completion of the captive flight, all seemed in readiness. The time for conversation about supersonic flight had come to an end. The program was ready to begin the journey for the transonic zone.

2

The Pinecastle Tests

The Trip to Pinecastle

On January 14, the first group of Langley personnel departed Virginia for Orlando.[1] After a stopover in Savannah, the Beechcraft C-45 (#47262) arrived at Pinecastle Field and the NACA team began setting up the telemetering and recording equipment. The group, comprised of Gerald M. Truszynski, Charles A. Taylor, and John Lekas, represented a diversity of talents. NACA selected Truszynski to handle the radar tracking chores for the tests. Truszynski graduated with a bachelor of science degree in electrical engineering from Rutgers University in 1944. Shortly thereafter, he started work for NACA on radio telemetry systems, radar trajectory measurements, and wind tunnel instrumentation. During the Pinecastle tests he would be responsible for the SCR-584 radar system.

Taylor would be in charge of the telemetry system at Pinecastle. A supervisory telemetry engineer in the instrument research division at Langley, Taylor was known as a responsible engineer and a good supervisor. He was present because of his familiarity with the instrument package aboard the XS-1. Lekas would be assisting Taylor as an instrument technician.

Located southeast of the city in a thickly wooded area surrounded by marshes and large lakes, Pinecastle Army Air Field in Orlando would serve as the home for the XS-1 tests over the next several weeks. The deed for the Pinecastle base had been signed on November 13, 1941, and construction work on the airfield started shortly thereafter. When finished in 1942, Pinecastle Field was officially declared a secondary airfield for the busier Orlando Army Air Base. Pinecastle's 10,000-foot runway was reportedly one of the longest in America. During World War II, the field was used by B-17s to perfect strategic bombing techniques. By January 1946 the 1,385-acre facility was scheduled for early deactivation.[2]

On Friday, January 18, the B-29 flew from Bell's Niagara airfield south to Pinecastle Field with the XS-1 strapped securely beneath it.[3] The 4-hour, 30-minute trip was tiring but uneventful. To ease the monotony the passengers tuned in various radio stations as they went south, as specially wired speakers in the cockpit enhanced the quality of the passing stations. Aboard the B-29 were Bob Stanley, Jack Woolams, and Doug Rumsey of the experimental group; Joe Cannon as pilot, Harold "Pappy" Dow as copilot, Bill Miller as flight engineer, and Ivan Hauptmann as B-29 crew chief. The rest of the team included Charles MacLean "Mac" Hamilton as XS-1 crew chief and William Means and Herman Schneider as the XS-1 ground crew.

During the week, other Bell technicians and support personnel drove from Niagara Falls to Florida. The weather contrast could not have been greater. In order to leave the airfield in New York, the B-29 group had to shovel several inches of snow off the runway. To arrive in Orlando was to journey to another planet, or so it seemed to people from New York in the dead of winter. Florida, even in January, was nothing like the rest of the United States. It was back to what seemed like early fall weather.

The XS-1 program at Pinecastle Field was supposed to be a joint project between Bell Aircraft and NACA; however, it was not off to an auspicious start. With the XS-1 deposited at the airfield, the Bell contingent wondered where the remaining members of the NACA team were. Although they were supposedly arriving at the same time as the Bell contingent, it was now obvious that the first NACA group was not ready and thus the second group would not be coming soon. The weekend was approaching and the Bell team, as private nongovernment employees, did not expect to work over a Saturday/Sunday without overtime compensation (something the Army had been reluctant to pay earlier in the program). In any event, one Bell employee summed up the delay as spotlighting the pattern for the entire project at Pinecastle—Bell waiting on the NACA. Or at least it must have seemed that way since the two groups came with different objectives.

Bell Aircraft was under contract to the ATSC to wring out a new experimental plane. Flights exposed the pilot to the intricacies of the aircraft's handling qualities and hence were the objective. On the other side of the same coin, the NACA wished to get data on the characteristics of the plane. Without data, the NACA believed there was little of interest to be learned from the flights. Each team viewed its requirements in a different light and each side felt the other was holding up the process.

In Virginia, the NACA crew prepared to depart Hampton on January 19.[4] The LMAL group aboard the C-45 consisted of Walt Williams, Norm Hayes, Elton Miller, and Maj. Don Eastman of the ATSC Langley Liaison Office. Joel "Bob" Baker would be piloting the C-45 aircraft. Each came with impressive technical credentials for this project.[5]

LMAL had selected Walter C. Williams to head up the NACA's XS-1 project team. Williams obtained a bachelor of science degree in aeronautical engineering in 1939 from Louisiana State University. After working a short time at the Glenn Martin Company in Baltimore, he moved over to the NACA at LMAL to investigate aircraft control and stability issues. Having worked with John Stack, Milton Davidson, and Harold Turner on the transonic research airplane concept studies, Williams was in an excellent position to head up the flight test program when the XS-1 finished its construction phase.

O. N. "Norm" Hayes was born in 1916. After serving three years in the Army (1934–37), Hayes went to work at the LMAL in February 1938. Shortly after arriving, he was assigned to the flight research department's instrument research division. He used his work situation to attend extension classes of the University of Virginia. During the war years, Hayes installed instruments and flew on the PBS Sikorsky flying boat. At various times he instrumented the Japanese Zero (Mitsubishi A6M2), the British Supermarine Spitfire and de Havilland Mosquito, and the American Boeing B-17 and Douglas B-19. Well before the XS-1 was finished, Hayes had been selected to instrument the aircraft for flight testing. On several occasions, as previously recounted, Hayes flew to Bell to supervise the installation of the NACA test instruments in the XS-1. By the trip to Florida, the Bell research plane was an old acquaintance.

Born in Pennsylvania in 1920, Joel Robert Baker graduated from Rensselaer Polytechnic Institute in New York in 1941. Since he was color blind, he was ineligible for wartime duty and ended up serving two years during the war as a test pilot for the Curtiss-Wright Company. During a presentation visit by Mel Gough from the NACA Langley Lab Baker became interested in becoming a test pilot for the government. Through a mutual friend, it was arranged for Baker to join Langley in the spring of 1945. During that summer he ran speed dives to obtain preliminary XS-1 data utilizing a P-51 with a special XS-1 wind tunnel model attached to the wing. In many respects, the air flow over the model simulated a regular flight at transonic speeds. However, Baker's assignment to Orlando came not from his XS-1 related test experience. None of the senior NACA pilots wanted to go to Orlando simply to watch Bell pilots fly the XS-1, thus Baker, the more junior pilot in time of service, drew the task. His role would be to observe and note the aircraft's characteristics in anticipation of the later NACA portion of the flight tests.

The fourth person on the C-45 was the ATSC liaison officer, Maj. Donald Eastman. Born in 1917, Eastman had come to work at the LMAL as a civilian employee in November 1940. In June 1941, he was mustered into the Army with the rank of second lieutenant and continued to serve as the MC/ATSC liaison officer at LMAL. He was involved with the XS-1 from its inception although the experimental craft was only one of the many projects he was associated with at Langley. Eastman replaced his boss who died in the summer of 1945. At the request of Captain Pearsall, the Wright Field XS-1 program officer, Eastman joined the NACA contingent for the trip to Pinecastle.

The NACA delegation departed Langley early in the morning of January 19, arrived at Pinecastle Army Air Field at 3:15 P.M.,[6] and immediately contacted Taylor, Truszynski, and Lekas. Now complete, the NACA group proceeded to the south end of the airfield to inspect the B-29 with the XS-1 mated to it. No Bell personnel were present and one observer noted the airfield itself exhibited an unnatural quiet as few other people were visible. Later, the group conducted a procedural review for their own anticipated activities during the tests. It quickly became apparent that work on the Taylor/Lekas portion of the preparations would take at least another 1½ days. Truszynski had the radar truck set up but was still awaiting the arrival of the Eastman-Kodak 16mm camera for installation on the radar parabola. Taylor had managed to obtain use of a large trailer for the telemetry receiving station, which Williams had designated as a makeshift command post. Tables and chairs would be set up to utilize the space as the on-site office for the NACA crew. By 6:00 P.M. the NACA crew had put in a full day and decided to call it quits. Major Horning, the local ATSC liaison, had provided an Army carryall for the use of the group. They all piled in to make the short ride to Orlando Army Air Base, where they would be staying in the bachelor officers quarters (BOQ).

Like many other military posts after World War II, Orlando Army Air Base was full of people and activity. However, the influx of jobs and personnel had not brought sufficient construction dollars, so the BOQ could not be considered luxury accommodations. Simple and stark, it exhibited all the best creature comforts of what it was—an Army barracks. Surprisingly, though, the food at the base was a strong point and the mess service made it doubly so, since the service was still provided by a large group of German prisoners of war. Although these waiters did not speak English very well, there were plenty of them. In any event, the overnight price was right—one dollar per day. To the NACA crew, on a government per diem of six dollars per day for room and food, it was definitely the way to go in accommodations. Besides, most of the NACA team were in their twenties and participating in one of the greatest adventures of their lives up to that point. A little discomfort could easily be overlooked.

On Sunday morning the NACA group proceeded with the work program as outlined during the previous day's review. At 9:00 A.M. part of the crew departed for Pinecastle Field while Baker, Williams, and Eastman went into downtown Orlando to meet with the Bell group. Stanley and Woolams were staying at the Angebilt Hotel on Orange Avenue, one of the finer hotels in Orlando. Its use by the Bell people clearly indicated the pay and per-diem differences between government and private industry employees. Upon their arrival the NACA group discovered that Woolams and Stanley had gone fishing. Likewise, none of the other Bell staff could be located. After waiting a short while, they left a message for Stanley and returned to Pinecastle Field at 1:00 P.M. to reunite with the other NACA personnel already hard at work.

The work progress at Pinecastle Field on that day was seemingly one step forward and two steps back. Although the radar work by Truszynski was just about complete, Lekas and Norm Hayes were having a problem with the telemetry equipment. With no Bell personnel on the scene, it was impossible to finish installing the recording instruments. Bell personnel were required to remove the panels from the fuselage and wing of the XS-1; no one present wanted to take responsibility for opening up the airplane. However, Hayes's checkout of the NACA instruments proceeded well and indicated that installation should be a straightforward matter at the appropriate time. By 6:00 P.M. the NACA group had finished the work on hand and departed for the base.

After dinner, Williams, Taylor, and Eastman again left for downtown Orlando and the Angebilt; this time the trio sat down with the Bell contingent. Present for Bell, in addition to Bob Stanley, were Jack Woolams and Stanley Smith. The purpose of the meeting involved more than idle conversation—significant details needed to be reviewed in order to prepare the two groups for the first flight. Also, work conditions and schedules needed to be established to prevent a repetition of the delays caused by the sequence of events on this day.

Taylor quickly went over the details of the problem with the telemetry equipment. Stanley agreed to allow Bell technicians to remove the panels so that the telemetry equipment could be installed. This work would have to be done on Monday since Bell had scheduled a drop test for Tuesday. A Bell technician was designated to man the telemetry panel in the B-29 drop plane during the test. As for working arrangements, Stanley said that from now on the Bell work schedule would be the same as the NACA schedule, that is, from 7:00 a.m. to 4:00 p.m.

With these details aside, Stanley proceeded to outline the Bell schedule for the flight tests. From the experiments at Niagara Falls, Bell had determined that a down-load pressure of 1,300 pounds on the wing surface of the XS-1 could be obtained when the B-29 speed was set at 180 mph and the inboard engines were feathered. With the flaps set down at 25 degrees, the total aerodynamic conditions allowed the airflow under the B-29 to smooth out and the specially installed pusher strut, exerting 3,000 pounds of downward pressure, to do its job. Stanley believed that taken together these factors ensured that the XS-1 would obtain a clean separation from the B-29. To observe the separation and flight, a P-51H aircraft would perform chase duties. Captain David Pearsall, the Wright Field project officer, was scheduled to arrive Monday with a specially prepared B-17 to do the photography work.[7]

On Monday, January 21, activities at Pinecastle started early with the NACA crew arriving at 7:00 a.m. However, once again there was confusion over who would do what and work progress was delayed. While the B-29 was positioned over the XS-1 in its loading pit, the previous night's conversation regarding the removal of the XS-1 panels caused further delays. Stan Smith did not know that permission had been granted to open up the aircraft. After the XS-1 was hoisted into the B-29 so that the landing gear on the experimental craft could be checked, Stanley finally arrived and gave the go-ahead, but added a word of caution. He intended to try a test-run flight at 2:30 P.M., thus placing the NACA crew under a very tight time frame.

By noon, the Bell team had finished its work on the XS-1, but the NACA crew was still working. Stanley advised them that 2:30 P.M. was the cutoff time, whether they were ready or not. By working continuously the NACA group finished by 2:30 P.M. and the XS-1 teams were ready to attempt a preliminary test. There was, however, still no ground communication station set up.

The purpose of the exercise was to go through all the steps to simulate a drop launch. At 2:30 P.M., Harold Dow guided the B-29 down the runway at Pinecastle with the XS-1 attached underneath.[8] The first simulation was to be done at 10,000 feet, so that it could be done without oxygen masks and the difficulty with speaking that the masks entailed. If satisfactory, then a second test would be made under actual drop conditions. That is, the bomber would climb to 25,000 feet and the simulation would proceed as before except this time under actual altitude conditions (with oxygen masks for the thin atmosphere).

The dry run at 10,000 feet was very unsatisfactory due to the number of unauthorized "sightseers" onboard the bomber without oxygen equipment.[9] Lacking the necessary equipment, it was impossible for the B-29 to go to the higher altitude. Normally, the B-29 was a pressurized aircraft, but the extensive modifications to allow the XS-1 to be carried by the bomber had eliminated the pressurization capability. As Stanley quickly discerned, there was no way to attempt the higher altitude test until the pressurization capability was restored. In fact, the entire flight was ragged, so Stanley scrubbed the mission and the B-29 returned to Pinecastle at 4:30 P.M. In the future, Stanley decreed, only necessary personnel would be on the bomber during the missions. At the end of the day, Stanley announced that another test had been scheduled for Tuesday. The actual drop flight was now planned for Wednesday.

The procedural foul up over the bomber was not the day's only controversy. Earlier, Stanley, Woolams, and Baker had huddled to discuss the landing procedure for the XS-1.[10] Several issues came up in that conversation that were to figure significantly in the first flight. Specifically, should the XS-1 land with the flaps up or down? If the flaps were up, the rollout on landing would be longer simply because the landing speed would be higher at touchdown. If Woolams set the flaps in the down position, then the XS-1 could land at a slower speed with a consequent shorter landing roll. The risk involved with the latter procedure, however, was that the XS-1 would exhibit a higher rate of descent (sink rate) at the slower speeds, which could cause a problem in an unpowered flight, especially in a plane with unfamiliar flight characteristics.

Upon hearing this debate, Baker referenced an NACA report on sink rates and the concurrent demands of higher stalling speeds per rate of drop. Stanley the engineer certainly understood sink-rate theory enough to surmise a rough lift-drag rate for the XS-1 even in the absence of hard data. But it was still a crap shoot as to proper procedure for the first flight. In any event, the best instructor was going to be experience. The end result was neither individual paid much attention to Baker's remarks. As usual, the presence of a young government employee at a private company "show" did little to enhance Baker's "outsider" opinions.

Aside from the issue of pilot visibility, two other questions received attention in the Stanley-Woolams-Baker conversation: the proper characteristics for the landing flare, and whether to aim the landing for the beginning of the runway or somewhere halfway along the strip. As in the sink-rate discussion, the questions were left open. Stanley was willing to let Woolams's excellent judgment be the guide.

The last major actor in the XS-1 drama also arrived that day; flying in from Wright Field was the especially configured B-17 (#44-85774) photographic airplane.[11] On board was a crew drafted to this mission. Also aboard was Capt. David Pearsall from the pilotless aircraft branch of the Aircraft Projects Division of Wright Field. Although an experienced pilot, Pearsall had chosen to avoid the long flight down alone and hitched a ride on the B-17 for convenience. As the Army's XS-1 project officer he was to observe the tests while trying to allow the contractor maximum flexibility of operations. After watching the afternoon procedures, Pearsall joined the NACA group at the Orlando BOQ.

On Tuesday, January 22, the Bell contingent determined they would get in one afternoon dry-run test of the launch procedures to correct the fiasco of the previous day. However, heavy rain and a very low cloud ceiling at Pinecastle Field greeted the test group and lingered throughout the day. As a result, no flights were made so Stanley finally ordered four complete simulated launch exercises to be conducted with the B-29 on the ground.

Concurrently, Williams and Baker called upon the Orlando Air Forces XS-1 liaison officer to request additional assistance from another ground vehicle for use while on the Pinecastle airfield. The other members of the NACA group had spent the morning calibrating the aileron, rudder, and elevator control position recorders as well as the telemetering equipment channels. When completed, they participated in the practice launch exercises of the afternoon by monitoring the delivery of the telemeter recording data.

On Wednesday, the scheduled first launch was scrubbed not only because of the continuing miserable weather, but also because a new complication arose. The B-29 was grounded in response to a technical order mandating the changing of its propeller governors. The local AAF safety board had used the weather delay to inspect the aircraft at Pinecastle Field. Between the weather and the technical order, there would be no flight.

Williams telephoned Mel Gough at Langley and advised him of the B-29's grounding. Williams stated that D. R. Shoults, a senior engineer at Bell Aircraft, was even then trying to get special permission from Wright Field to make at least one more flight with the B-29. Gough informed Williams he already knew of the B-29 problems and was concerned. Gough had a reputation at NACA as a stickler for safety; his rule in experimental flight was "Don't take chances, fly safe." Tech orders were nothing to ignore. Gough finally decided that he would call at 9:00 A.M. the following day to reevaluate the situation.

The morning of January 24 again demonstrated the different schedules for the two participating teams. Promptly at 9:00 A.M., Gough called Williams about the status of the B-29 tech order and informed him that Hoover was making arrangements for a quick service of the B-29 at the Langley Laboratory. He further stated that Wright Field had advised not to fly the B-29 until the tech order had been fulfilled. This suited Gough, who did not want to risk the whole project for this type of problem. Williams indicated that, at this time, the Bell group had not advised him of anything regarding the tech order. Unstated was the fact that it was now past their normal start time for the daily schedule, which in itself indicated continuing problems.

After hanging up with Gough, Williams called to the Bell crew area. Jack Woolams said that Bell Aircraft had received permission for one additional flight, without fulfillment of the tech order, but was still negotiating for further flights. Since the possibility existed that only this one flight would be approved before the continued grounding of the B-29, Bell was determined to make it the first drop test for the XS-1. This news caused a flurry of activity among the NACA people, since they were not ready to proceed with the launch. After calling Gough to advise him that Bell was determined to push on, the NACA crew raced to Pinecastle Field to load the film drums for the XS-1 recording instruments. Likewise, the NACA telemetering equipment was again rechecked.

After all preparations were complete, the NACA group moved to their stations and took up the watch, as the Bell contingent finished loading the XS-1 into the bomber. However, the weather, which had recently proven so unreliable, once again caused problems. Although the sky above Pinecastle Field was clear for the first time in several days, the 10,000-foot runway was experiencing a tricky 90-degree crosswind of about 30 miles per hour. With no weather letup in sight, the Bell group once again made the decision to scrub the flight for the day.

While the weather continued to make for rough going in the program timetable at Pinecastle, bureaucratic events in Washington revealed little that was any smoother. On January 8, the Washington NACA HQ finally responded to the ATSC letter of November 23 regarding the future of the XS-1 program.[12] The NACA letter provided ample proof that the two agencies were not of the same mind regarding the requirements for the program. In notifying Wright Field of its "agreement" on the outlines of the program, the NACA set itself up for a serious misunderstanding. NACA reiterated the "agreed" program:

1. Final Army acceptance will occur after the rocket motor is demonstrated and without the necessity for a mother ship.
2. Pinecastle flights are to determine Langley Field operational feasibility.
3. Bell will supply a pilot until the NACA decides to take over the program.

However, the "agreed" program was not the one in the Army letter of November 23. Indeed, a close review of the Army letter reveals the NACA was responding to the letter that seemingly they wished they had received. The misunderstanding on these points was to continue throughout the program. In short, the XS-1 project was still an unfocused effort. Aside from agreeing to attempt to breach the sound barrier, little else of the program seemed in agreement among the participants.

The First Flight

Friday morning, January 25, dawned without the weather problems that had hampered the previous two days' efforts.[13] After all the set-backs, everyone was ready for some progress on this so-far-unlucky project. Williams and the NACA group arrived at Pinecastle at approximately 9:00 A.M., as did the Bell group. Both the XS-1 and the NACA instruments were ready. The weather cooperated with clear skies and unlimited visibility. The cool temperature, however, was decidedly un-Florida-like; the high for the day was only expected to reach the mid-60s.

Among the many last-minute preparations, the Bell technicians at the pit smeared fresh red paint on the B-29 wooden guide rails installed under the bomber and on the XS-1 on the fuselage tube. If the XS-1 drifted on separation and struck the bomber, a visible record would be created of the contact. At the NACA camp difficulties arose when during the preflight final checkout the NACA group discovered that the ground radar equipment was not working. In his first inspection, Gerry Truszynski speculated that the drive motor was burned out. Without radar, proceeding with the flight test was seriously called into question for the NACA. However, the Bell group huddled and decided to go ahead without the radar. As Truszynski continued working desperately on the radar system, the B-17 photo plane proceeded with takeoff at 10:15 A.M. At 10:40 A.M. the mated B-29/XS-1, with Joe Cannon and Pappy Dow at the bomber controls, and Woolams riding along, took off and began the long, slow (about 20 minutes) climb to altitude. Bob Stanley concurrently took off in the P-51H to perform chase functions.[14] On the ground, Doug Rumsey remained in the control tower to monitor the XS-1's radio transmissions. Meanwhile, Joel Baker positioned himself alongside the telemetering truck to listen in over the radio and to monitor the landing of the XS-1.

The earlier B-29 flight test at Bell Aircraft had already determined that the mated B-29/XS-1 was solid enough to allow the bomber to perform normal operating maneuvers on both takeoffs and landings, as well as flights at normal cruising speeds. During this first operational flight the earlier experience was reconfirmed. In fact, the actual performance of the bomber was found to be only slightly reduced and the B-29 ride exhibited neither uncomfortable buffeting nor impaired handling. Aside from the rather high elevator loads required to hold the nose of the B-29 angled up for the final stages after landing (i.e., after touchdown and during rollout) the plane's flight characteristics were quite predictable.

As the bomber began to climb above 5,000 feet, Woolams initiated the previously tested launch procedure. Equipped with his familiar white-faced helmet, Mae West, and a parachute (for luck if no other utility), Woolams moved from the forward crew compartment into the forward bomb bay and along the catwalk to start the descent down the shielded ladder beside the XS-1.[15] The idea of crawling down a ladder suspended over a open hole in the belly of a bomber was unsettling enough, but the cantilevered sliding ladder required that the pilot bounce on the platform to force it to drop. With the addition of the roaring noise, the frigid air, and the constant buffeting from the wind howling through the bomb bay, it was truly a bracing experience. After crewmember Mac Hamilton assisted with lowering the adjustable ladder to a position even with the entry hatch of the XS-1, Woolams initiated the cockpit set-up procedures that in the future would become standard practice.

Easing through the entry hatch, Woolams quickly began the task of buttoning up the aircraft for launch. It was eerie to be alone in the XS-1; it was not like normal test flying from the ground, surrounded by a crew and technicians, with the sun beating down from a clear sky. Now, although daylight streamed in the bomb bay along the bottom edges of the XS-1, in the cockpit on the top of the experimental craft the light was much more shrouded. While communications with the bomber still existed through the intercom, Woolams was now alone in his element. He first slipped on the shoulder straps and the lap belt to bind him to his seat, then connected the oxygen mask and the dehydrator and checked for leaks. Finally, Woolams plugged in the headphones and attached his microphone. When Hamilton lowered the XS-1 door to a position on his right alongside the entry hatch opening, Woolams pulled on the handle with his right hand (the cockpit was too narrow to allow a pilot to get leverage with his left hand) to close the lever and create an airtight compartment. Or so it was supposed to do; for it was at this point that the XS-1 experienced its first serious problem.

Since the XS-1 would be operating above 10,000 feet, oxygen and heat for the pilot were a necessary part of the overall plan. In the thin atmosphere of higher altitudes, where the XS-1 would perform best, without a source of heat the extremely cold outside temperature would affect the pilot very quickly. When Woolams pulled the door of the XS-1 shut, it did not seal tightly. Woolams urgently indicated to Mac Hamilton, who was in the B-29 monitoring the system that pumped heat from the B-29 to the XS-1, that the thermal conditions in the XS-1 were uncomfortable and were rapidly getting worse. Contrary to plans, the heat from the B-29 was cooling off in the passage line. Woolams was now sitting in a dark icebox with only his flight jacket, coveralls, helmet, and gloves for warmth.

Meanwhile, on the ground, Truszynski had been attempting to repair the radar system in anticipation of the launch. The XS-1's cockpit heating problems, however, put an end to any potential remedies. Once the B-29 reached drop altitude, there would be no waiting for Truszynski to finish. Time finally ran out on the NACA ground crew.

With the bomber at 25,000 feet, and Woolams reporting very cold temperatures in the XS-1 cockpit, Woolams decided to initiate the launch procedure. In the B-29, Dow began to count off the last minute before drop at 10-second intervals. During that time, Cannon feathered the B-29 inboard engines and set the flaps positioned down 25 degrees. With these steps, the B-29's speed fell off sharply. Upon reaching the desired cruising speed of 180 MPH, Cannon put the B-29 into a shallow dive and then leveled the plane. Similarly, Woolams went through the final checkout for the XS-1. Landing gear up. Flaps up. Spoilers closed. Horizontal stabilizer one degree up. Controls neutral. Check. The B-29 was ready to execute the launch.

Ticking off the seconds verbally, Dow began the countdown prior to release of the plane. Counting backward from ten, the tension began to build. 10—9—8 . . . Woolams switched on the telemetry equipment. 7—6—5. The initial major test of the XS-1 shakeout would come in the first few seconds after B-29 separation. The XS-1 was hung on the B-29 at an angle such that at launching conditions the airload exerted a downward force of 1,500 pounds on the research plane. However, to assist nature in separating the two planes, a yoke had been fitted to the bomber just forward of its center of gravity. At the moment of separation, a sphere would be activated containing compressed air to exert an additional 3,000 pounds of downward pressure through the yoke. If the Bell engineers were correct, this additional assist would ensure a clean break with the mother craft. 4—3—2—1. It was 11:15 A.M., January 25, 1946. The journey to supersonic flight was about to begin.

Pressing the solenoid switch in the bomber, Dow activated the shackle release for the XS-1. At slightly below 22,600 feet, he pulled the shackle release and the XS-1 separated from the mother ship.[16] The little saffron-colored plane dropped away with an initial force of −1 g and the tail slightly low, a very clean break. In the B-17 the cameras recorded the historic first moments in slow motion.[17] The bomber crew barely felt any reaction by the B-29 to the release; Woolams noted no discomfort as well. Stanley, in the P-51 flying chase, agreed that the drop characteristics were ideal. The XS-1 had not yawed into the bomber. (Nor had it floated backwards into the B-29 as the lack of red paint stains would later confirm.) It was apparent that the pilot had not lost control of the experimental craft. Quietly, almost anticlimactically, the XS-1 had passed its first major test.

Two seconds after the drop, Woolams put the nose of the XS-1 slightly down and began to pick up speed. From an initial 177 MPH, the XS-1 began a slow, steady acceleration. Within four seconds the plane had only increased its speed to 182 MPH. Immediately after clearing the bomber area, Woolams executed several turns to check out aircraft handling. At speeds up to an observed 275 MPH, he found the XS-1 rock solid, experiencing neither vibration nor significant noise.

At 18,000 feet, Woolams tried out the spoilers and again registered no problems in handling, even while executing a stall with the airplane clean (wheels up and flaps retracted). The XS-1 controls felt as light as a feather, mostly due to the balance of the aircraft in maneuvers. Longitudinal stability was quite positive and stick force versus g force was satisfactory up to the 3g level (which was as high as Woolams attained on the initial flight). Directional stability was also positive with air dampening and lateral stability noted as about neutral, although satisfactory for normal conditions. However, at the observed airspeed of 120 MPH during the stall test, Woolams noted center-section buffeting. No tendency to drop off on a wing was found, and some aileron was maintained to the end, giving Woolams the feeling that the wing did not completely stall.

Five minutes and twenty-eight seconds into the flight, Woolams began to execute the XS-1's second set of handling tests. At just over 12,000 feet and with an observed airspeed of 110 MPH, Woolams lowered the landing gear, then the flaps. While attempting to stall the aircraft again, Woolams noted additional but clearly less severe buffeting. The following stall was complete and satisfactory. Although the plane exhibited a slight tendency to drop off on one wing or the other, Woolams readily corrected the condition by the application of rudder. The only problem the pilot noted was a slight amount of lateral instability at the moment of stall. He managed to correct this condition with the ailerons and did not believe it to be serious.

Woolams felt that the XS-1 airplane was simply a joy to experience. Of all the aircraft he had flown over the years, only the Bell XP-77 and the German Heinkel 162 compared to the XS-1 in maneuverability, control relationship, responsiveness to control movements, and lightness of control forces. Sitting in the tiny cockpit, Woolams felt that the ruggedness, silence, and smoothness of operation made it the "most delightful one to fly of them all."[18]

The laws of nature say that no flight lasts forever, which is especially true in unpowered flight. Reality quickly intruded upon the joys of initial discovery. The lowering of the landing gear continued to cut into Woolams's speed. Since the XS-1's landing gear and flaps were initially designed as one-shot deals—that is, there was no system to redeploy either—it was time to land the XS-1. Woolams noted that visibility from the XS-1's high cockpit while the plane was in flight was only adequate. This problem, of course, had been anticipated as far back as the summer of 1945 when Bell engineers Benson Hamlin and Paul Emmons had asked Jack to sit in a mockup of the cockpit to determine visibility. Woolams had suggested changes in the cockpit canopy design. While these changes had improved the view from the XS-1, Woolams still had to bank the plane to see landmarks within a five-mile radius beneath him. However, Woolams noted, by coming down at a steep glide angle the landing visibility improved significantly. With Pinecastle's long runway clearly in view, Woolams executed a sharp left turn toward the strip.

Dropping toward the concrete strip, Woolams focused on the cockpit instruments as he tried to observe all aspects of the experimental craft's landing characteristics. However, he miscalculated the steepness of the XS-1's rapid descent. As speed fell off quickly, he realized the XS-1 was not going to make it to the runway. Losing speed at an alarming rate, Woolams faced possibly missing the first landing. In fact, the ground observers at Pinecastle Field were sure the XS-1 was not going to clear the trees that lined the far west end of the strip preceding the runway. While managing to coax the XS-1 over the trees, Woolams recognized the hopelessness of trying for a straight-on landing. Dropping toward the ground, Woolams slammed full elevator and tried to flair the nose to pull a little extra distance out of the aircraft. His efforts were to no avail as he undershot the end of the runway. At 9 minutes and 42 seconds into the flight, the XS-1 touched down at 105 MPH on Pinecastle Field. Landing 400 feet short of the runway on the hard grass shoulder, the XS-1 shot across the tip of the runway at a 45-degree angle. No damage was sustained and ground roll-off into the grass median and toward the taxiway went about 900 additional feet. The plane came to rest with its nose over the taxiway and its tail still in the grass median.[19] The first flight of the first XS-1 was now history. When the XS-1 finally stopped rolling, the necessity for 10,000 feet of runway remained an issue not yet answered; the requirement for some additional width of runway, however, seemingly loomed larger.

Down by the telemeter equipment truck, the NACA contingent watched the landing with a mixture of concern and joy. While the XS-1 finished its rollout, they climbed in the carryall and set off on the short run to where Woolams had brought the XS-1 to a stop. By the time they arrived at the aircraft, Jack was standing by the aircraft absorbed in thought. The NACA group dismounted to check the equipment and the aircraft.

Stanley landed the P-51 and taxied over to Woolams as the Pinecastle Field fire units arrived as a precaution against possible damage to the XS-1. After dismounting, Stanley had a detailed discussion with Woolams on the reasons for the unusual landing. The XS-1's rather unorthodox flight conclusion, while frustrating to a skilled pilot, changed little in terms of testing the plane. Woolams and Stanley were professional enough to know that first flights in experimental aircraft were always learning experiences. Woolams told Stanley he had no doubt that he could put the XS-1 on the runway the next time. Stanley later called Lawrence Bell, president of Bell Aircraft, back in Buffalo to convey the good news: "Larry, we're in business. We've got a flying machine here."[20]

For the NACA group, although the flight was somewhat marred by unexpected hangups and minor glitches it was still a success. The recording instruments worked fine with the exception of the accelerometer film drum that stuck halfway through the flight and the elevator component of the wheel force recorder that did not function at all. Otherwise, the data, once analyzed, would provide useful insights into what was happening to the airplane. Similarly, the communications system between the B-29 and the ground station did not fulfill expectations. Using the Army communications truck for signaling the B-29 when conversation was desired seemed entirely too cumbersome. In the future a radio would be installed in the telemetry truck to facilitate direct communications with the airborne platform.

Major Donald Eastman, the AMC liaison, called Brig. Gen. Laurence Craigie, the head of the Engineering Division at Wright Field, to inform him that the XS-1 was down and the flight had been a success.[21] Although the flight did not go completely as planned, Eastman told Craigie, the Bell team "ought to be congratulated." Craigie simply expressed satisfaction that the program was underway.

Later that afternoon, the Bell team retired to the officers' club to discuss the day's events, but conversation turned away from business when Jack Woolams suggested that they attend a rodeo in town that evening. In the freezing chill of a cold Friday night, the Bell group went to the rodeo, where Woolams quietly disappeared. The Bell engineers finally realized his absence when someone saw him climb over the chute and onto the back of the upcoming bull. Before his companions could utter a word, Woolams was out of the gate on the second wild ride of his day. Although it was a short trip, the crowd enjoyed it. Back at the hotel, the story of the evening's adventure spread quickly. It was another landmark in the saga of Woolams's wild times.[22]

Although success was the mood of the moment, Bell Aircraft negotiations for additional flights before the tech order enforced grounding of the B-29 proved fruitless. The delay meant that the B-29 would be out of service for at least four or five days. For the Bell group, the lack of the B-29 meant a "long weekend." Williams called Gough at Langley to inform him of the first flight and to advise him of the now definite grounding of the B-29. After discussion, it was decided that the NACA group should return to Virginia. After the NACA group loaded the aircraft, Baker piloted the C-45 back to Orlando Army Air Base. The NACA crew took leave of Orlando at 8:45 A.M. the morning after the flight to return to Langley. Aboard the plane were Williams, Truszynski, Taylor, Miller, Eastman, and Hayes. Baker was at the controls of the C-45. (Lekas would remain behind.) They arrived back at Langley Laboratory at 2:30 P.M. the same day.[23]

The Flights Continue

On Saturday, February 2, the Langley group returned to Orlando.[24] Aboard the C-45 were Baker, Williams, Hayes, Truszynski, and John H. Householder as crew chief for the NACA shuttle aircraft. The group departed Virginia at 9:30 A.M. and, after a stopover at Savannah, arrived at Orlando Army Air Base at 3:30 P.M. John Lekas was on hand to greet the NACA technicians. He quickly brought the group up to date on the Bell-instituted changes since the first flight and the upcoming plans. Although the latter were of importance, they imposed no restrictions on the flight schedule. However, the engine governors for the B-29 had only just arrived and would not be installed until Monday morning. An XS-1 test flight was scheduled by Bell for later that afternoon. Again, no work was seemingly being done on the weekend by Bell Aircraft.

Early on Sunday morning, the LMAL crew departed Orlando and went to Pinecastle. A full eight-hour workday was planned. While Hayes and Williams checked the recording instruments and calibrated the control position recorder and the yaw angle recorder, Lekas checked out the telemeter and Truszynski and Householder worked on the radar data box. Problems cropped up on every project. Hayes discovered that the wheel force recorder was not working and thus he could not balance the electrical system of the wheel. While the problem was simple enough (being found in the strain gauges), it was not possible to repair the equipment at Pinecastle and the wheel would have to be returned to Langley. Similarly, Truszynski found that the selsyn in the new radar box had no dampers and thus was unusable. Either new dampers would have to be made in Orlando or selsyn flown in from Langley.

With one success under their belts, the combined Bell/NACA team was ready to proceed with the continued wringing out of the new aircraft. It was normal procedure to put any new plane through a standard series of tests to determine initial flight characteristics. With the apparent success of separation and airdrop seemingly behind it, the XS-1 group was ready to commence work on other aspects of the test.

Since the previous flight, the Bell crew had made one change to the XS-1. To make it easier to examine the handling qualities of the XS-1, the landing gear/flap system had been reworked to provide independent operation for both. This would allow Woolams to test the flaps without the inherent degradation of flight time and quality due to the deployment of the landing gear. Center of gravity remained at 24.9 percent mean aerodynamic chord but gross weight had been trimmed slightly to 4,230 pounds. The purpose of the flight would be to continue checkout of the XS-1's handling characteristics.

Monday promised to be a busy day. Truszynski met with base personnel to arrange for dampers to be made in the machine shop in Orlando. While awaiting Truszynski's return, the NACA group discussed plans to contact Langley from the radio in the B-17. With the powerful sending unit in the B-17, the airborne aircraft could theoretically contact Virginia. By 10:30 A.M. the Langley group reached Pinecastle. Unfortunately, they were late and the B-17 had already taken off at 10:25 A.M. Efforts to contact Langley from the ground using the B-29 radio failed and the NACA group proceeded with their prelaunch work. The instruments in the XS-1 were again checked in anticipation of an afternoon flight. At 2:00 P.M., however, the afternoon flight was canceled due to low cloud cover and the stiff crosswind blowing at a 90-degree angle across the long runway at Pinecastle.

Two flights were scheduled for Tuesday, February 5. For the first test, takeoff was initiated at 9:15 A.M.[25] The weather was clear and visibility was unlimited as the B-29 began its taxi run for takeoff. During this flight Joel Baker would fly chase in the newest addition to the drop team, a P-47N (#488903). The P-51 previously utilized as the chase plane was released from the program. Bob Stanley now joined Doug Rumsey in the control tower to listen in on Jack Woolams's conversations.

Once airborne, with Dow at the controls and Cannon as copilot, the B-29 crew and Woolams went through the same procedure as for the first launch. However, one significant difference was quickly apparent to Woolams once he was in the cockpit and the XS-1 door was levered into place. This time the heat and pressurizing output from one B-29 engine were diverted to the external XS-1 heating tube. As a result, the heating in the cockpit was now adequate for Woolams to work while attached to the bomber. In addition, a cabin pressure differential of 2,000 feet was obtained at a pressure altitude of 20,000 feet. Although neither issue presented problems on the current short flights, the leaky door seal remained a serious concern. It was imperative for solutions to be found for the later high-altitude flight work.

The drop was accompanied by the same set of external characteristics as in the first flight. The B-29 speed was set at 180 MPH, with the inboard engines feathered, the flaps set down 25 degrees. However, this time the ejector was only charged to 600 psi. The drop occurred near 22,000 feet at approximately 10:16 A.M. During the flight, Woolams was scheduled to check out XS-1 handling characteristics.

After separation from the area of the B-29, Woolams checked out XS-1 stability on all three axes. He found directional stability to be positive and satisfactory and longitudinal stability to be quite positive and probably excellent for high-altitude operations. Stick force versus g force was also satisfactory for as far as Woolams pushed the aircraft. Finally, he rated lateral stability as only barely positive, but satisfactory.

In diving to pick up speed, Woolams noted that handling qualities of the aircraft were quite good. Again, he detected little wind noise due to the XS-1's clean aerodynamic shape. By 13,000 feet, the XS-1 had reached an observed speed of 350 MPH. This was the fastest speed obtained on the flight. Leveling off and slowing to an observed airspeed of 250 MPH, Woolams trimmed the plane and discovered it required an up horizontal stabilizer angle of 1⅓ degrees: that is, the aircraft trimmed out slightly nose down. The awkward positioning of the horizontal stabilizer control, behind and above the pilot's left hand, received negative comments from Woolams.

As the glide flight began its descent for landing approach, Woolams checked out the other characteristics of the experimental craft. The ailerons were found to be very light and quite effective, but the XS-1 exhibited poor centering characteristics, especially at slow speeds. He did not believe this situation would constitute a problem during powered flights, since the XS-1 was soon to be equipped with an irreversible mechanism. The action of the spoilers for glide angle control was excellent in the clean configuration, but less so when the flaps and landing gear were extended. In fact, Woolams estimated the XS-1 spoilers were only half as effective in the latter configuration as the average sailplane equipped with high aspect ratio wings. When using maximum spoiler deflection at speeds below 150 MPH, Woolams discovered frequent lateral correction by the ailerons was necessary since partial wing stall accompanied by somewhat uneven buffeting would occur.

As discovered during the first flight, the gliding angle of the XS-1 when in the clean configuration was very shallow. However, once flaps and landing gear were extended, it became quite steep. In fact, when spoilers were extended the glide angle became extremely steep, which

Woolams believed would allow the XS-1 the ability, should it be necessary, to clear obstacles during landing. Since the high cabin structure presented a problem in viewing the ground in the immediate vicinity of the plane, Woolams recommended the XS-1 be leveled off with a bit of excess speed. The actual approach itself was no problem since the steep angle of landing, necessitated by the nose-down attitude necessary to keep speed above 150 MPH, provided a clear view of the runway. The trick was to let the XS-1 settle down on the runway gradually since its landing distance from the ground was more difficult to judge than for most aircraft.

Ground handling and braking of the XS-1 appeared to be normal and forward visibility during rollout was adequate. After landing, the NACA undertook a postflight inspection of the XS-1. The examination of the telemetering instruments revealed that the yaw angle recorder had failed on the flight. Hayes initiated a hasty field repair in preparation for the afternoon flight.

Although scattered clouds had appeared at about 6,000 feet, the weather conditions were still excellent for an afternoon test.[26] Takeoff for the third XS-1 flight was at 2:30 P.M., with Cannon piloting the B-29 and Dow serving as copilot. Joel Baker once again did chase work in the P-47N. However, this time the conditions of drop, while not radically different from previous missions, were configured to initiate subtle changes in future launch procedures. While the B-29 speed was again at 180 MPH and inboard engines were feathered, the flaps were set down only 15 degrees, instead of the previous setting of 25 degrees. Drop launch of the XS-1 was accomplished at 3:40 P.M. and, despite the altered drop characteristic, XS-1 departure was visibly the same as on the first two flights.

After separation from the B-29, Woolams quickly wrung out the XS-1 by putting it through a series of handling tests and dove the plane to an observed airspeed of 400 MPH. During the descent sequence he examined the landing approach characteristics using the spoilers at interval. Again the landing proved no problem and Woolams indicated that he now felt comfortable on the approach pattern. The only major postflight observation that Woolams made was that there probably would not be enough oxygen in the airplane to sustain the pilot during the high-altitude flights. Woolams suggested either increasing the onboard oxygen supply or installing a filler neck outside the pressure compartment, accessible for refilling just prior to B-29 drop launching.

Bell Aircraft planned to complete three flights on February 6 and as a result the work pace quickened to meet the ambitious schedule. Even with the visible portions of the flight completed, the logistical aspects of the program continued. Late into the night at Pinecastle, Lekas, Truszynski, and Hayes continued to labor to pull together the data and photographic aspects of the permanent record. All photographs were developed and Hayes and Truszynski assisted Lekas with the telemetering equipment. During the second test flight of the day, the NACA telemeter discriminator had failed to function. But repair work was completed before the group left that evening for the Orlando BOQ.

The Program Suffers a Setback

Wednesday morning, February 6, the Bell ground crew had the XS-1 loaded and ready to go by 8:00 A.M.[27] However, bad weather would again sabotage the day's flight plans. Likewise, since it was raining at dawn on Thursday and the weather was bad for most of the day, the test personnel spent the time doing routine preparations for another shot at flying on Friday. The NACA crew checked and loaded the telemeter equipment and the recording instruments. Joel Baker took the opportunity to pilot the B-17 photographic plane as they rehearsed their mission.

Friday morning the XS-1 teams arrived early to start the test procedure.[28] The weather looked passable at that moment with scattered clouds at 5,000 feet, but in Florida, with its quickly changing meteorological conditions, this meant nothing. Setbacks began almost immediately. While Williams and Hayes attempted to ground-check the horizontal stabilizer, the unit's limit switch failed to shut off the pneumatic actuating motor during the stabilizer's travel to the extreme position, resulting in the stabilizer actuating screw breaking off at the bell crank. Observing the work, Woolams wandered over to the aircraft to inquire about the damage. He coolly remarked he was glad it chose to fail on the ground rather than in the air.[29] To Hayes, the remark was typical of Jack Woolams—nothing seemed to rattle him, but his intelligence and inquisitiveness were always apparent. Corrective measures were undertaken to fix the damage and prevent a recurrence. As a result, takeoff for the test flight was delayed several hours.

Once airborne, the B-29, with Dow at the controls and Cannon as copilot, climbed to altitude to initiate the drop procedure. Joel Baker was again in the P-47N doing chase duties. However, on this mission he departed early so that he could conduct some research and perform additional photographic chores for the program. The B-29 conditions for dropping the XS-1 were the same as in the past except this time the bomber's flaps were not extended for the release. XS-1 separation from the B-29 occurred as in the previous drops and was once again considered excellent.

The purpose of the fourth drop test was to examine glide characteristics and communications. Woolams contacted two different ground stations with the radio transmitter on board the XS-1. Both tests were executed above 20,000 feet and both ground stations were contacted successfully. However, Woolams later noted that power input to the unit was considerably less than would be necessary for the follow-on higher altitude and greater distance tests that would come when the plane utilized its proposed rocket engines.

Next, Woolams attempted to test the XS-1 under speed and dive conditions. When he pushed the XS-1 over into a steep dive, the plane quickly built up speed to approximately 400 MPH. Capping the dive, Woolams executed a 4g pullout. Once again the XS-1 fulfilled Woolams's early expectations as he found the plane quite responsive. Stability and handling were excellent and full aileron deflection could be obtained up to the maximum obtained speed. In pulling out of the dive, Woolams noted that stick force gradient to g force was satisfactory. The only complaint was that wind noise picked up noticeably above 300 MPH.

At 10,000 feet Woolams lowered the landing gear and watched the indicator light show green signifying the gear was locked into place. Lining up for final approach, Woolams (as was his custom) rescanned the gauges and then proceeded to bring the XS-1 in for a landing. After initial touchdown the plane briefly skipped up about one foot and then settled back onto the strip. However, as rollout proceeded and the weight of the airplane settled back onto the landing gear, the left main gear wheel started to retract. Realizing that the left main landing gear was folding up underneath him, Woolams attempted to hold the left wing up as long as possible while the XS-1 cut its speed. Gradually he let the left wing settle toward the ground. The trick was to let the plane simply slide down the runway on its wheel and wing tip while minimizing the damage to the latter. Rolling down the runway at a high rate of speed, Woolams fought to hold the XS-1 on the strip. At approximately 80 MPH the left wing touched the ground. To prevent the wingtip from acting like a skeg in the water and causing the plane to veer to the left, Woolams applied the right brakes in an effort to keep the plane somewhat centered on the runway. But the XS-1 began a gradual swerve toward the left in spite of Woolams's continuing efforts.

About halfway down the strip, with speed dropping off rapidly, Woolams met the final surprise of his flight—a runway boundary light standing firmly in the path of his dangling wingtip. This permanent fixture, approximately eighteen inches in height and eight inches in diameter, posed a formidable barrier to the left wing. Striking the light fixture with the leading edge of the wing, the XS-1 spun 90 degrees to the left and skidded off the strip. The plane came to rest in the grass field between the runway and the taxi strip. Woolams was unhurt in the crackup and quickly crawled out to inspect the damage. He was soon joined by other personnel who had raced to the crash site from all directions.

While damage assessments were held, Woolams received another task to perform.[30] Arriving with the ground personnel were Bell subcontractors responsible for providing a pressure suit for the later high-altitude flights. Although the suit would not be needed on the Pinecastle flights, the subcontractors were already working on it and needed Woolams to try on the suit before they could proceed. While everyone huddled around the damaged XS-1 as it lay on its side, Stanley instructed Woolams to try out the new pressure suit. Pulling an impromptu dress rehearsal, Woolams changed into the flight suit while standing beside the downed aircraft.

Fortunately, damage to the XS-1 was confined to the wing area. The leading edge of the wing was dented from hitting the landing light, the wing tip and outboard section of the left aileron were damaged by the ground, and the ventral fin was worn away in several places. No definitive reason for the failure of the left landing gear was readily apparent. The visual inspection, plus the review of the landing sequence, supported the idea that the slight skip on landing had caused an interaction on impact between the landing gear retraction cylinder and the landing gear downlock, just sufficient to cause the lock to move out of position, allowing the gear to begin retracting as soon as pressure was applied.

Although the damage was minor, once again demonstrating the ruggedness of the airplane, it rendered field repairs impossible. After making an estimate of the parts needed, Woolams called Bell Aircraft Corporation to discuss possible options. From this conversation, a decision was made to bring the XS-1 back to Niagara Falls to perform the repairs. The mated B-29/XS-1 departed on February 9 for the flight back to Bell Aircraft.[31] The NACA crew also prepared to return to Langley Laboratory. Packing the film and the inoperative recording instruments, the NACA group (minus Truszynski, who remained behind) also flew out of Orlando on February 9.

The completion of the fourth flight marked the last involvement for Capt. David Pearsall.[32] Early in February, Pearsall was notified he had been accepted at graduate school. As a result, with the enforced delay due to the XS-1 wing damage, Pearsall notified Wright Field that this event marked a convenient time to turn over his duties to Capt. George Colchagoff as the new XS-1 project officer.

Return to Pinecastle

On February 14, the B-29, with the repaired XS-1, departed Niagara Falls with Dow at the controls. As part of the B-29's return journey to Orlando, Stanley had ordered it to stop off at Langley Field to allow the AMC and NACA personnel who had not been present at Pinecastle to inspect the experimental aircraft. The Buffalo contingent also brought the new XS-1 model that Bell had constructed for Langley to use in the eight-foot high-speed wind tunnel tests. On February 15, after picking up Stanley and Smith, the B-29 continued its flight to Orlando.[33]

The NACA contingent also left Langley Laboratory on February 15 to return to Orlando.[34] Aboard the C-45 were Williams, Hayes, Lekas, Abraham E. "Ace" Ruvin to help with the telemetry data, and H. N. Hinman as crew chief for the shuttle aircraft. Joel Baker was again the pilot for the trip south. Arriving late in the afternoon, the group met with Gerry Truszynski who had remained behind in Orlando.

The entire group loaded up and traveled to Pinecastle to begin work on the XS-1. The telemeter transmitter was reinstalled in the experimental craft; the NACA recording instruments were checked out and calibrated. The team worked until early morning to finish the tasks.

On February 18, the planned morning flight had to be scrubbed since the Bell preparations were incomplete. The weather turned bad in the afternoon, finishing any possibility for flights that day. However, both teams continued to prepare the craft for another try the next day. Ruvin worked on the telemetry equipment and had to make a cutter since neither base had such a tool to produce a new gear. When complete, the new gear had to do the work of three units. This promised to be only a temporary solution, and a new set of gears was ordered from Langley.

By early morning, February 19 looked to be a promising day for further flights.[35] Visibility was good in spite of scattered clouds around 10,000 feet. The purpose of the fifth flight was airspeed calibration. The XS-1 would still follow the same drop procedure as in the other tests. The center of gravity would still be 24.9 percent mean aerodynamic chord and gross weight remained at 4,230 pounds. Although the P-47 would handle the chase duties for this flight, Baker refused to fly it because the Bell inspector would not release the plane for flight due to its supposed poor mechanical condition. Baker's boss, Mel Gough, had strict rules about flying uncleared aircraft. Baker wasn't about to fly an unauthorized mission. As a result, Captain Colchagoff flew the P-47 during the test flight.

The B-29 left the ground at 9:00 A.M. to begin the long climb to drop altitude. Dow flew the B-29 as pilot with Cannon as copilot. Once down in the cockpit, Woolams noted the same cold draft as before seeping into the XS-1. It was a problem that had to be fixed since air temperature dropped rapidly in the compartment once the B-29 hose was disconnected. Woolams also noted that any fix on the door seal (the easiest way to fix the leak) could not be done at the cost of ease of door closure. Closing the cabin door was already difficult enough in the cramped confines of the cockpit. Nor was cold the only distraction. Once the pilot was in the XS-1, the ladder was normally withdrawn away from the experimental ship. Without that additional bracing, the XS-1 developed a very noticeable vibration when still attached to the B-29, so intense that Woolams literally felt his eyelids fluttering. While no ill-effects to the XS-1 could be observed, it was something to be watched.

The drop from the B-29 was made with the flaps fully up, the inboard propellers feathered, and the observed airspeed at 200 MPH instead of the previous 180 MPH. Separation time was noted by Woolams as 10:17 A.M. Once again the drop separation was perfect.

After slipping away from the bomber, Woolams steadied the XS-1 and the P-47 pulled alongside to provide a rough airspeed calibration.

This simple test was to be done using the airspeed calibration for the P-47 as given in the Pilot's Operating Instructions, because the Bell engineers felt that the stalling speeds, as recorded on the first flight and reported in the Preliminary Data Report #1 issued by Langley Laboratory, were simply too high.

It was discovered that the P-47 and the XS-1 shared the same position error in regard to speed. (This was to be expected due to the similarity of airspeed systems: a boom-type static pickup ahead of the leading edge of the wing.) Woolams noted that the XS-1 speed read about 1.5 MPH slow at stalling speed. In fact he calculated the XS-1 system about 1–2 MPH more negative than the P-47, which meant that the stalling speed was actually about 2–3 MPH faster than actually indicated.

At approximately 10,000 feet Woolams extended the landing gear. However, he immediately noted that the red light on the instrument panel monitoring the nosewheel did not go out. This meant that the nosewheel was apparently not locked in the down position. In the XS-1's current configuration, this situation could have been a problem. The XS-1's main landing gear could not be retracted once it was lowered except by releasing the downlocks. Unfortunately, the release procedure could only be done from outside the airplane. Luckily the nosewheel was not a part of that system and Woolams had a chance to attempt a midair fix of the problem—quickly, as the XS-1 was rapidly descending and he was running out of time.

Woolams first tried to lock the nosewheel by turning on the auxiliary landing gear pressure system. No luck. Next he turned off the auxiliary system and tried step 2—retract the nosewheel and lower it again. Then turn on the auxiliary pressure system again. No go. The red light was still on and time was rapidly running out. As Woolams made the turn to begin the final descent, he tried one more sequence to lock the nosewheel into place. Following the previously performed steps, he raised and lowered the nosewheel and then turned on the auxiliary boost. Success! The red light disappeared and the instrument panel flashed green, indicating that the nosewheel and the main landing gear were now locked down into place.

Woolams lined up the long runway and cruised on in to Pinecastle Field to attempt an uneventful landing. Without touching the brakes, Woolams let the XS-1 roll out along the runway until the plane's speed slowed to about 10 MPH. At that speed more of the plane's weight was being borne by the nosewheel. Suddenly, the nosewheel retracted and the XS-1 settled on to the edge of the nosewheel door. Accompanied by an awful noise, the XS-1 skidded down the runway on the door edge. When the plane finally came to a stop in the nose-down position, the crew rushed over to inspect the plane. Fortunately, the damage was confined to the door, which was badly worn along its lower edge.

In the control tower, Bob Stanley and Doug Rumsey watched the landing with intense interest.[36] Upon seeing the landing gear difficulties, a worried Stanley raced down from the tower with Rumsey in hot pursuit. When Stanley reached the ground, without pausing a moment he leaped into Rumsey's rental car and took off—without Doug Rumsey. Rumsey was left standing at the tower with only Stanley's motorcycle, something Rumsey had never learned to ride. "Oh well," thought Doug, "no time like the present to learn," and off he went in pursuit of Stanley.

By the time Rumsey reached the Woolams-Stanley meeting, the flight review was complete. Preliminary examination of the XS-1 led to the conclusion that the plane's nosewheel actuating cylinder, which engaged/disengaged the nosewheel downlock, had disconnected from the downlock mechanism.[37] The cylinder had moved to the retracted position with considerable force, indicating that it had disengaged after the landing. The XS-1's landing gear retraction system was a closed hydraulic system that could be actuated by the passage of pneumatic pressure through a diaphragm. Thus, the postlanding investigation postulated, air must have somehow leaked into the hydraulic system and increased the pressure to such an extent that the actuating strut could not be extended without creating a back pressure, thus causing the gear to disengage.

To prevent a recurrence of the nosewheel difficulties, the Bell technicians altered the downlock actuating cam so that the nosewheel actuating cylinder engaged the downlock during extension, but could not disengage the downlock for retraction. In short, once lowered the nosewheel could not be retracted except manually from outside the aircraft. This now made the entire landing gear system identical since the main landing gear had the same characteristics. The Bell technicians also increased the diameter of the locking hole into which the nosewheel downlock pin deployed so that it could be accomplished more easily.

In reality, although this fact was not discovered until much later in the XS-1 program, nose-gear problems would remain a frequent occurrence. The failure of the nosewheel actually resulted from the rather limited elevator control available at the XS-1's normal stall speed. Pilots oftentimes discovered that sufficient elevator responsiveness was not available on landing with the result that main and nose gear wheels contacted the ground almost simultaneously. When that situation occurred, the XS-1 was liable to overload the nose-gear cylinder assembly and, because of the short travel of the shock in the extremely small nosewheel compartment, stimulate a gear failure.

The nose-gear door was removed for repair and the decision was made to continue the flight tests in its absence. Repairs were made at Pinecastle Field on February 20. During the downtime, the NACA group gave Woolams a new program of flight tests. Within the constraints of the XS-1's short flying time, Woolams agreed to attempt to obtain the requested data. The NACA and Bell groups utilized the period for general maintenance and operational work. Since there were no chase duties to perform, Joel Baker flew the Beechcraft C-45 to assist Truszynski with his radar checkout.[38]

Although the XS-1's repairs were finished by the afternoon of the twenty-first, bad weather forced the scheduled mission to be scrubbed. The necessity to work within the system now cropped up as Pinecastle Field was closed Friday, February 22, for a federal holiday. Next it was again the weather as clouds, wind, and rain intervened over the weekend. Thus, it was not until Monday that the experiments could continue.

February 25 promised to be a perfect day for the XS-1's sixth and seventh flight tests.[39] The purpose of the first flight of the day was to examine the XS-1's static directional stability. For the second flight, Woolams would investigate the longitudinal and directional stability. The weather was clear and visibility around Pinecastle Field unlimited. Aside from the changes to the nosewheel mechanism, the plane was the same as in previous drops. The weight remained at 4,230 pounds, and the center of gravity was still 24.9 percent mean aerodynamic chord. Baker would fly chase in the P-47.

Following takeoff, the bomber began the now-familiar slow climb to altitude. Consistent with past procedure, the two B-29 pilots alternated their respective roles—for this mission, Cannon flew as pilot and Dow as copilot. Woolams also followed the now "standard" XS-1 entry and readiness procedures. The drop from the B-29 was made under the same conditions as before, except that the technicians now felt secure enough from examining the films and data to recommend that the inboard B-29 engines no longer be feathered. The instability of the air flow under the bomber did not seem a problem and the next logical step was to see what would happen if the inboard engines were idling rather than feathered. Observed B-29 airspeed was again 200 MPH.

The drop launch was executed at 10:37 A.M. The negative acceleration was −2.5 g's. As Woolams noted, the previous launches at 180 MPH had produced a g force of −2. Thus it was quite apparent that because of the angle at which the XS-1 hung from the B-29 a small increase in bomber speed meant an appreciable increase in the aerodynamic download.

To obtain static directional stability data, Woolams put the XS-1 through full rudder deflection sideslips to the left and right while holding the plane on a constant heading. These experiments were conducted under three settings. The settings were at an observed airspeed of 150 MPH with the landing gear and flaps up, at 250 MPH in the same flight posture, and at 150 MPH with the landing gear and flaps down. Throughout the tests, Woolams noted no buffeting or impending fin stall. As usual, the XS-1 exhibited excellent handling qualities. Landing this time was, thankfully, uneventful, indicating that the nose-wheel fix was suitable.

By mid-afternoon, the crews had recovered the data and serviced the various aircraft. Another flight, number 7 in the test series, was ready to go. No changes were made from the previous flight configuration and takeoff conditions for the XS-1 remained as before. For this mission, the B-29 crew again alternated pilot duties. Once airborne, the B-29 climbed to altitude, but this time the drop conditions were altered. Instead of 200 MPH, the B-29 left the inboard engines running at cruising speed and increased its launch speed to 220 MPH. At 3:37 P.M. the XS-1 began its seventh flight. Baker again handled chase duties in the P-47 and the B-17 was ready for filming of the flight.

The drop was perfect with no tendency to yaw or roll noted in the launch. The speed increase seemingly had no effect on drop conditions. However, negative acceleration had increased to about −3 g's with the increase in speed. Since the acceleration was only momentary, coinciding with the drop, Woolams noted little particular discomfort. It would be possible, he believed, to relieve the magnitude of these accelerations by reducing the ejector launching pressure to lessen downward force.

Once free of the B-29, Woolams put the XS-1 through a series of maneuvers to examine longitudinal and directional stability. At 200 MPH he executed a series of steady turns as he attempted to measure stick force versus the g data. Stabilized data was taken at 2 g's, but attempts to attain 3 g's were discontinued when the XS-1 experienced buffeting at 2.2 g's. After lowering the landing gear, and with the flaps in the down position, he executed a series of full rudder sideslips. Maintaining a constant heading but with speed dropping, Woolams sideslipped both right and left, first at 175 MPH, and then at 125–130 MPH. Landing was uneventful.

On February 26, the Bell Aircraft group prepared to conduct two additional flight tests.[40] The nose-gear door was still under repair and the flights were made without it. Flight conditions were excellent and no changes were made in the XS-1 from the previous day's tests. Takeoff was early in the morning with Cannon (pilot) and Dow (copilot) at the controls of the B-29, and Joel Baker in the P-47. The purpose of the flight was to examine the XS-1's dynamic stability.

Once airborne for the eighth glide test, the crew went through drop procedure. The B-29 inboard engines were set at cruising speed and the observed airspeed at drop was 220 MPH. To curb the sharp negative g force that Woolams had encountered on separation during the last two drops, the ejector pressure was reduced from 600 psi to 400 psi. In effect, this would lower the downward force (from 3,000 to 2,000 pounds) being exerted on the XS-1 at the moment of separation.

The drop occurred at 9:44 A.M. and was as clean as previous efforts. Even more important, the negative acceleration at the moment of separation from the B-29 was reduced, dropping from the more than -3 g's to -2.7 g's as a result of lowering the ejector pressure. During the flight, dynamic directional, lateral, and longitudinal stability data was taken at observed airspeeds of 250 MPH and 150 MPH. While performing these tests, Woolams noted that centering of all three flight controls was poor. He attributed this to the light aerodynamic forces in the low-deflection range and the relative high static frictional forces. However, he indicated in the pilot report that this characteristic was not objectionable for an aircraft of this type, since centering at high speeds remained adequate and it was "considered desirable" to equip supersonic test aircraft with irreversible controls. This latter item on the typed pilot report brought forth an exclamatory handwritten notation from Bob Stanley stating for the record that not everyone was in agreement on this statement.

By late afternoon, the XS-1 was ready for another test. The weather remained good with only scattered clouds at 4,000 feet. No changes had been made to the XS-1 and the purpose of the flight would be to investigate the experimental craft's rate of roll. However, Joel Baker had been sent to the maintenance facilities at Robbins Air Base in Georgia to pick up some equipment and was unavailable to perform his standard chase duties.[41]

The pilot duties for the B-29 continued the alternating pattern.[42] The B-29 conditions for the drop were as previously utilized, that is, inboard engines cruising with speed at 220 MPH. However, for this test the ejector pressure was lowered once again and set at 200 psi. As a result, the XS-1 separation was again clean and negative g force fell to −2 g. It was obvious by now that the XS-1 launch did not require elaborate precautions to obtain a clean separation. The remaining question concerned the correct mix of precautions needed to ensure safety but allow reduced B-29 preparations. The ninth flight of the XS-1 commenced with separation from the B-29 at 4:54 P.M.

After clearing the bomber, Woolams began the scheduled tests for the flight. At an observed airspeed of 150 MPH he put the XS-1 through a set of abrupt rudder, fixed aileron rolls. The rate of roll versus aileron deflection data were obtained for one-quarter, one-half, three-quarters, and full aileron deflection under the following conditions. First, Woolams examined the aileron deflection with the rudder fixed and the flaps and landing gear up. Next, he applied excessive rudder, but with the aircraft still clean, to overcome what Woolams found to be excessive yaw. Finally, with the rudder fixed, Woolams lowered the flaps and landing gear.

Upon completion of the tests Woolams prepared to land. Suddenly, the attention dedicated to wringing out a fine aircraft gave way to the dangers always inherent in flying. At about 6,000 feet and well into the landing pattern, the windshield glazed over, making it impossible for the pilot to see. Only a strip about four inches wide along the lower portion (of all the luck, it would have to be the bottom rather than the top, given the poor visibility out the XS-1's windscreen to begin with) was clear enough to provide vision for the pilot. With the slant of the XS-1's canopy, the bottom of the glass gave an excellent view of the nose rather than parallel to the ground as in other aircraft. Woolams had faced this type of situation once before in his career, at Muroc while testing the XP-63 Kingcobra. The XP-63 engine had developed an oil leak that covered the windshield with black fluid. Witnesses present that day later testified to his skill in belly-landing the aircraft and walking away as cool as if on an afternoon stroll.[43] Now he was in similar trouble.

Large quantities of glycol had siphoned out of the windshield de-icing tube and covered the windscreen, a condition similar to someone spraying a garden hose on your car's windshield on a cold day. The pressure of the airflow over the canopy insured a crazy-quilt vista and made judging distances impossible. The lack of depth perception was a serious situation both because Woolams did not have enough instruments in the experimental craft to fly blind and because he was on final—indeed only—approach to land. With no time to lose, Woolams tried to clear the windshield, but the problem was external and Woolams had few remedies at hand. Nothing helped and as he passed through 4,000 feet the problem became even worse. The glycol kept draining and the thin strip through which Woolams was viewing the outside slowly closed down to almost nothing. It is at times like these that experienced test pilots reveal their skills. With steps 1, 2, and 3 failing to clear the windshield, Woolams reached for the XS-1 door lever. By jettisoning the door, at worst Woolams could look out the open hatch and try to fly the XS-1 down using his sidelong view to judge distance to the ground. Certainly there was no thought of leaving, since no one was sure bailing out was entirely feasible. However, just as Woolams was on the verge of popping the XS-1 door, the glycol siphoning stopped and the outer windshield suddenly cleared.

Touchdown was normal and Woolams breathed easier. Preliminary investigation failed to determine the reason for the siphoning. However, a decision was made to substitute alcohol for glycol in the windscreen de-icing system to prevent a recurrence of that extremely dangerous situation.

During the four flights on February 25 and 26, the nosewheel door remained off the XS-1. A new door arrived on February 26, but Stanley made the decision not to delay the flight tests simply for the door installation. It would not be possible to install the door on the twenty-seventh because Pinecastle Field would be closed for an Army parachute team demonstration.[44] On the morning of February 27, not only was it very cloudy, but the temperature dipped unexpectedly into the 40-degree range. The Army demonstration was postponed until February 28. But, as anyone by now could predict, on February 28 Pinecastle Field again experienced very bad weather with scattered showers and thunderstorms. With that sort of weather, the Army simply gave up and canceled the planned demonstration. Bell Aircraft was notified that the field would again be available for the XS-1 tests on March 1.

Between Friday, March 1, and Tuesday, March 5, the weather at Pinecastle was very mixed.[45] No flights were attempted and the Bell team spent the time working on modifications to the XS-1 and B-29. Because of the successes with lowering the ejector pressure for XS-1 separation, it was decided to make all subsequent flights with the ejector removed. Similarly, the XS-1 compartments designed for the fuel tanks were modified to create bins to store ballast. Shot-bag ballast, equal to the weight of the absent rocket engines, was added to the experimental craft. However, to check handling and separation characteristics, the ballast would be added in two stages. First, for the tenth glide flight, only one-half the ballast to equal the weight of the engines would be added. When added to the weight of the material used to fabricate the storage bins, the XS-1 would increase in weight 580 pounds. Care was taken by the Bell team to add the weight in such a manner that it did not affect the center of gravity of the aircraft. It was planned to add the additional ballast on subsequent flights. This would allow the plane to be tested at the true weight of the plane minus the weight of the fuel.

While work at Pinecastle slowed due to the weather, new names and problems appeared at Buffalo. In February, Bell test pilot Richard H. "Dick" Frost was named as the new XS-1 project engineer.[46] (This change allowed Stan Smith to move over to the new XS-2 project.) Born in 1917, Frost received his bachelor of science in aeronautical engineering from Rensselaer Polytechnic Institute in 1940. After working in the aviation industry for three years, he joined Bell Aircraft as a test pilot in late July 1943. Frost's extensive mechanical systems background made him a natural for the upcoming rocket-powered portion of the XS-1 project.

Also during February, technicians at Bell Aircraft continued to move forward with finishing the two other XS-1 aircraft. Work on the XS-1 numbers 2 and 3 revealed remarkable progress. While the number 2 aircraft continued under assembly, the components for the number 3 plane were nearly finished. A third 10 percent wing was almost ready and the first of the new thin 8 percent wing was under component construction. Bell engineers were currently conducting tests on the new high-pressure nitrogen evaporator. Pressure-testing on February 27 revealed leaks in the system. Repairs were anticipated to delay further testing until early March.

At RMI, the progress was not so satisfactory. The long-anticipated rocket engine was still being delayed by the failure of the igniter system. At General Electric, which had taken over work on the turbine pump from RMI, the news was equally bad. No appreciable progress had been achieved on the turbine pump for the number 3 airplane. On February 25, a conference at Buffalo with Bell and GE personnel indicated the latter company's efforts to move ahead in spite of a strike at the GE plant. While Buffalo continued to experience delays, the work at Pinecastle received some added help.

On the afternoon of March 5 Alvin "Tex" Johnston, Bell's assistant chief test pilot, flew down to Pinecastle Field to check out in the XS-1. During a leisurely dinner that evening, Woolams briefed Johnston on the handling and flight characteristics of the XS-1. Years later, Johnston still remembered how pleased Woolams was with the response to controls and handling of the plane.[47]

On the morning of March 6, the weather improved but not to the level of conditions in the early flights. The sky above Pinecastle exhibited broken clouds at 4,000 feet, but the Bell contingent determined to go ahead with the scheduled flight. Tests were planned on the XS-1's static longitudinal stability characteristics. Woolams would fly the morning flight and Johnston would handle the afternoon sortie.[48]

Changes made during the period of inaction had been numerous, including the addition of the above-mentioned 580 pounds of ballast. This figure brought the XS-1 launch weight up to 4,764 pounds. The nosewheel door had been reinstalled and a new gasket put in the defroster system along with the planned fill-up of the de-icing unit with alcohol. Finally, the static pickup for measuring the XS-1's airspeed had been removed from the left wing pitot boom and hooked to the nose pitot boom.

Takeoff of the B-29 was scheduled early in the morning in order to improve the chances of avoiding any bad weather. Cannon flew as pilot and Dow handled the copilot duties. Climbing to 25,000 feet, the B-29 prepared to release the XS-1 on its first weighted flight. The B-29 flaps were set in the up position, the inboard engines were set on cruising, and the ejector strut had been removed. Bomber speed was raised to 240 MPH. At 8:32 A.M. Dow pulled the shackle release and the XS-1 dropped away from the B-29. In the P-47, Joel Baker gave chase to the orange experimental craft. As before, the launch was perfect. Notably, in the absence of the ejector, the negative acceleration on separation was reduced to -1.9 g's.

The reason for the relocation of the pitot boom was quickly confirmed. The position error of the pitot boom in the nose configuration was discovered to be slightly more negative than in the wing location, a fact revealed by putting the aircraft into stalls and noting the observed airspeed. For flaps up with the ballast, it was seen as 120 MPH. For flaps down, stall was noted at 110 MPH. Again Woolams observed that the outer wing sections did not stall with the elevator in the full up position. His reasoning for that phenomenon was that the XS-1's center section would buffet and stall before a high enough angle of attack could be reached to stall the outboard wing sections.

Static longitudinal measurements were taken from stall speed up to 385 MPH. This exercise was conducted with the horizontal stabilizer trimmed for zero elevator force at approximately 300 MPH. Conditions set for this required the XS-1's nose to be trimmed down, with the stabilizer set two degrees up from neutral.

The ground technicians began preparations for another flight as soon as the plane landed. While loading the XS-1 back aboard the B-29, the nose airspeed boom that also functioned as the telemeter antenna was broken. The break occurred right at the micarta insulating block that divided the boom into two pieces. Work was undertaken to fabricate a new micarta block in the Pinecastle Field maintenance shop.

While awaiting the boom repair, Woolams and Johnston went over the details for the latter's afternoon checkout flight. At noon, Woolams received an unexpected call from Bell HQ. There was to be an inspection visit to Buffalo by Maj. Gen. Benjamin Chidlaw, the deputy commander of the Air Materiel Command at Wright Field. During his tour of the Bell plant, Chidlaw wanted to observe the aircraft firsthand. Since the ten drop tests had successfully demonstrated the feasibility of releasing the XS-1 from the B-29, one of the major purposes of the flight tests had been accomplished. As a result, Bell did not wish to risk the aircraft on any further tests without the engine and tanks installed. Also, Bell wished to rework the landing gear in light of the experiences gained at Pinecastle. As a result, Woolams was instructed to load the XS-1 and return to the plant. On March 7, Dow took the B-29 controls for the return flight to Niagara Falls.[49]

After the call detailing the end of the Pinecastle tests, the NACA crew busied themselves with final calibration of the test instruments. Then, the recording instruments and telemeter transmitter were removed from the XS-1 for transport back to Langley. On March 7, Baker, Hinman, and Hayes took the C-45 on a direct flight back to Langley Field. Walt Williams left Orlando that same afternoon on an extended annual leave. Ace Ruvin and Gerry Truszynski remained at

Pinecastle to pack the telemeter equipment and turn in the borrowed Army equipment. On March 8, a C-47 from Langley arrived to transport Truszynski, Ruvin, and the NACA equipment back to Virginia.[50]

The Florida phase of the program had come to an end. In spite of the many past program difficulties, the project was finally underway. It had proven to be an outstanding success from the standpoint of the flights. The ten glide tests had demonstrated the feasibility of the B-29 mother ship air-launch procedure. The issues relative to separation and drop had proven satisfactory. The tests had revealed excellent XS-1 flight characteristics and provided no major surprises. The NACA had obtained promising data to form the basis for analysis of the later flight tests. Finally, the difficulties regarding landing space and the frequency of accidents had virtually eliminated any thought by Bell or AMC of using Niagara Falls or Langley Field as a test site. The only philosophical drawback, cooperation between the three participants, remained unresolved. From the perspective of the teamwork of the three (Bell/NACA/AMC) players, the Pinecastle tests could only be labeled a qualified success. But more tests were to come and this issue would be revisited.

3

The Program Takes a Pause

Politics

The completion of the Pinecastle tests marked a respite in the program. The work to finish the number 2 airplane for the next phase of flying promised to keep the project at a standstill for several months. During that time, the AAF and NACA continued a debate on two subjects that would weigh heavily on the XS-1 program over the next year: the control of the program and the cost of research flying.

Bureaucratic imperialism is a fact of life in Washington politics.[1] This condition applies when two or more agencies seek to apply permanent control over the same area of policy. In 1945, the politics of policymaking had created a collision course between two old allies: the NACA and the recently invigorated Army Air Forces. The postwar allocation of aviation resources surfaced as the primary area of disagreement between the two agencies. As a result, the politics of budgetary allocation framed the early debate over the MX-653 program. Now with the initial Pinecastle tests of the XS-1 finished, the larger issues of the supersonic program—who was to lead the way, NACA or the AAF, in aviation research and development, and what was to be the degree of independence between the two efforts—came back to the forefront.

During the war, the significant increase in all research and development expenditures propelled the United States to technological supremacy in many areas.[2] Aviation research was particularly enhanced as annual expenditures rose from $250 million in 1939 to over $800 million in 1945. AAF research expenditures increased from $10 million in 1940 to $121.6 million in 1944. (The Navy received almost $80 million and the NACA about $45 million.) Perhaps one-half of the total federal government research and development dollars spent between 1939 and

1945 were spent on aviation. Most of these contract dollars ($337.4 million out of $418.7 million) went to the nation's booming aviation industry. Conversely, most of the $95 million expended by the NACA during the war was directed to support research within its own facilities. Both the AAF and NACA managed to vastly expand their facilities and capabilities.

The new partnerships fostered by the war and the vastly increased technological capabilities, when coupled to the experience of the war years, provided fertile grounds for new directions in aviation research by both agencies. The NACA recognized the fundamental difference in the way they conducted basic science research before the war and their role in applied research problems during the war. The postwar AAF wanted merely to control its own aviation research destiny.

As discussed earlier, after the American discovery of the extent and significance of Germany's high-speed research efforts, both the NACA and the AAF moved to carve out independent programs in the emerging research area. By December 1945, each agency had developed a new and very expensive wind tunnel program that offered significant bureaucratic policy advantages to the parent organization, but each also required an enormous commitment of scarce new program dollars. When the NACA discovered the outlines of the AAF program, they publicly urged coordination (under NACA direction) between the two agencies to ensure no costly duplication of effort.

The AAF seemingly offered little enthusiasm for the NACA "coordination," feeling it was another attempt to maintain the status quo in aviation research—a status quo that senior AAF officers believed had failed the service during the war. Pearl Harbor had changed the illusion of the safety of American isolation. In future wars, airpower would initiate hostilities on the first day, and America had to be ready *before* that day. The only way to ensure that readiness, the AAF believed, was to have a ongoing long-range research and development program under military control.

With that divergence of opinion, the ground was set for a policy collision. As a result, an intense behind-the-scenes competition grew during early 1946. Both agencies moved vigorously in an effort to find allies in Congress and industry, while at the same time attempting to neutralize the other's efforts. This sparring continued throughout the spring and summer as industry, the services, and the NACA competed for political support. Each group proceeded from a different perspective: the NACA wanted to continue its role in basic research for the nation and to establish dominance in the emerging area of high-speed research; the military services were concerned that the NACA would not move fast enough or in great enough depth to ensure continued American aviation supremacy; and industry was concerned that government agencies would continue to encroach into development work that traditionally had been the private companies' milieu.

All participants proceeded from the recognition that bureaucratic power was necessary to create the foundations for the new program. That power would be organized around a constituency, which in turn centered on the jurisdiction of the new program. Lose out initially and any hope of recovery would be difficult, based on the simple lack of power to influence the system. In such a zero-sum arena, it is little surprise that no compromise could initially be found. However, the need for some degree of consensus among the three participants in order to produce a congressional agreement finally forced a degree of cooperation. Thus in April 1946, in an attempt to find a middle ground, the NACA, the AAF, and industry agreed to form a review panel under the chairmanship of Arthur E. Raymond of Douglas Aircraft. By June, the Raymond panel had merged the various competing proposals. However, as the policy compromises were made by the panel, the initial program costs of slightly under $1 billion in new facilities grew to over $3 billion. The proposed "combined" wind tunnel program for supersonic research seemingly had everything for everyone in a willful attempt to find the necessary congressional and industry support.

The federal Bureau of the Budget, worried over the competing agency claims and costly programmatic overlap, struggled to find another way to meet the needs of supersonic research. Collectively, Congress expressed considerable dismay over the bureaucratic struggle for supremacy over the wind tunnel program, but at the same time its individual members sent contradicting signals to their allies. Throughout the summer of 1946, the struggle continued with no resolution in sight, but with profound implications for the overall supersonic program.[3] At the same time as the policy struggles for the soul of the research program were under review, the tactical issues of research flying were receiving renewed attention.

Late in 1945, a series of memoranda surfaced concerning the financing of the NACA research projects.[4] Although the stated purpose of the initial memo (by Maj. Don Eastman, AMC liaison at LMAL) was to outline the large stake the AMC had in Langley projects, no one at Wright Field was quite sure what the end result of such information would be. The Eastman memo revealed that the cost of research at Langley Field from 1940 to 1945 totaled about $35.5 million. Of that total, $22 million were the NACA-related overhead costs and thus not directly applicable to any research project. Of the remaining amount, Eastman estimated that the AAF had provided $8.5 million and the NACA had spent $5 million on direct costs of research. As a result, Eastman indicated that during the war the AAF had contributed 50 percent of the cost of operating the LMAL research program. The news, while welcome in revealing the AAF's investment in the NACA's research program, stirred the Engineering Division at Wright Field to ask Eastman why such a memo was prepared. If the purpose of the memo was to suggest the establishment of a directing board to handle these matters, then the AMC considered such action "detrimental to the amicable relations between the Army Air Forces and the NACA."

After further discussions, the AMC reached a better understanding as to the purpose of the Eastman letter. Two intentions were cited for the correspondence. First, the Engineering Division's personnel were reminded of the large wartime participation of the AAF in LMAL research. It was AAF funds, not NACA money, that drove much of this research. At budget time in Washington, with postwar defense dollars so scarce, this was an important consideration. Second, the memo served to point out to Engineering Division personnel on the NACA committees that their participation was not merely a theoretical exercise in research management but rather served to define the AAF's aviation future. But the new directions for flight research did not end with a discussion of costs.

In January 1946, in response to the flurry of cost estimates and memos on cooperation, Brig. Gen. Laurence Craigie outlined the new policy on how research was to be conducted. In the past, Craigie reminded his personnel, the NACA conducted its research projects based on Army/Navy requests for assistance. In the postwar period, the NACA would continue to be responsible for ongoing research programs. If the AMC were to derive any benefits from this basic research, it was imperative that the AMC representatives be aware of that research and how it fit Wright Field's needs, which the AMC needed to define for NACA. Conceding projects to the NACA would weaken the AAF's ability to set its own research agenda. The AAF would have to increase its reliance on NACA goodwill and cooperation. The discussion, while straightforward in purpose, obscured a bigger issue. The NACA would soon have large amounts of new congressional funds. To compete for research issues priorities, the AAF must do a better job identifying and promoting its own research agenda. Craigie was ensuring that his troops knew the score on this issue.

It was not just NACA research that AMC wanted to adjust. On January 24, Craigie took the additional step of outlining a new policy regarding research and development aircraft in a letter to all aircraft manufacturers,[5] indicating that the AMC expected that aircraft manufacturers would deliver experimental aircraft to the AAF after not more than 30 hours of flight testing. If the (conventional) plane failed to meet manufacturer speed guarantees, then AMC would extend "overriding protection." However, no safety net would be provided for experimental aircraft. Other performance deficiencies would be subject to negotiation. Additionally, all components were expected to operate at a reasonable design efficiency (expressed as 80 percent of design). If the airplane was suitable for production, the AMC would issue a contract for any additional redesign, modification, installation, and flight testing necessary to produce a 100 percent specification production prototype. The results of this policy, Craigie hoped, would be that aircraft manufacturers could substantially reduce the amount of contingency funds necessary to achieve a 100 percent experimental aircraft. All existing contracts were to be reviewed with this procedure in mind and resubmissions revising those contracts were "invited." The AMC had made it easier to develop aircraft, but how much easier would it be?

On August 28, 1946, Bell Aircraft replied to a AMC inquiry regarding how the new policy might fit the most unusual of Wright Field's stable of projects. While reminding the AMC that the XS-1 contract originated before the new policy, Bell Aircraft indicated it was striving to follow the policy in the "interest of conserving public funds" and would work to limit its acceptance tests to those required to demonstrate that the plane functioned "in a satisfactory manner and is safe for flight." While this statement would be welcome at Wright Field, the cross purposes of the AMC policy and NACA interests in supersonic research would quickly surface.

Bell Continues the Program

On March 25 Dow and Woolams piloted the B-29 on a 5-hour, 10-minute flight to Oklahoma City for a routine maintenance overhaul, including an engine change and cowling modifications. The maintenance work continued until April 28. Cannon and Dow piloted the bomber for the return flight to Niagara Falls.[6] Woolams could not be along for the return flight because he had disappeared to an faraway "exotic" location.

Muroc Dry Lake in Southern California's high desert had received considerable attention during the initial discussions for a place to test the XS-1. At that time, the primary concern for the project revolved around the annual bad weather due to strike the Muroc area in December/January. The resulting flooding of the dry lakebed rendered testing impossible until the evaporation of the water. The ATSC/Bell participants agreed that any postponement for better weather would simply delay a program already significantly behind schedule. But after the return from Pinecastle Muroc received renewed attention. With the difficulties he had experienced with the XS-1's high rate of descent and the numerous clouds in Florida, Woolams recommended testing in California. Bell Aircraft based their recommitment to Muroc on several circumstances. First, a flight test base was already present (as at Pinecastle) complete with facilities and equipment; military personnel were available for assistance and security; the weather was usually excellent (unlike the unpredictable Pinecastle); the dry lakebed was available for emergency landings; and the remote site promoted secrecy and provided distance from populated areas. Bell Aircraft had previously tested at Muroc in 1942–44 with the XP-59 jet and was aware of the advantages of the AAF base.[7]

Muroc Army Air Field was the end of the universe. Even the official history described it as "located in the middle of a sandy waste dotted with sage brush and Joshua trees."[8] The base was first used in 1933 as a bombing and gunnery range. Primitive conditions for personnel existed from the first days. The original Army Air Corps (AAC) units landed on the vast Rogers Dry Lake, and the crews bivouacked nearby. In 1937, the entire AAC participated in extensive maneuvers at the base. The attack on Pearl Harbor triggered an increased interest in the use of Muroc Army Air Base (so designated until November 8, 1943, when it changed to Field). In late 1941/early 1942, the U.S. government spent approximately $2 million to upgrade the facilities.

The result was a concrete runway, new sewage and water systems, and "temporary-type" buildings. Throughout the war Muroc Army Air Field trained significant numbers of U.S. fliers. By April 1946 all training had ceased at Muroc and it was designated as an AAF flight test center for experimental research and engineering development work.

All those who served at Muroc endured the extremely primitive conditions. Many learned a valuable lesson of service life—in the military, temporary means forever, including the wooden buildings at Muroc. Many of the same buildings that were erected in 1941 greeted the XS-1 team in 1946. For all the upgrades, Muroc was still a surprise to the unsuspecting. Warm sun and beautiful skies masked windswept desert flats; dust and dirt were everywhere. The temperature could fry a body at 115 degrees Fahrenheit and then freeze that same soul with chilling cold, easily within days of each other. The offbase rents were high, the distance to most homes long, and the temporary base structures barely adequate. But Muroc had an abundance of that which the XS-1 program desired: wide-open spaces, a vast dry lakebed, and nearly year-round flying weather.

In late March, Woolams made the journey to "exotic" Muroc to prepare the Bell flight facility for future XS-1 tests,[9] but this was not the first time he had seen the dusty landscape. During 1943–44, Woolams was the principal test pilot on the XP-59A, America's first jet aircraft. He eventually became responsible for the Bell Aircraft flight station at Muroc. During his time in California, Woolams added substantially to his legend as a flier on the edge: a wild individual, fun-loving and unpredictable, yet one balanced by extreme professionalism.

Since this program required specialized facilities, Woolams traveled to Muroc to ensure there would be no delays when the XS-1 number 2 was ready to test. Preparations proved to be minimal. Woolams contracted to have a large tank constructed to hold the liquid oxygen (lox) and a smaller tank to contain liquid nitrogen erected at the west end of the flight-line apron. Additionally, a mixing tank for alcohol and water was created by modifying a small Army fuel trailer. Finally, a pit similar to the ones already constructed at Niagara Falls and Pinecastle was prepared to handle B-29/XS-1 loading. A spur railroad track was run to the pit for a liquid oxygen tank car to be brought to the large lox tank for periodic refills. Office facilities were obtained for Bell personnel, although space was extremely cramped. Woolams managed to rent six houses at Willow Springs (about 25 to 30 miles from the base) for Bell personnel as well as finding some quarters on base. Ground transportation consisted of renting a car since the military had no vehicles to spare. Little progress, however, was made on securing the continued use of the necessary radio frequency. By the time the modifications and new equipment were ready, Woolams had been at Muroc over a month. He finally returned to Bell Aircraft during the second week of May. Just as he was leaving, the new base commander for Muroc arrived on the scene. Colonel Signa A. Gilkey was destined to play a recurring role in the events at the base over the next several years.

While Woolams prepared the new site for additional XS-1 testing, work on the two aircraft at Buffalo continued unabated. During April, the fuselage on the number 1 airplane was prepared for installation of the nitrogen and fuel tanks. A fin cap antenna, similar to one already on the number 2 and 3 airplanes, was installed. The major XS-1 structural changes at Buffalo centered around the construction and retrofit of the new 8 percent wing and 6 percent tail for the number 1 airplane. The change would not affect the number 2 plane, which would continue to utilize a 10 percent wing and 8 percent tail assembly. By mid-May, two sets of 6 percent tail assemblies and one 8 percent wing were nearing completion. By the end of the month, the fuselage center sections that coupled the wings to the body of the aircraft were ready to be riveted into place. The thin wing sections were, however, still receiving their outer skins. Bell engineers completed installation of the number 2 plane's nitrogen and fuel tanks and pressure-tested the cockpit. Final mounting provisions for the NACA equipment were completed and the ARC-5 cockpit radio installation was finished. All structural components of the number 3 airplane were complete by mid-month. Mock-up work on the number 3 airplane's anticipated General Electric turbine pump was also progressing satisfactorily, although no progress reports from GE on the item had been recently received.

At RMI, the rocket motor passed its acceptance tests on March 23. An engine delivered to Bell Aircraft on March 26 was used for mock-up purposes to complete the number 2 airplane's plumbing and wiring requirements. The only delay in moving the RMI rocket motor to the Bell powerplant test facility was caused by the fire damage to the building from Bell Aircraft's own single-cylinder engine test on March 4. After completing the mock-up, the RMI engine was to be used for indoctrination of the Bell Aircraft personnel. RMI would deliver another engine for the number 2 airplane. However, at the end of the month, Bell Aircraft was advised that RMI intended to move its entire plant to a new location. The result was to once again delay testing and delivery of the number 2 airplane's motor.[10]

In spite of the delays by Bell for its engineering work, the XS-1 program bosses at Wright Field wanted a closer look at their new research airplane.[11] On April 25, the AMC requested by wire that Bell Aircraft send the XS-1 number 1 to Wright Field by May 10 for exhibition purposes. On May 8, the NACA forwarded to Wright Field the still photos and film of the tests at Pinecastle. While the AMC wanted to see the real airplane at Wright Field by May 10, like everything else in the program this date quickly slipped to May 15, and then finally to May 17. Prior to that date, Bell Aircraft continued to make improvements to the two airplanes and the mother ship. During this delay, the B-29 received installation of a loxygen tank for topping off the XS-1 system, an additional catwalk, and an emergency XS-1 jettisoning control. When that work was finished, the B-29 took the number 1 airplane to Wright Field for an extended stay as part of an AMC open-house publicity effort. (Seemingly, everyone was willing to ignore the classification of this program.) The publicity "coincidentally" followed the April appointment of the Raymond panel selected to sort out the competing NACA/AAF claims for supersonic research facility dollars.

The special Wright Field extravaganza featured ten exotic X-craft, including new jet bombers and prototype jet fighters.[12] Jet bombers displayed for the press included the North American XB-45, the Consolidated Vultee XB-46, the Boeing XB-47, Martin XB-48, and Northrop's XB-49 Flying Wing. Fighters included the Northrop XP-79B with its prone pilot, the North American XP-86, and the Curtiss XP-87. But in spite of the impressive hardware present, the XS-1 received top billing in much of the written press coverage. During the open house, the AAF revealed the previous XS-1 glider tests (without mentioning Pinecastle), the extensive new NACA instrument package for identifying the characteristics of the transonic zone, and the recently developed flight pressure suit to allow the pilot to reach 80,000 feet. Also unveiled for the public was the new XS-1 ejection seat. Supposedly, this new equipment would allow a pilot disabled by passage through the transonic barrier to be blasted free from the XS-1 by a special charge and then allow his parachute to open automatically.

The wonderful news story certainly preceded reality as no ejection seat was ever fitted to the original XS-1.

The massive air show was intended to support the military need for the substantial federal investment dollars required to provide an expanded AAF research and development program. During the press tour, both Maj. Gen. Curtis E. LeMay, deputy chief of the Air Staff for research and development, and General Craigie of the AMC publicly reiterated the call for an annual expenditure of $250 million per year for five years and $600 million for the proposed Air Engineering Development Center. Dr. Vannevar Bush, director of the secretary of defense's research and development board, perhaps unintentionally outlined the real reason for the publicity campaign when he stated that the supersonic unitary wind tunnel program must be coordinated, but "it must preserve the vigor and resourcefulness of highly autonomous groups pursuing actively the ideas which they believe most important." In his call for friendly competition, Bush further stated, "It is a good thing for this country that we should have vigorous groups pursuing their several parts somewhat along parallel lines at times, and with emulation and strong competition to keep them on their toes." Thus, while the Raymond panel looked for a compromise, the AAF pursued its independent objectives even though the program was couched in cooperative terms.

On June 5, Dow and Woolams piloted the B-29 with the XS-1 attached back to Bell Aircraft.[13] During the number 1 plane's absence, the number 2 aircraft entered its final assembly stage. All fuselage sections of the plane were bolted together (utilizing the 10 percent wing and 8 percent tail assembly), the nitrogen system was completed, and the rocket motor, hookups, and landing gear were installed. As soon as the XS-1 number 1 returned to Niagara Falls, Bell engineers began the laborious task of disassembling the plane and preparing for the installation of the powerplant system. The disassembly would also allow for the addition of the new thin wing and tail. By the end of June, the first 8 percent wing and 6 percent tail were almost ready. At the same time, the number 2 airplane was finished except for the installation of the rocket motor and pressure regulators. Other Bell Aircraft technicians finished a temporary assembly of the number 3 airplane so that engineers could run a series of cabin-pressure tests. The cabin sealing problem remained difficult to solve. As a result, the modifications to permit cabin-pressure sealing on the number 1 and 2 planes was postponed until a suitable solution could be found through the tests on the number 3 airplane.

While progress was made on one front—the cabin-pressure sealing—difficulties appeared from another direction. In early June, scientists at the University of Southern California notified Bell Aircraft that a cabin pressure of 5 psi could be extremely hazardous to the pilot in worst-case conditions. They recommended a maximum of 2 to 3 psi for the tests. Woolams consulted with the scientists at the AeroMedical Laboratory at Wright Field, and those from USC. In the end, the 3-psi differential became the adopted target although leakage rates not to accede 1 psi per hour took many weeks to achieve. A request to the AMC for expedited approval of this change did not meet with rapid response.

At RMI, the second rocket engine passed its acceptance tests, and the corporation began work on the third engine. During the wait in June for the RMI engine, Bell Aircraft engineers tested the nitrogen, loxygen, and fuel systems and propellant jettisoning systems for the number 2 plane. They also checked out the landing gear and flaps. All systems passed Bell Aircraft and AAF inspection. In anticipation of the number 2 plane becoming ready for captive flight testing, Bell Aircraft technicians prepared the B-29 for a practice flight. A ground test for the lox topping-off and practice procedures was arranged and the B-29/XS-1 number 2 passed the test. Also, the nitrogen evaporator that would be used to generate the 4,500-psi gaseous pressure needed for flight testing performed successfully on numerous occasions. Seemingly, the mechanical equipment and the procedures were slowly coming together in a workable package. In spite of all the past delays, Bell Aircraft finally advised the NACA in late June that the XS-1 number 2 would travel to Muroc during the last week of August.

On July 17, RMI delivered the second rocket engine to Bell Aircraft.[14] After the number 2 plane installation preparations were verified as satisfactory, the engine was removed to the rocket test cell. From July 26 through August 12, the Bell Aircraft Niagara Falls facility was closed for vacation. Thus the first test of the engine did not occur until August 13. The test uncovered another failed part (propellant valve), and spare parts were ordered. This failure had also occurred with the first engine, which had been returned to RMI on July 29 due to the burnout of one cylinder. Even with the failure taken into account, that RMI engine had proven remarkably durable. The Bell Aircraft engineers had operated the engine several times a day every working day for three weeks. A total of 1½ hours of operating time had been gained on the engine. While Bell Aircraft expected the first engine's return on August 12, it was not until August 23 that the work at RMI was finished and the motor returned to Bell. When the engine was installed in the rocket test facility and test-fired, the results were disappointing. When Bell engineers discovered that the number 1 cylinder was almost burnt out, they proceeded with tests using just the other three cylinders. The problems with the rocket engine seemingly confirmed the Stack/Gough concerns about rocket propulsion systems being the XS-1's Achilles' heel.

Work continued at an accelerated pace on the other components of the XS-1 program. During the early summer, Bell Aircraft secured the use of an AAF North American P-51D (#44-13131) and Douglas C-47 (#349010) for use during the tests. After reopening the plant, Bell technicians fit the new thin 8 percent wing and 6 percent tail to the number 1 aircraft and calibrated the strain gauge installation on the number 2 airplane, work that had originally been scheduled during the July shutdown of the Bell plant. Bell Aircraft notified the NACA that the LMAL group should stand by to travel to Niagara Falls on September 16 to complete the NACA instrument calibration for the number 2 plane. All work was stopped on the number 3 plane to concentrate on the other two, ostensibly because little progress was in evidence at GE on the turbine pump so there seemed no need to hurry the number 3 plane to completion. Only the components for the second set of thin wing and tail, seemingly for the number 3 airplane, proceeded. As a result of this decision, the primary NACA requirement for the original program—the longer flight time provided by a turbine pump–equipped plane—slowly receded from consideration. At the same time, the leakage problems in the nitrogen evaporator caused a continuous round of inspections and reassembly. The good news on this vital piece of equipment was that the installation of larger liquid nitrogen spheres provided excellent results and cut the time necessary for the production operation. Likewise, efforts involving the B-29 revealed more success. On August 15, Dow and Woolams took the B-29 aloft for a 1-hour, 40-minute instructional flight. Procedural matters were simulated. By late August the B-29 was declared ready for the XS-1 tests and was standing by for the trip to Muroc. On September 4, Dow and Cannon spent 1 hour and 45 minutes in the B-29 rehearsing takeoffs and landings with simulated cargo. The B-29 was marking time until the XS-1 was ready.

Although the reported June deadline for an August trip to Muroc had slipped by, Bell Aircraft remained confident that the existing problems were manageable ones. By mid-September, the number 2 airplane was in flight configuration. All Bell Aircraft and NACA equipment had been installed and checked. Engine ground tests had also proven quite satisfactory. The B-29 was required to undergo several mandatory technical orders, but after completion of the maintenance, it also was declared ready once again.

In the meantime, work on the number 1 airplane proceeded rapidly. Cabin pressure tests had proven satisfactory and cockpit instruments and equipment were now being installed. The new thin wing/tail components were fitted to the plane and the additional rocket engine was ready for installation in the fuselage even though it would have to be overhauled before that step could be completed. The overhaul was necessary because during the tests in the rocket test shack, a second cylinder had burned out. But the need to allow the XS-1 pilot time to train on the engine did not permit repair work. Work on the number 3 plane remained suspended until the other program problems could be resolved. On September 5, Dick Frost took the P-51 chase plane up for a 40-minute familiarization flight. On September 17, the AeroMedical Lab finally delivered an anti-g pressure suit for use in the tests. By this time, the number 2 plane was scheduled to leave for Muroc on September 25. Before that departure, the program's politics were addressed one more time.[15]

On August 28, Dick Frost sent a detailed memorandum to the AMC at Wright Field outlining the Bell test program, in response to AMC inquiries regarding Wright Field's letter of February 7, 1946, entitled "Policy regarding research and development aircraft."[16] Frost indicated the impracticality of the recent AMC request in regard to the XS-1. Research flying by its very nature was unpredictable, Frost reminded the AMC. Successive flights are influenced by the events and learning curve of previous ones. As a result, Frost stated, Bell Aircraft could discuss its plans only in broad terms. Once again, he indicated the importance of the results to date. The Pinecastle experiments clearly revealed the XS-1 had "no sufficient deficiencies relative to power-off flight and landing characteristics." Further, the performance of the B-29 and the air-launch technique had seemingly been proven satisfactory. As a result of these successes, Bell Aircraft would now concentrate on the Mach 0.80 speed and controllable 8g test. Regarding the controllability of the plane up to Mach 0.80, Frost indicated that Bell Aircraft would accept its pilot's opinion on this issue. Bell Aircraft could not "envision any lengthy series of scientific tests to investigate all the byroads of stability in its various forms." Further, Frost indicated Bell's continued interest in cooperating with the NACA and its intention to make every effort to ensure that the NACA would be able to collect data. "We trust that good mutual judgement will prevail in our relations with the NACA during the *brief period* during which we will attempt compliance with [the specified goals]" (emphasis added). However, after clearly outlining Bell's intention of a quick flight program, Frost reinforced the picture by reminding the AMC (really NACA, since AMC knew the budget situation) that Bell Aircraft engineering funds for the XS-1 program were constrained. Any NACA changes, or delays promoted by them for the purpose of data or because of instrument delays, would have to be taken under advisement. For Bell Aircraft, the impending flights were not the first phase of the research program. Rather, the contractor flights were an acceptance test for a new aircraft:

> We do not foresee the need for delaying any flight test, for instance, to permit detailed analysis of numerous data which the automatic instrumentation may have recorded the previous flight, nor delaying a flight because radar, or telemetering, or say, a multiple manometer were not functioning 100 percent since *none of those items have any bearing on our contractual commitments* [emphasis added].

The importance of the XS-1 program was supposedly in phase 2—the supersonic test phase—not in the acceptance trials. Delays cost time and money and Bell Aircraft did not intend to tolerate either circumstance. Frost had plainly stated the current Bell Aircraft view of the XS-1 program. If the AMC had a different opinion, they needed to express it. Recognizably, this flight program was clearly not the understanding by which the NACA had agreed to cooperate on this unique research airplane project. It was, however, standard operating procedure for AMC acceptance of new potential *production* aircraft. But this was not a production aircraft. The confusion over roles and objectives, ways and means, was emerging rapidly and no one seemed certain of their part. The Bell Aircraft letter appears as a case of all agencies charting new ground in the manner in which the cooperative effort on research flying would continue, but constrained by the practice of continuing to utilize existing procedures in many instances.

With the physical components of the program almost ready, Bell Aircraft took time to reorganize its personnel efforts for the project.[17]

Earlier in the summer, Frank Nicholas succeeded Ivan Hauptmann as B-29 crew chief. This allowed Jack Russell to take Nicholas's vacant B-29 rear scanner position in the program. Russell was to have a long career in the XS-1 program and at Muroc. On September 18, Mark Heaney took his first flight as copilot of the B-29, replacing Joe Cannon, who was tabbed for Bell's helicopter test flight duties. (Eventually, Cannon would become the prime pilot for XS-1 number 3.) Heaney was a popular Bell employee who had been manager of flight testing at the Bell B-29 Marietta facility. He was also expected to serve as project engineer on the P-80 radio-control program at Muroc. The September 18 1-hour, 30-minute local flight was designed to familiarize Heaney with the characteristics of the loaded B-29 and to rehearse drop procedures.

On September 26, the Bell engineers arranged for a captive practice flight for the B-29/XS-1.[18] With Harold Dow in the pilot seat and Mark Heaney in his new role flying copilot, the two aircraft executed a 2-hour, 30-minute local flight. Project engineer Dick Frost practiced his anticipated role of chase pilot by conducting a 2-hour rehearsal flight in the P-51. The checkout proved satisfactory. Frost notified the AMC on September 30 that the B-29 only needed some last-minute modifications before it was ready to travel to Muroc (the changes were to enable the B-29 to later transport the nitrogen evaporator to Muroc for the powered flights).

Bureaucratic Struggles

While Bell Aircraft proceeded with the development of the two XS-1 aircraft, the NACA and the AMC continued to revise their plans for the forthcoming Muroc tests. The debate revolved around three issues: 1) the proper role for the NACA in the final phase of the acceptance tests; 2) the organization of the NACA team for Muroc; and 3) the results of the NACA wind tunnel tests at Langley on the MX-653 model and other high-speed research issues. As to the wind-tunnel test results, the NACA notified AMC on March 14 that MX-653 wind tunnel testing on the XS-1 model would commence on March 15.[19] The purpose of the initial eight-foot wind tunnel tests would be general issues of high-speed flight. Earlier results from the 7′ × 10′ wind tunnel tests had focused on questions of stability and control. These tests were completed and a final report was due April 30. The problems of the proper role of NACA and the organization of the NACA team were not so easily solved.

On April 11, Col. George F. Smith of the Aircraft Projects Section of the Engineering Division at Wright Field wrote to Dr. G. W. Lewis at the NACA to set forth the AMC position for the XS-1 program, a follow-up to the NACA letter of January 8, 1946. Specifically, Smith indicated that, first, Bell Aircraft would fly the XS-1 until the plane had demonstrated a capability for 8 g at 500 MPH. This was the minimum structural requirement. Second, the NACA would operate the aircraft for the duration of the high-speed tests. This included furnishing a pilot and collecting data. The NACA would be expected to fly beyond the range of safety. If the NACA believed the flight regime unsafe, they could return the plane to AAF jurisdiction. Finally, the air-launch practice would be continued as the most practical and safe method of flight testing the XS-1. The AMC desired that these criteria be adopted and a "firm understanding" be reached before the conclusion of the initial powered tests.[20] In summary, Smith rejected the previous NACA letter. The Langley reply would set no speed records.

On April 24, the NACA Washington office wrote to Langley to indicate the AMC's continuing differences in interpretation in regard to the MX-653 program. Specifically, NACA HQ was concerned about the AMC's objections to flight testing at Langley Field and the continued insistence on air-launching the XS-1 from the B-29. On April 26, Walt Williams prepared a detailed memorandum for Mel Gough to outline the NACA's view of the situation concerning the XS-1 program. The memo seemed a repeat of the January 8 letter. In the new memo, Williams again stressed that the NACA would not accept the airplane until the XS-1 was flight-demonstrated with the turbine pump because the NACA did not like the B-29 drop technique. Further, Williams reiterated that Bell Aircraft had no intentions of testing the airplane at Langley—a prime NACA consideration. Muroc Field was already being investigated as the site for the additional contractor tests. On March 29, the NACA was informed that the first rocket motor was ready to be installed in the number 2 XS-1. On April 24, NACA personnel learned that Bell Aircraft would begin XS-1 testing at Muroc on July 15. As a result, Langley staff convened a meeting to outline further NACA participation in the project. Attending the meeting were Hartley Soule and Milton Davidson of the compressibility research division, R. W. Rhode and Henry A. Pearson, and W. S.

Aiken of the aircraft loads division, W. H. Phillips and Walt Williams of the flight research division. The group recommended that a team of approximately twelve NACA Langley personnel be sent to Muroc for the tests.[21]

On April 30, Williams amplified his concerns regarding the future of the XS-1 program.[22] Discussing a recent visit by Bob Stanley, Williams indicated that Bell Aircraft still desired to operate the XS-1 *under* NACA supervision. Stanley advised the NACA that "some definite policy" would need to be decided on the testing of the XS-1 after the completion of the upcoming acceptance trials. Not only did Bell Aircraft desire to continue with the supersonic testing phase of the program, the decision to do so would substantially affect the size and scope of the test station that the corporation was activating at Muroc. Gough responded to Stanley that any such "definite policy" would be extremely difficult to establish. The NACA wanted to obtain the XS-1 for testing—after suitable acceptance tests—and was prepared to fly the aircraft until such time in the test program that the "airplane evidenced characteristics which would make the airplane uncontrollable and extremely hazardous to operate." At that time, Gough indicated, "a contract *may be negotiated to fly the plane at higher speeds*" (emphasis added). In reality, the door for Bell Aircraft to continue the supersonic flights was still open. However, Gough had other requirements for acceptance by the NACA of the XS-1 aircraft far beyond the simple standards outlined by the AMC:

1. Specified endurance at rated thrust;
2. An XS-1 capability for takeoff and climb under its own power to 35,000 feet;
3. Satisfactory XS-1 control at Mach 0.80;
4. A demonstration of an 8g acceleration at an indicated airspeed not exceeding 500 MPH. This included an expectation of Mach 0.80 in level flight;
5. An 8g acceleration at minimum speed.

All these conditions referred to the NACA stipulations outlined as far back as the Langley letter to the ATSC (AMC) of March 30, 1945. Finally, Gough indicated that the NACA might waive the requirement that the XS-1 demonstrate takeoff capability as the B-29 operation gathered experience. However, no NACA acceptance of the XS-1 should be anticipated without the turbine pump and fuel capacity for emergency landings. Neither of the two requirements were anywhere within reach at that moment.

Williams concluded his memo by again stating that if the AMC decided to accept the plane without meeting the above conditions, the NACA did not feel obligated to supply a pilot. The AMC could work out arrangements with Bell Aircraft for an additional contract *under* NACA supervision as envisioned in the ATSC (AMC) letter to Dr. Lewis of May 31, 1945. The NACA would continue in that case to collect, analyze, and disseminate the research test data. In summary, the NACA did not like the direction the program seemed to be taking based on the AMC letter. It was not the technical flight research program that the NACA believed was necessary to answer the engineering questions about supersonic flight. Worse, Bell Aircraft indicated that it was not ready to bow out of the program at the end of the acceptance tests.

While Williams argued the basic foundations of the program, Langley's engineer-in-charge, Henry Reid, informed the NACA HQ just how wide the gap was over the differing interpretations between the NACA and the AMC. On April 29, Reid advised the NACA that no one at Langley had yet seen the Army/Bell XS-1 contract. Shockingly, the lack of first-hand information on what constituted "acceptance" remained unclear although Reid outlined the general consensus. Specifically, Reid advised Washington that the NACA "understanding" was that Langley would take possession only after the turbine pump with reserve fuel supply could be demonstrated. Without that agreement, Reid stated, Langley "will not undertake to supply a pilot and operate the airplane." However, to avoid being totally left out of the project, Reid stated the NACA would still be happy to take over direction of the research test and data collection program. In any event, Reid urged the NACA HQ to press the AMC on the necessity—due to the AMC's desire to keep the time between the acceptance test flights and the transonic research flights as short as possible—to begin construction of Muroc-type XS-1 fueling and handling facilities at Langley's airfield as soon as practical.[23] Seemingly, the AMC, Bell Aircraft, and the NACA were discussing different programs, and just how different became apparent in late May.

After analyzing the AAF response to the April 11 letter, on May 28 Reid again advised the NACA HQ that he assumed that the acceptance tests only referred to the number 2 aircraft since it was the only one near ready for tests. If so, Langley should withhold an answer on the conditions for the NACA acceptance. Langley was willing to wait for the number 3 plane with the turbine-driven pump. If the AMC wanted an answer immediately, it was Langley's opinion that the AMC be notified that NACA pilots would *not* fly the plane with the high-pressure tanks. The AMC should make arrangements with Bell Aircraft to continue the flights. Langley would continue as the party responsible for the research program and would accept responsibility for planning the flight program, instrumentation, and data collection and dissemination. If the AMC did give a contract to Bell Aircraft, Reid stated, the NACA should insist on some role in the preparation of that contract to ensure proper research needs were met.[24]

The NACA wanted to fly the XS-1 after the acceptance flights. The problem was the conditions under which to take possession of the plane. If the XS-1 was in the proper posture—turbine pump, reserve fuel, and ground-capable—then the NACA was ready to fly its supersonic research phase. To that end, the NACA had continued its own preparations. As part of the program for the XS-1, LMAL had over the summer been developing a flight program for the Muroc tests. On June 7, Walt Williams outlined the major points of the program for Mel Gough. The program would be designed to make measurements on stability, control, aerodynamic loads, and on drag and performance characteristics of the XS-1, in two phases. The first phase would determine stability and control at high Mach numbers. The flights would progress in defined Mach-number increments up to the "operational limits" of the airplane. It was estimated that it would take eight flights for each speed increment and a total of forty-eight flights for the first phase. Any changes in wing thickness would require another forty-eight flights. Complete instrumentation and radar would be available at all times. The second phase would measure aerodynamic loads on the wing and tail, additional stability and control data, and drag/performance data. Only sixteen flights were anticipated for the second phase for each configuration. On June 24, Reid forwarded the program to NACA HQ.[25] The lengthy program outlined in the NACA memo was to have long-range implications at the AMC and Bell Aircraft. The number of flights would pinch Wright Field's schedule for results. The costs for both Bell Aircraft and the AMC would be difficult; and the test program did not say the plane would go supersonic, merely that it would be tested up to the operational limits. This NACA program could only be a disappointment for those desiring a more rapid program. But while flight programs were important, the overall issue of the direction of the XS-1 project continued to dominate center stage.

On June 13, the NACA HQ finally provided a detailed response to the AMC's April 11 policy letter on flight requirements. In summary, the NACA stated its belief that Bell Aircraft should meet the *original* requirements of the program, which did not seem to be the current contractual requirements. The NACA stressed it wanted to see a program consistent with the one that they agreed to participate in in the beginning. Again the NACA reiterated that its pilots would only fly the plane after the (original) acceptance requirements were completed and that the NACA would only fly the plane up to operationally safe speeds. Beyond that range, the AMC and Bell Aircraft could negotiate a separate contract. In that eventuality, the NACA only asked that it be a participant in that contract.[26] The letter sent a clear signal to the AMC—unfortunately, the signal was that Bell Aircraft and the AMC should continue to assume that Bell had a role in the follow-on flight program. That discrepancy would lead to many complications over the next several months, not only in publicity for the program, but also for the eventual ownership of the XS-1 planes.

As to the remaining issue—the construction of an NACA Muroc team—while the AMC and NACA bosses argued policy, the engineers at Langley attempted to prepare the program for the new remote location. On May 8, Walt Williams directed a report to Mel Gough outlining the composition of the proposed NACA Muroc team.[27] The list included twelve people: one project engineer; one assistant project engineer (stability and control); two assistant engineers (stability and control); three aircraft loads engineers; one instrument technician; one radar technician; one telemeter technician; one pilot-observer; and one crew chief for the NACA plane.

Not everyone was ready for the great challenge. Several NACA personnel decided against uprooting their lives to move to nowhere. On July 25 the NACA notified Muroc of the names and personnel to accompany the by-then thirteen-member NACA staff. Several new names appeared on the list. Some positions were not yet finalized.

Equipment procurement at Muroc seemed to proceed much more smoothly than personnel decisions. Williams had indicated a significant need for equipment, including a modified SCR-584 radar, photographic darkroom facilities, a building/trailer for use as a telemeter receiving station, a building/trailer for technical evaluation, and facilities for battery charging for several government vehicles. On May 7, Major

Eastman wrote to Wright Field to request assistance in procuring a modified SCR-584 radar to track the XS-1 tests. On May 28, the AMC replied to Langley that a radar would be made available within 36 days. (On July 25, the AMC amended the radar request to notify the NACA that it would be an unmodified radar.) On July 10, Langley's Reid asked the AMC for an M-2 optical tracker for the program. On August 2, Wright Field again replied in the affirmative. During an early July visit to Muroc by Eastman and Williams, the NACA found the desert base personnel ready to help (within their limited means) with all requests for assistance. Housing was promised and dorms were indicated for the visiting personnel. The housing was requested to be ready to occupy by September 1. Office space and equipment was assigned to the NACA. Finally, several transportation vehicles were procured as well as maintenance and power trucks. But no AAF personnel would be available to participate in the initial flight tests, which caused Eastman to indicate to Wright Field on September 11 the very real possibility that the first flight would go off without tracking data. In the final analysis, much had been assembled for the tests. Missing still were the radar, radio equipment, and operators.[28]

The remaining NACA issue was the cost for personnel to travel and be stationed at Muroc for the next year. On May 23 and again on June 5, Dr. Reid wrote to the NACA to remind HQ that the intended operations and equipment for the Muroc station would not be inexpensive. Reid anticipated that travel expenses alone could run $30,000. This was an expense that no one at the lab had anticipated when the 1947 budget was prepared since they hoped that the tests would be at Langley. The estimated figure would be 75 percent of the lab's annual travel budget. Reid requested that HQ explore alternate sources for the funding, possibly the AMC or Bell Aircraft, and indicated that $25,000 would be a fair offset for the costs. Backstage conversation on this issue continued throughout July, and on July 30 Reid decided to force the issue. He wrote a letter asking the NACA HQ to approve the travel for the Muroc station team. But Dr. Lewis had already beaten Reid to the punch and on August 1 formally requested that the AMC cover the expenses for the XS-1 tests at Muroc. On August 16, Williams traveled to Wright Field to discuss the flight test program and needs with Capt. George Colchagoff and civilian James Voyles, the two AMC representatives for the XS-1 program. Fortuitously, on September 10, the NACA learned that the AMC was going to provide the full $30,000 cost to cover travel and per diem for the NACA Muroc station. Part of that cost was covered on September 11 when the AMC authorized the NACA to use and maintain at AMC expense three AMC aircraft for travel and transportation on the tests: a C-47B (#43-49526); a C-45F (#44-47110); and a UC-45 (#44-47264). Five pilots (Mel Gough, Joel Baker, Herbert Hoover, W. E. Gray, and Stefan R. Cavallo) were certified to operate these aircraft for the NACA. The other costs were more difficult to deal with, as quickly became obvious through the continuing correspondence. While the previous notification allowed approvals for NACA travel to be executed expeditiously, the official wheels of progress did not turn as fast or as completely. It was only on September 27 that Langley was formally notified that approval had been received for the NACA to determine its own travel orders and to submit supplemental appropriation requests to the Bureau of the Budget in Washington. In conclusion, the latter procedure was a long way from the initial request.[29]

The Torch Passes

Over the summer, the Bell Aircraft Corporation public relations team did its best to promote the visibility of the secret XS-1 program. Typical of that effort was the NEA wire story in mid-June about the XS-1's visit to AMC. Written from Wright Field, the story focused on Woolams and the effort to break the mysterious sound barrier. Quoting the Air Force, Woolams's goal was described as "the most important mission an airman has undertaken since man's first flight." The plane was a unique vehicle. The flier was equal parts test pilot and scientist with a full measure of bravery thrown in. His final goal was to break the sound barrier and pave the way for future endeavors by other pilots. His ultimate objective was to lead the way in outer space exploration and travel between the planets!

Supporting this story was the AAF's own publicity as represented by a United Press wire story of early summer. Here the focus was on the AAF's new project and Woolams's efforts to break the sound barrier. The key issue, quoting Lt. Gen. Nathan Twining, chief of the AMC, was "whether the plane will remain intact and can be controlled by its human pilot at speeds exceeding sound." The hope was that the XS-1's power would allow Woolams to smash through the sound barrier at speeds up to 1,500 MPH. Every precaution had been taken to make it safe, but the inherent risk was palpable in the tone of the article.[30]

The initial burst of publicity did not sit well with all involved parties. The coverage of Bell Aircraft and its feats, and Woolams's connection to Bell did not endear the manufacturer to its NACA partners. On June 27, Major Eastman wrote to Wright Field explaining that the slant of the articles had "caused considerable unofficial comments by representatives of the N.A.C.A." Their displeasure centered on the fact that no mention of the NACA or its portion of the team effort had been discussed in the publicity. Eastman stressed the confusion at Langley over the substance of the high-speed and high-altitude tests that Woolams was supposedly going to fly. The NACA believed that would be left to their portion of the research effort. Eastman again stated this fact as if to challenge Wright Field to deny that this was indeed the situation. His conclusion was to request the Engineering Division to arrange an additional press release to cover the ruffled feathers at Langley. On July 16, Wright Field notified Eastman that discussions had occurred with XS-1 project officer James Voyles over the nature of the Bell publicity. Voyles was asked to "arrange a War Department publicity release" to cover the role of the NACA. If Wright Field was serious about changing the publicity coverage on the XS-1, this was certainly a straightforward approach to ask the AMC project officer to "arrange" the publicity.[31] Yet it would be hard to compete with the glamour of a private company's airplane and its pilot.

In any event, nothing Bell Aircraft, the AMC, or the NACA did slowed the public fascination with the new experimental plane and its fearless pilot. Growing press coverage of Woolams seemed to culminate in late summer. *Collier's* magazine dedicated a two-part series to the XS-1 and Woolams in August which outlined the existing nature of this great adventure into the unknown.[32] According to *Collier's*, late that summer Woolams would fire the XS-1 up to 75,000 feet and over 1,000 MPH. The risks were severe. "This sounds like murder," the magazine stated but then indicated that Woolams had volunteered. Everyone knew the risks, but Woolams had decided to take them in the quest to move aviation to the next stage of development. The *Collier's* article revealed painstaking interviews and significant background information; not only would the tests be at Muroc, but the article also indicated the XS-1's cost at over $1 million. Further, the magazine mentioned the tests at Pinecastle and the drop-test procedure for the Muroc tests. For a nonaviation journal, it was very creditable reporting. But the concluding piece of journalistic coverage came in an article by Woolams for *Aviation* magazine.[33] In the article, Woolams set forth his goals for the program and the dangers he faced in making the effort. It was a no-nonsense analysis of the risks and the rewards. However, as with any test pilot, Woolams attempted to put the risks in perspective:

> I do not, however, look on the forthcoming flights as anything particularly sensational; they will be the logical result of a great deal of hard, cooperative work. There will, of course, be some danger, as there must be in flying at new altitudes and speeds. But we are well prepared. I have complete confidence in the airplane and the men who have worked so long and hard on the project.

The XS-1 would go up to 1,500 MPH and would explore flight characteristics as high as 80,000 feet. When completed, the flights would shatter the myth of a barrier in the sky. In the end, Woolams was sure the tests would justify the efforts. Short and sweet, Woolams knew what he was doing in the XS-1's efforts to break the sound barrier. If it could be done, this plane and this pilot were going to do it. But once again that magic speed number of 1,500 MPH received prominence in the coverage. However, the NACA had already determined that *this* plane was not going to achieve 1,500 MPH.

The NACA continued to watch the press buildup of the XS-1 with professional skepticism. The published numbers for speed and altitude certainly raised a few official eyebrows. The NACA's estimation of the maximum velocity of the XS-1 in level flight after discharge at 28,000 feet from the B-29 was around 950 to 1,000 MPH. This speed would occur near 60,000 feet after taking into account fuel consumption, weight of the plane, and wing loading. In actuality, above 35,000–45,000 feet, the XS-1 performance would be limited by the amount of fuel and not the available thrust. Without larger fuel supplies, the XS-1 as currently configured (i.e., without the turbine pump) was a limited airplane. With the addition of the theoretically more efficient turbine pump, the XS-1 could achieve estimated higher velocities on the order of 1,500 MPH at 70,000–80,000 feet.[34] The NACA hoped that the correct information could dispel all of the myths currently gaining strength and perhaps dispel false notions of the plane's abilities.

The AMC public relations repair efforts were not the only attempts to gain the NACA recognition from the program. When the NACA witnessed the continuing Bell Aircraft XS-1 coverage, they decided to tell their own side of the story. On August 11, the NACA HQ released a press bulletin on their involvement in the XS-1.[35] The story was the same, but the thrust of the report was that "the XS-1 is the result of high-speed research carried on by the NACA since 1939." Thus the plane was built by Bell Aircraft using NACA specifications. It was built for the NACA. It was instrumented by the NACA. It was to be flown supersonic by the NACA. Once the research data had been gathered and analyzed, the plane would be turned over to the AAF "for further experiment as they see fit." Only at the bottom of the second page did the NACA mention that there would be three aircraft and the AAF and Bell Aircraft would each possess one plane.

No press release seemed to satisfy everyone involved with the program. The different angles to the press stories highlighted the continuing questions about the responsibility for the XS-1 program. The NACA saw the flights in one light. The Bell team viewed the program in another. The AMC, supposedly in charge of the effort and the paymaster for the program, was charting new ground in this project. Usually the contractor developed the project and flew the plane until acceptance. But this program was not for a production airplane. It was a research plane designed to provide data on aerodynamic characteristics to incorporate into later production aircraft. The limits of responsibility and the demands of research were still not fully defined. The publicity matter continued to play out until finally subsumed by the larger issue of whose program the XS-1 really followed.

During the long delay between the end of the Pinecastle flights and the initiation of the Muroc tests, Jack Woolams remained busy on several other Bell Aircraft programs. Two projects received particular attention. In addition to his XS-1 duties, Woolams became involved in the XS-2 program then in its initial stages. During 1945, the AMC, the NACA, and Bell Aircraft had discussed a swept-wing follow-on for the XS-1 program. This project would incorporate the latest AMC, NACA, and captured German research information. On December 14, 1945, the AMC and Bell Aircraft signed a preliminary contract to develop a swept-wing model for flight testing. The program was built on previous research conducted by Bell Aircraft under a Navy contract. Two P-63As were modified under the direction of Stan Smith and received the in-house Bell designation of XP-63Ns. They were quickly redesignated by the Navy as the Bell L-39-1 and L-39-2. Tex Johnston made the initial flight while Woolams was at Muroc, but on July 20 Woolams took off in the L-39-2 to conduct a series of tests with the new bi-

convex wing section. It was almost his last test. The wing modifications had relocated the center of gravity of the airplane too far aft of
the wings' center of pressure, thus making the L-39 extremely tail-
heavy. Bell engineers had missed this point. Woolams discovered the
serious control problem shortly after reaching flying speed. He mas-
tered the problem with a magnificent display of flying skills and
brought the L-39 safely back to the Niagara Falls airfield.[36] Lady Luck
was still with him.

The other time-consuming project for Woolams was not an official
Bell Aircraft program. The National Air Races postwar resumption at
Cleveland on Labor Day 1946 would culminate with the 300-mile
closed-course Thompson Trophy Race—always a most spectacular
event. To promote the reputation of Bell Aircraft products, Larry Bell
and Jack Woolams devised a subterfuge to allow a P-39 to compete
without appearing to have factory sponsorship. At that time, two Bell
employees, Chalmers Goodlin and Dick Frost, had initiated a sideline
business, Skylanes Unlimited, to cover private flying opportunities.
The new plan was to obtain several surplus P-39 aircraft, prepare them
for racing, and pay for it through Skylanes Unlimited. Two aircraft
were purchased for $1,500 at the surplus depot at Walnut Ridge.
Painted in dazzling red and yellow colors and named Cobra I and Cobra
II, the two extensively modified planes were expected to provide an
excellent opportunity to win against much more famous planes such as
the P-38, F4U, F8F, and P-51. A coin was flipped among Woolams,
Johnston, and Goodlin. The agreement was that the two winners would
fly the planes and split the money between the three pilots. Woolams
and Johnston won the toss. However, tragedy struck before the race
could occur. Although both planes performed well at the trials three
days before the race, Woolams decided to return to Niagara Falls to
install a new engine in Cobra I. While testing Cobra I at very low
altitude over Lake Ontario late on Friday before the Labor Day race,
Woolams's plane dove into the water, killing the gifted and fun-loving
aviator. Tex Johnston in Cobra II won the Cleveland Air Race on
Labor Day. As per the coin-toss agreement, Johnston, Goodlin, and
Woolams's widow split the $19,200 prize money.[37]

Although a search of the lake was instituted as quickly as possible,
no definitive account of the accident ever emerged. It was assumed
that the recently installed plexiglass canopy on Woolams's aircraft had
imploded during flight, killing the pilot. Tributes were quick in coming.

On September 4, General Craigie wrote a personal letter to Larry Bell expressing his regrets. Woolams's death, Craigie said, was a tragedy for Bell Aircraft, but it was

> an equally severe blow to us in the Air Forces. Jack has, for a number of years, had the highest regard and respect of all of us in the Engineering Division . . . and we all share with you this loss of one of the finest and most promising individuals engaged in the development of military aircraft.

Larry Bell's reply of September 6 clearly indicated the depth of his personal feelings for Woolams.

> Jack's death was a great loss to aviation; one of the most capable test pilots I have ever seen and a fine boy of great charm and personality to whom we were all deeply devoted. I felt closer to Jack than any pilot we ever had. His willingness to explore the unknown was always a great inspiration to me.

On the evening of December 7, 1948, just after the award of the Collier Trophy to Larry Bell, Chuck Yeager, and John Stack for the XS-1 achievements, Bell spoke at the Aero Club of Washington, D.C. In his remarks, Bell returned to the subject of his dear friend Jack Woolams and paid a special tribute to his memory.[38] His place in aviation history was secure.

With the death of Jack Woolams, Bell Aircraft was left with a vacancy in its most prestigious flight program. To fill that gap, it could logically be assumed that Tex Johnston, now Bell Aircraft's chief test pilot, would assume the role previously handled by Woolams. However, Larry Bell had other plans. Even before the end of the Cleveland Air Show, Bell casually informed Goodlin while introducing him to an associate that Goodlin would be the new XS-1 pilot. Goodlin said nothing at that time to respond to the honor. In retrospect, the unmarried Goodlin believed the tragic death of the married Woolams weighed heavily on Bell's mind. Indeed, Bell had almost canceled Johnston's participation in the Cleveland race because Tex was a husband and father. Now a new era would be written in the XS-1 program.

Chalmers "Slick" Goodlin was born in Greensburg, Pennsylvania, in 1923.[39] When he was five, his family moved to New Alexandria, Pennsylvania, where he grew up. Goodlin led an active early adult life. He was especially fond of speedy machines, which he exhibited through his use of two motorcycles and his interest in airplanes. The airport at New Alexandria regularly featured several air shows. After six hours of lessons from the airport's operator, Goodlin made his solo flight in a Piper J-3 at the age of 16.

While after two years of high school Goodlin had intended to become a farmer, as happened so often at that time the stirring emotions aroused by the Battle of Britain changed his career plans. Since he lacked both the age and the college education required by the U.S. Army Air Corps, he went to Canada in February 1941 to enlist in the Royal Canadian Air Force. During flight training, Goodlin's classmates noted his excellent instrument flying skills and nicknamed him "Slick" in tribute to his ability to handle blind flying techniques.

On December 5, 1941, Goodlin was commissioned as probably one of the RCAF's youngest officers. He returned home on leave just in time to hear of Pearl Harbor. After many months as a flight instructor in Canada, and a combat tour in England where his Spitfire squadron flew with RAF units, he transferred to the U.S. Navy in December 1942. His duties with the Navy included performance acceptance test and ferry flights for newly delivered aircraft. With this experience, Goodlin was released from active duty in December 1943. He begin work for Bell Aircraft on January 28, 1944, in flight test operations. On March 5, 1945, Bob Stanley transferred Goodlin to experimental test flight work. During the first week of September 1946, the NACA and the AMC were notified that Goodlin would be the new XS-1 pilot. At the time of his assumption of the role as Bell's prime pilot on the XS-1 program he was 23 years old.

Preparing to Move to Muroc

On September 16 and 17, a team from Langley, including Walt Williams, John Gardner, Norm Hayes, Warren Wallis, Paul Harper, and Stefan Cavallo, traveled to Niagara Falls to visit the Bell Aircraft plant.[40] Gardner was along to see the rocket motor, Hayes, Wallis, and Harper (strain gauge specialist) to assist in instrument installation, and Cavallo as pilot. The purpose of the meetings was the final installation of the NACA instrumentation in the XS-1 number 2 and a discussion of Bell Aircraft's flight test plans. The NACA team arrived at approximately 10:00 A.M. on September 16.

Frost met the group in the experimental hangar where the Bell technicians were revamping several engine control switches that had proven faulty in the recently concluded first engine test. Notation was made by the NACA group of the change in layout of a potential XS-1 instrument panel. A new attitude gyro and compass had been added during the makeover. Since the Bell Aircraft work precluded the NACA technicians' entry to the cockpit area, Hayes, Wallis, and Harper began work on the center section instrument bay. At the same time, Williams, Cavallo, and Gardner proceeded with Frost to the rocket test shack where Frost reviewed the engine installation and tests. To date, two engines had been received from RMI; one engine was in the test shack and the other in the XS-1. Bell engineers had already run the test shack engine for about 30 minutes, Frost reported, and were very pleased with the results.

After moving to the engineering department, Frost outlined Bell's flight plans. First, Bell Aircraft intended to conduct additional engine ground tests on the XS-1 that should be completed by September 20. A captive flight under the B-29 was scheduled for the week of September 23. Second, final Bell XS-1 cleanup and preparations at Niagara Falls would be completed between September 23 and 30. Third, the XS-1 would travel to Muroc on September 30. Because of the tight time schedule, Frost emphasized, any work or items that caused delays would be postponed until arrival in California. Upon hearing this cautionary note, Williams launched into a discussion of issues for which he was afraid he already knew the answers.

First, Williams asked about the pedal-force recorders that the NACA wished installed in the XS-1. Frost said there were no plans to do that at this time although Williams was free to further discuss this matter with Bob Stanley when he arrived later in the day. Second, Frost asked exactly what data the NACA expected from the XS-1 in the acceptance tests. Williams replied that the NACA desired to obtain complete stability and control data. This data would also furnish the required aerodynamic load information. Frost said that Bell Aircraft anticipated no special flights to obtain that data since Bell would accept their pilot's opinion on the control and stability issues for the satisfaction of the AMC contract requirements. Further, Frost revealed to Williams the Stanley letter that Frost had sent to Wright Field on August 28: the letter that discussed Bell Aircraft's test plane and financial restrictions.[41]

During the afternoon, Cavallo and Williams met with Frost and Bob Stanley to discuss the XS-1 test program. Stanley again brought up Frost's letter to the AMC. Specifically, Stanley referenced the AAF directive that limited manufacturers of research aircraft to a minimum of flight tests only to prove an experimental craft safe within prescribed limits. Total testing time was limited to 30 hours by the Air Force policy. With that schedule, Stanley indicated, Bell Aircraft would try to assist the NACA in gathering the data it desired, but no delay would be allowed in the Bell flight tests. With that statement, Williams saw another Pinecastle exercise in frustration on the horizon. He repeatedly asked Stanley and Frost to outline in a written statement exactly what Bell Aircraft planned to do in the tests. Without that statement, Williams indicated, it would be hard for the NACA to get the essential data it required to take over the program. Williams said specifically that data on longitudinal stability, control in steady and accelerated flight conditions, and the definition of the buffet boundary were required by the NACA. To prevent any technical delays, the NACA promised to duplicate all important XS-1 instrumentation. He again brought up the issue of the rudder pedal-force recorders and Stanley finally agreed that Bell technicians would install them.

The discussion then moved on to the exact intentions of the Bell Aircraft Corporation during the remainder of the acceptance flight tests. Frost advised Williams that Bell planned to make an empty first flight and then add 1,000 pounds of water ballast each flight until a full fuel-load level was achieved. Bell Aircraft estimated that this process and the powered flights would require about sixteen to eighteen flights. Stanley would not provide any additional written statements on the Bell flight schedule. Unspoken but understood by Bell was the attitude that flight research was a building block. Each new flight test built on the results of the preceding one. Commitments to follow a set research pattern might mean unnecessary flights. In keeping with the AAF requirements, it was better to trust the pilot's judgment and feel of the plane. Upon this uncooperative note, the meeting concluded with Stanley reiterating Frost's estimate of the schedule for the XS-1 departure to Muroc.

Since the primary purpose for the NACA trip was to ensure proper instrumentation, Hayes, Wallis, and Harper were hard at work during the Williams-Stanley meetings. As usual, the installed instruments exhibited several problems and work was not finished by the scheduled NACA departure time. As a result, Williams made the decision that the NACA personnel would remain until Tuesday so that they could all return on the same plane. The NACA technicians worked until very late into the night correcting the instrumentation. The delay also afforded Williams another opportunity to meet with Stanley on the NACA difficulties in preparing for the flight test program. These were covered on Tuesday in a lengthy meeting with Williams, Frost, and A. Joseph "Joe" Marchese (XS-1 assistant project engineer). Williams again explained the preparations necessary before the first flight: final instrument checkout, instrument calibration of some equipment, and installation of the telemetering equipment. No final decisions were reached.

One other issue loomed in the background of the Bell/NACA disagreement. Unstated in the larger debate over the speed of the XS-1 program were the ongoing financial problems at Bell Aircraft. The postwar lack of business was not unique to Bell Aircraft. Every major aircraft company, many of them heavily involved in war production, now faced the immediate task of how to generate new business in the face of their simultaneous loss of the world's biggest customer. For Larry Bell, a man for whom intimate government contacts and successful contracts were long a part of his life, it was doubly painful. In early 1945 Bell Aircraft's facility at Niagara Falls employed 19,264 people. By the end of the year that number had fallen to 5,326. Similarly, the Marietta Georgia B-29 production facility fell from 26,514 people on the payroll in August 1945 to only 92 people by October 1 of that year. (The plant was mothballed soon thereafter.) In 1945–46, Bell Aircraft was asset-rich but contract-poor. The corporation had plenty of cash on hand but no future potential sales, with the exception of its newly developed helicopter. By year end, Bell Aircraft would have losses totaling $657,900 on sales of $11.5 million, and it would be worse in 1947, even with the start of sales of the Model 47 helicopter. In addition, throughout late 1946 Larry Bell and his board of directors were often distracted by the actions of a group of Wall Street investors who since early 1946 had been making a move to take over and liquidate Bell Aircraft for its lucrative physical assets. While much of this was public knowledge, the specifics remained secret. All of the takeover talk, however, generated enormous pressure on Bell Aircraft both to show financial restraint and to develop new products.[42]

The NACA response to their treatment at Niagara Falls was quick in coming. On September 26, LMAL's engineer-in-charge, Henry Reid, registered a strong letter of "serious concern" to the NACA HQ.[43] Reid pulled no punches with HQ when he stated that the Bell Aircraft flight program as outlined to Williams would make it very difficult for Langley to determine the reliability and safety of the XS-1. Without an assurance that the XS-1 was safe below Mach 0.80, Reid stated, LMAL engineers would have no satisfactory basis for understanding of the aircraft's performance above Mach 0.80 and in the transonic zone. The clear implication from Reid's remarks was that he expected NACA HQ to act to correct this situation. Unfortunately, Reid reminded HQ, an examination by LMAL of the AMC–Bell Aircraft contract had already confirmed that the Bell position was technically legal. The AMC had only required Bell to demonstrate the limiting conditions (Mach 0.80, and an 8g pullout). These conditions did not, from Langley's viewpoint, constitute a "suitable preparation for the research flying."

Reid stressed to the NACA HQ that the LMAL position was that the XS-1 program required quantitative data collection on the airplane's subcritical behavior and loadings. Additionally, Langley desired to see further experience with the mechanical features of the plane, including all aspects of the rocket powerplant operation. Reid took this opportunity to remind the NACA HQ of the physical difficulties the plane experienced in Florida (two landing gear failures and a damaged wing). Without further information, and with the foregoing concerns, Reid speculated that Langley would have to provide an additional twenty people and two pilots to its personnel at Muroc. This programmatic increase did not seem likely, given the budgetary situation. Thus, in summation, Reid put the problem squarely before HQ. Langley *could not* operate the planes and *would not* allow its pilots to fly the XS-1 aircraft if Bell Aircraft's limited flight program was allowed to be the operative plan.

Having defined the issue, Reid now proposed the only solution that Langley saw for the foregoing circumstances. He urged HQ to contact the AMC and advise them to begin immediate negotiations with Bell Aircraft for the operation, maintenance, repair, modification, and research flight testing of the XS-1 program *under NACA supervision.* Part of the acceptance flight program would be rolled into the new research flight contract. The NACA would then not object to an abbreviated Bell Aircraft contractor acceptance program and would urge the AMC to quickly pay Bell Aircraft for the airplane tests. In short, Langley wanted to redo the program with the NACA clearly in charge. If Bell accepted these conditions, then the NACA did not mind their continued participation. If they refused, . . . well, that was not yet clear. The September 26 letter cleared the way for Bell to retain a primary role in the program. But the Reid letter was also a veiled threat: make changes in the program or it will be difficult for NACA to continue.

The Reid letter finally brought the XS-1 policy questions to a head at NACA HQ. As a result, on September 27, Hartley Soule and Milton Davidson traveled to Washington to meet with E. J. Ames and Clotaire Wood. Both the XS-1 and the new Navy D-558 airplane projects were discussed. However, the D-558 project received far more attention than the discussions on the XS-1. HQ informed Langley that the first D-558 would be ready by January and the second aircraft—to be turned over to NACA—would be ready by May 1947. Soule and Davidson urged HQ to contact the Navy and ask that the timetable be expedited as "the utility of the phase I airplanes tends to decrease with time." This bold admission that the D-558-I might produce fewer results than NACA originally desired elicited no further comments. LMAL also inquired whether the Navy could continue to retain Douglas Aircraft for assistance during its research flying. Once the discussions on the D-558 were complete, the HQ personnel asked the LMAL staff why Reid had sent the September 26 XS-1 letter. The existing Soule memorandum of the meeting does not explain in detail the "background and reasons" for the LMAL request to reorient the XS-1 contract. But HQ realized that the LMAL contract revisions would effectively put Langley out of the supersonic program with the XS-1, regardless of the legal niceties of "under NACA supervision." Ames and Wood agreed to discuss with Dr. Lewis and John Crowley (now on loan from LMAL to NACA HQ) the Langley request for AMC to contract with Bell for the research flying. If HQ agreed, the AMC would be notified to proceed with a separate contract for Bell Aircraft. But HQ did not agree. The continued push for congressional funding of the National Supersonic Research Center logically required a heavy NACA involvement in supersonic research. While from a technical standpoint the frustration of LMAL with Bell was certainly understandable, the politics of supersonic flight quickly killed this proposal.[44]

To underscore the seriousness of Langley's feelings on this issue, Major Eastman took the additional step of formally notifying Wright Field of his concerns about the ongoing problem. In reply, Col. George Patterson, chief of the Engineering Division Liaison Section at Wright Field, sent a teletype on September 26 saying that the Aircraft Projects Section would take immediate steps to clarify the issues concerning the XS-1 program. This clarification came in the form of a letter from General Craigie to Larry Bell on September 30. Bell replied on October 4 that he agreed with Craigie "thoroughly that nothing should interfere with the scheduled tests of the XS-1 to get the information for which it was produced." Although seemingly straight-forward, the ambiguous reply could not end the issue.[45]

By the end of September, Bell Aircraft was seemingly ready to move the components of the XS-1 to California. First, Slick Goodlin flew to Wright Field to be interviewed by Col. Albert Boyd, the head of Flight Test Division. During the visit, Goodlin had met with the staff of the AeroMedical Laboratory and was later introduced to the tortures of the gravity machine. Second, the XS-1 number 2 airplane was ready. The number 1 airplane had received the 8 percent wing/6 percent tail combination and the fuselage was undergoing installation of the gaseous nitrogen containers. Even the second set of thin wing/tail components were nearing completion, although no work was currently being done on the number 3 airplane. RMI had almost completed work on the third and fourth rocket motors and Bell Aircraft anticipated another engine within a week. Since the B-29 dress rehearsal on September 26 had gone smoothly, no problems were envisioned for the trip to Muroc. Last, on September 20 Gen. Carl Spaatz, commander of the AAF, had visited Bell Aircraft to see the XS-1. As part of the performance, the rocket engine in the number 2 was ignited and the engine run for 2 minutes and 20 seconds. All systems functioned smoothly.[46] Finally, the program appeared to be moving ahead. Similar progress occurred at LMAL in readying its team.

In spite of the failure to clarify the issue of program control, the XS-1 program appeared ready to proceed and the NACA decided to follow through with its part of the project. On September 20, LMAL notified Muroc Field that Williams would be the NACA representative in charge of their project team. Williams had complete authority on any questions regarding the NACA's participation on the project at Muroc. On September 30 Williams, George Minalga, Aiken, and C. E. Matheny arrived at Muroc. They found to their surprise a telegram from the NACA advising them that the XS-1 would be delayed a week due to problems with the B-29. Although this issue was serious enough, the NACA team discovered more pressing problems—no one at Muroc seemed to remember the assurances provided in July about housing for the NACA personnel and their wives. It was not an auspicious beginning, although within a day better news on the housing issue and the vast amount of set-up work required to get ready for XS-1 testing changed the focus of the discussions between Langley and Muroc.[47]

On October 4, Williams reported to Mel Gough at Langley on the progress in setting up the new NACA facility at Muroc. Minalga and

Donald Youngblood had already moved into the Muroc barracks space formerly occupied by personnel from Republic Aircraft. The NACA office and its furniture were installed and the NACA personnel put in overtime in anticipation of the arrival of the Bell Aircraft contingent. With the mundane details out of the way, Williams again reminded Gough that the continuing AMC indecision regarding the Frost (Stanley) letter of August 28 concerning Bell's flight plans was a disagreement waiting to happen. Williams gently urged Gough to not let this point continue unanswered. The decision on this issue would "directly affect our plans." Unfortunately, no decision had been made, and the lack of an answer from the AMC would soon cause the testing issue problem to rear its head again.[48]

One other matter remained to be finished before the Bell team made the journey to Muroc. Research flying was a dangerous business and test pilots were a superstitious group. Within a three-week span, Bell test pilots Milton Carlson and Jack Woolams had been killed in crashes. Many pilots believed that disasters came in groups of three. The question was who was next. On the morning of September 4, the recently designated Bell XS-1 pilot Slick Goodlin was on a "routine" test flight in the new Bell P-83 jet fighter, with Charles Fay along as project engineer, when the plane suddenly burst into flames. Goodlin radioed for Fay to jump out while he tried to guide the plane away from populated areas. By the time Fay had cleared the craft and Goodlin felt it was safe to jump, the right wing had virtually disintegrated. Barely clearing the plane before it plunged out of control, Goodlin floated down into a buckwheat field near the village of Tonawanda. The curse of three deaths passed by on this occasion.[49] Yet far away it visited a fellow pilot with tragic results.

In 1946 other nations were also pursuing the quest for supersonic speeds.[50] During the summer, Great Britain had a promising supersonic research program under contract. Abruptly, however, the Ministry of Supply announced that Britain's postwar financial crises would preclude a more extensive manned research program. Other authorities cited the persistent climate of danger associated with the program. Regardless, as a consequence of the decision British efforts would focus on unmanned rockets for research testing. The British decision seemed to leave the U.S. program as the only serious manned effort still being attempted. But one individual, 36-year-old Geoffrey de Havilland, chief test pilot of the de Havilland Aircraft Company, was set to pursue the dream of supersonic flight in spite of the Ministry of Supply's decision. The son of the company's owner, de Havilland was an excellent pilot with a distinguished record. He had flown the first Mosquito bomber, the first Vampire jet fighter, and the first D.H. Hornet long-range fighter. Now he was ready to pursue his greatest challenge.

The de Havilland Company actually had two aircraft ready for flight trials. Both had been built to analyze swept-wing research issues and stability and control problems. As a result, they did not fall under the Ministry decision regarding supersonic flight testing. The D.H. 108 Swallow was a swept-wing, flying wing with a vertical stabilizer. Possessing neither a horizontal stabilizer nor a tail (similar to the later American X-4), it was built around a standard Vampire fuselage and contained a Goblin turbojet powerplant. It lacked the power necessary to achieve supersonic speeds in level flight and for that reason it was anticipated that flight testing would be conducted in a dive. Throughout mid-September de Havilland practiced for a new speed record run to demonstrate the capabilities of the D.H. 108. On September 27, he was ready for one last test before a speed record run. At 5:30 P.M., he taxied out of the company's Hatfield aerodrome, cleared the runway, and prepared to execute the designated flight plan. The plan called for a low-level maximum speed run followed by a maximum speed dive. Within 30 minutes word reached Hatfield that an accident had occurred. Wreckage spotted in the Thames River estuary had been identified as the Swallow. Ten days later, de Havilland's body washed ashore. The D.H. 108 had fallen victim to compressibility effects in its high-speed dive. Great Britain was definitely out of the supersonic race.

After recovering from slight burns from the near-fatal crash of September 4, Slick Goodlin met with Bob Stanley in mid-September to discuss the flight contract for the XS-1 work. Stanley indicated that the program was behind schedule and asked that Goodlin leave immediately for the West Coast. The discussion quickly turned to money. The issue of a risk bonus was standard procedure for test flying. Woolams had been promised $10,000 for the acceptance tests on the XS-1. After discussing the problems with the rocket engine and the inherent dangers in the test program (notably the lack of a proven escape method), Slick Goodlin decided that a $10,000 (minus $2,000 to Mrs. Woolams) bonus to fly the experimental program was not adequate compensation. Stanley concurred but indicated the acceptance tests were covered by an existing AMC contract. Stanley believed this situation could not be changed.

The two men discussed a larger pay package for the remainder of the supersonic test flights and additional work over the next several years. As a result of the anticipated size and length of the program, Goodlin requested that Stanley consent to a large bonus and a personal services contract from Bell Aircraft. A figure of $150,000 (minus the existing $10,000), paid over five years, was discussed as compensation to fly the XS-1 *through* the sound barrier and perform additional test work. This contract would be executed by Bell Aircraft (not AMC) to Goodlin, either through a new (Goodlin) corporation or directly for personal services. Stanley produced a graph that supposedly revealed how the payment schedule would be paid (bonus per Mach number flown) to Goodlin. The graph was to play a prominent role in the program at a later date. The details seemed agreeable to both men. Under a handshake agreement with Stanley to work out the specifics with Goodlin's lawyers, the rocket pilot departed for Muroc in his new 1946 Ford. The result of that "uncompleted" deal was to haunt the program, Slick Goodlin, and the history of the XS-1 for four decades.[51]

4

The Goodlin Era

The Glide Flights

On Monday, October 7, Bell Aircraft was ready to move the XS-1 number 2 to Muroc.[1] Dow took the pilot chores on the B-29 and Bob Stanley flew as copilot. In a 10-hour and 20-minute flight (including two hours of instrument flying), the mated B-29/XS-1 traveled to Muroc with one stopover. The Bell team arrived around 5:00 P.M. and was greeted by Walt Williams and the small NACA group already on the scene. Since they arrived so late, no operations were attempted that evening.

The next day Bell's C-47 arrived with the rest of the contractor contingent, and Dick Frost arrived in the Bell P-51 that was to be used for chase duties. He had made the two-day flight from Niagara Falls in 10 hours and 10 minutes. Work on the XS-1 had already begun. Goodlin immediately took the P-51 up for an inspection of the Muroc area. Upon his return, he departed by car for a ground view of the lakebed. During the inspection, Goodlin found bomb craters remained on the lakebed from the wartime practice missions.

Since Stanley and Williams were together again, the next round of their disagreements quickly broke out. Stanley indicated that Bell Aircraft intended to fly the XS-1 on the morning of October 9. Williams said that his team of NACA technicians would not arrive at Muroc until later that same day or perhaps not until Wednesday, but in either case the flight should wait for their arrival. A meeting was set up in Maj. Clarence A. Shoop's office to reach an agreement on the flight time. (Shoop was chief of technical engineering at Muroc and the AMC's representative for the XS-1 project.) At the meeting, Williams repeated his objections about conducting a flight without the NACA instruments being able to record data. Since the whole point of the

Bell/NACA cooperation was to provide data for determination of the proper path to supersonic flight, Williams stated, it was pointless to fly without instrumentation. Stanley objected to the delay, noting that since this first flight was merely a glide flight to familiarize the pilot with the new aircraft, it was senseless to force a delay. Any data collected on this flight would be similar to that obtained at Pinecastle and therefore redundant. But then Stanley couched the debate in different terms, reiterating his belief that the real issue was who—the NACA or Bell Aircraft—was in charge of the contractor flight program. Sensing an explosive issue, Major Shoop declined to referee the disagreement and deferred the matter to Wright Field for guidance.

The Shoop conference call reached James Voyles and Charles Hall at AMC. Williams and Stanley stated their positions: Williams that it was the AMC's desire to obtain data for each flight regardless of the purpose of the flight, and Stanley that Bell Aircraft was in charge of the contractor program and, in any event, the flight was a repeat of those conducted at Pinecastle. Referring to the previous Bell letter to the AMC outlining the proposed flight program, Stanley asked if the AMC disagreed with that plan, or if they did not understand it. Stanley's heavy-handed effort at coercion produced no immediate results. After hearing the details of the dispute, Voyles said AMC "generally" agreed with the Bell letter but thought "some points should be clarified." In any event, Voyles told Stanley to proceed with the immediate flight test as he thought best. But Voyles added a parting warning: this decision was not a clear statement of the AMC policy, and "if something did happen to the XS-1 without the telemetering installed it would be very embarrassing for the AMC as well as the Bell [Corporation]." In other words, Wright Field would not decide the policy issue in this call. On that note the conversation was terminated.

Even with all of the verbal maneuvering, the NACA was left with the immediate problem—Bell Aircraft intended to conduct a flight the following day whether or not the recording instruments were installed on the plane to gather data. Stanley finally relented somewhat by saying that while a flight was scheduled for the next day, he was willing to base his decision about the flight on how things progressed. Frost quietly advised Williams that he personally could be more cooperative than Stanley with the NACA after Stanley had departed for Buffalo (and the pace had consequently slowed). Late that same day, the rest of the NACA technicians group arrived at Muroc aboard the Langley's C-45.

On October 9, the XS-1 number 2 was loaded aboard the B-29 for its first glide test.[2] Stanley agreed to allow the morning for the NACA technicians to finish their work. Although time expired before the NACA personnel could verify instrument functioning, the equipment was installed and loaded. No effort was made to record the flight by telemeter or radar since the B-29 radar control panel had not yet been installed and there was no communication for timing the records. The radar was still awaiting the arrival of Wright Field personnel and some equipment. Nor was there any optical tracker available for the program and no personnel to man it even if it had been available. In short, Bell Aircraft was in a hurry.

At Muroc, the B-29 was a busy aircraft. For the drop launches of the number 2 XS-1 Dow would fly as pilot and Heaney as copilot (not alternating duties as at Pinecastle), with Dow handling all the Bell contractor tests. Mac Hamilton would serve as XS-1 crew chief. Throughout the tests, he and Goodlin would sit behind the two pilots on the short climb to XS-1 entry altitude. In the B-29 rear compartment, Bill Miller served as B-29 flight engineer, Frank Nicholas performed assistant flight engineer and B-29 crew-chief duties, and Jack Russell served as rear compartment scanner.

On each flight, every person performed specific functions. While the two pilots handled the B-29 flight duties, Miller and Nicholas monitored the bomber's engines, Russell observed the XS-1 controls and engine prelaunch conditions, and Hamilton preceded Goodlin into the bomb bay to assist with pilot entry into the XS-1. Once Goodlin entered the XS-1, Hamilton assisted with closing and buttoning up the cockpit door, which was no mean feat in the turbulent conditions of the bomb bay. Finally, as drop altitude was reached, Dow would check off the time remaining in one-minute intervals. At the moment of drop, Heaney would reach back to trip the pull lever and release the shackle. Through practice, the seemingly complex procedure never presented a problem.

At 2:10 P.M. the B-29 lifted off from the high desert at Muroc and began the long climb to altitude. Bob Stanley followed in the P-51, although in the future Frost was expected to have that duty. At a pressure altitude of 7,000 feet, Goodlin departed the B-29 cockpit and climbed into the bomb bay to enter the XS-1. However, 50 minutes into the flight, Goodlin became cognizant of a serious problem. At 20,000 feet, after the heat from the B-29 was disconnected, Hamilton started charging the pressure domes in the nitrogen regulators. The

XS-1 cabin pressurization system was designed to ensure a safe interior pressure for the higher altitude flights. Nitrogen was used to drive the attitude gyro, as well as being exhausted into the cabin. As Goodlin activated the nitrogen to power the gyro, he noticed an ominous rise in cabin pressure. He set the gyro discharge valve to atmosphere, but cabin pressure continued to rise. Next, he pulled the manual pressure release; it failed to operate (the trigger mechanism was later found to have sheared off). Cabin altitude passed below sea level as the aircraft passed 24,000 feet. Concerned that the continuing increase in pressure could cause him to black out, Goodlin decided to jettison the XS-1 cabin door. Hamilton quickly secured the outside of the door with the chains used to lower the door after the pilot's cockpit entrance. By the time this maneuver was complete, the cockpit altitude was 1,000 feet below sea level; Goodlin could delay no longer. When the door handle was pulled, the cabin pressure seal was broken. The cockpit pressure buildup literally caused the door to explode outward, hitting the ladder with such force that the door was bent and the ladder damaged. The door ended up wedged between the XS-1 and the ladder.

When the seal was broken, Goodlin experienced an immediate pressure change from $-1,000$ feet to the real altitude of 23,000 feet. He instantly noticed a clearing of his ears and a momentary stomachache. However, what he saw was worse than what he felt. Peering out from the cockpit, Goodlin surveyed the ladder damage and the yawning opening below the bomber. Deciding that it was not possible to climb back into the B-29, Goodlin informed Dow that he would ride out the return journey in the XS-1! After the events of the last few minutes, no one wanted to argue an alternative. The B-29 return to Muroc was mercifully uneventful, in spite of the tight XS-1 ground clearance on landing. But for the XS-1 number 2 it was certainly a shaky beginning to what was recognized as a dangerous project.

On the ground, the Bell technicians surveyed the damage to both planes. The XS-1 door was dented, the aircraft's door frame was warped, and the B-29 ladder was damaged. None of these were judged as major repairs. Indeed, Stanley quickly repaired the door by himself with a sledgehammer. The ladder was straightened and reinstalled, and the manual release handle and its attachment to the emergency dump plug fixed. Surprisingly, the gyro exhaust valve had not failed. After reviewing the facts, the technicians initially pinned the cause on the exhaust from the pilot's oxygen breathing system. But the real culprit turned out to be a rubber gasket used to help pressurize the cockpit during tests at Bell. Someone had forgotten to remove it after the tests at Buffalo. Only after a thorough inspection was the gasket spotted and removed. The emergency dump relief valve was not fixed, but they armed Goodlin with a screwdriver to pry the valve open if necessary. As Joel Baker wryly noted, Goodlin was apparently supposed to hold the valve open with one hand, fly with one hand, and work the controls with another hand.

After several days at Muroc in the company of the entire Bell/ NACA/AMC team, the personnel involved made several interesting observations. The NACA personnel were frustrated by Bell's drive to get the aircraft ready as quickly as possible, which prevented the NACA people from working on their instruments. (NACA needed to work when Bell technicians were not around the small aircraft.) But the priority had to be getting the XS-1 ready to fly, because absent this condition there was no need for the instruments. And, of course, once the XS-1 was ready, Bell wanted to fly it. Frustrating as this was, it was the usual order of business on contractor planes.

Procedure was not the only point of friction. Walt Williams initially perceived Goodlin to be less than enthusiastic in his new role as lead pilot. While Woolams had been excited by the prospect of flying the XS-1 as fast as possible, Goodlin seemed to Williams to be only interested in getting the demonstration flights over as quickly as possible. Goodlin expressed concern about the ladder, the location of the cabin door for emergency exit, and other flight procedures. It was only after a personal meeting between the NACA team and Goodlin that Williams altered his opinion. Goodlin would do the NACA program, Williams believed, *if* he was away from Frost and Stanley. Joel Baker, however, felt that Goodlin had little direct interest in the mundane details of the aircraft and its readiness, but that he did have an intense interest in piloting the aircraft. In Baker's view, Goodlin did not make as much of a "fuss" as Woolams did about flying the airplane. Goodlin counted on Frost to ensure the other details. Baker also observed that Goodlin was not in favor of wasting all the scheduled time to prove the XS-1 could carry and jettison water ballast simply to learn some theoretical handling properties. Slick told Baker that he was paid to fly to Mach 0.80 and he wanted to get there as soon as possible. Baker did not remember ever hearing him mention 1,500 MPH or even 1,000 MPH. But Baker didn't know about the Goodlin/Stanley handshake agreement or the Bell conversations with the AMC for a follow-on contract.

Fortunately, everyone agreed that Goodlin's trust in Dick Frost was well placed, as he proved to be a superb program manager. As one observer put it, "The only person who knows all the answers is Frost." It was clear that Frost not only knew the technical details, but he was also not afraid to make a decision. If there was disagreement on the attitude of the XS-1 pilot and unanimity on Frost, there was also no dispute as to who was initially in charge at Muroc. It was clear that Bob Stanley was making the early decisions and running the show. Williams believed Stanley was trying very hard to get through the acceptance tests quickly in order to save money, perhaps in 10 days, which would result in the NACA getting very little data. Nothing in the past indicated the tests would go that smoothly, but Stanley seemed ever the optimist in setting deadlines. To Williams, it was another case of Bell Aircraft cutting corners a little too close. Many hoped that when Stanley returned to Buffalo things could assume a more routine pace.[3]

In spite of all the hurry and excitement centered on the XS-1 it was still very difficult to ignore the fact that the experimental Bell aircraft was not the only advanced airplane undergoing tests around Muroc. Sometimes the base seemed like a giant repair yard as many of the new models spent more time in maintenance than flying. The NACA personnel attempted to keep LMAL advised of the other work in progress, but it was an extremely difficult task. Lockheed Aircraft was trying to establish a new speed record with a P-80R. (Although several attempts were made, the best the P-80 could initially achieve was 617 MPH.) Meanwhile, Douglas Aircraft had run into money problems and their work on the XB-43 was proceeding slowly. The XP-84 did achieve 611 MPH in September, but burned out three engines in quick succession, which delayed that program. Vought Aircraft was conducting tests on the XF-5U (which experienced wing skin problems), and the XF-6U. The latter's flight program consisted of one test before the ever-present refrain of engine problems. The Northrop XB-35 Flying Wing had managed only three flights and continued to be plagued by mechanical leaks. Likewise, the North American XFJ-1 had so many engine problems in its half-dozen flights that some observers commented that the engine was in and out of the plane as often as the pilot.[4]

These recurring flight problems were serious but not unexpected setbacks for any new aircraft testing program. If the supply system had been adequate, all of these problems would have been manageable. Unfortunately, Muroc sometimes seemed at the end of the maintenance universe; the desert test facility was notorious for supply failures. The postwar need had simply overwhelmed the services' ability to provide spare parts. Even on the high-profile XS-1, Wright Field and Bell Aircraft were perennially late in forwarding parts. Add in problems with sand, winter rains, lack of money, distance, and inertia, and each Muroc program seemed plagued by delays.

Slow or not, however, progress on the XS-1 continued. On October 11, all was judged reasonably ready for another attempt at a first shakedown flight for the XS-1 number 2.[5] In preparation, the NACA group rushed to finish their instrumentation work, but "finished" was a loose description of the results. While the strain gauge and recording instruments were checked and loaded, the telemeter installation was still incomplete. Worse, the M-2 optical tracker had disappeared in the mail. A trace was issued from Wright Field and Muroc to locate it. Finally, even though the transmitter communications had not been repaired, Stanley was determined to proceed with the next flight.

By the time of takeoff, XS-1 gross weight was 6,831 pounds and center of gravity would be at 25.3 percent mean aerodynamic chord. The purpose of the initial #6063 flight was to let Goodlin familiarize himself with the basic handling characteristics of the new plane. The weather conditions at Muroc cooperated, so unlike the rainy and cloudy conditions prevalent at Pinecastle. During late morning, with a clear sky and unlimited visibility, the B-29 lifted off from Muroc airfield and began the familiar climb to drop altitude. Stanley followed in the P-51 as chase. After reaching an altitude of 5,000 feet, Goodlin climbed into the B-29 bomb bay and was lowered on the ladder to the XS-1 cockpit. After entering the cockpit, Goodlin quickly went through the standard flight procedure by securing the safety belt and accessing the hose connections for the oxygen supply and the mask dehydrator. After lowering and securing the XS-1 cockpit door, the umbilical cabin pressurization heater hose was connected from the B-29. A full XS-1 cockpit checkout was conducted as the B-29 continued its climb to launch altitude.

As with the difficulties on October 9, XS-1 cabin pressurization quickly proved faulty. Goodlin observed that the cabin could not maintain the proper differential pressure. At a pressure altitude of 20,000 feet, the cabin altitude was in the vicinity of 5,000 feet (the correct differential at that altitude should have been 7,000 feet). After Mac Hamilton disconnected the pressurization and heater hoses, Goodlin noted the cabin pressure rose to an equivalent 12,000 feet at which time the cabin-pressure relief valve began operating. Throughout the B-29 climb to launch altitude, the XS-1 cabin pressure continued to fluctuate between 12,000 and 13,000 feet. This annoying distraction finally caused Goodlin to place the pressure relief valve in the dump position prior to the drop.

After the first stage and lox dome-loading operations were complete, the oxygen supply hose from the B-29 was disconnected. Goodlin immediately noticed that the amount of breathing oxygen in the XS-1 system was extremely limited. As the anticipated flight time was also limited, this would not seriously impact the current flight. However, for any longer duration flights, Goodlin believed the present system would have to be upgraded.

At noon, about 25,000 feet over Muroc Field, and at an indicated airspeed of 230 mph, the B-29 released the XS-1 number 2 on its first mission. As at Pinecastle, the XS-1 separated cleanly from the bomber and Goodlin noted none of the g-force discomfort that Woolams had experienced from the use of the ejector. Once away from the bomber, Goodlin quickly found the XS-1 controls to be quite light and, as a result, had to fight an initial tendency to overcontrol the aircraft. Goodlin specifically noted a lack of feel in the aileron control system and recommended reducing the friction. Rate of roll was extremely quick.

To test out the handling characteristics, Goodlin put the XS-1 through a series of stall tests. At 15,000 feet he conducted a stall with the airplane in a clean configuration. Stall was first noted at an IAS of 140 mph. After a shallow dive to gain an IAS of 170 mph, Goodlin extended wheels and flaps and conducted another stall test. The aircraft exhibited similar characteristics to that of the clean configuration. More important, Goodlin found the XS-1 required minimum time for stall recovery.

Goodlin brought the XS-1 in on the northeast runway at Muroc. He lined up the landing at approximately 180 mph with touchdown at 140 mph. In the interest of safety, approach was made rather high and at a greater speed than necessary. As a result, contact was made with the runway about 3,000 feet down the 6,500-foot track. The XS-1 continued off the runway and out 4,500 feet over the dry lakebed. Rollout was inordinately long due to malfunctioning brakes. Goodlin estimated that due to the clean lines of the XS-1 7,000 to 8,000 feet of space would always be needed for the landing in case the brakes ever again failed.

This distance once again indicated the desirability of using the Muroc site over Langley.

Three days later, on Monday, October 14, Goodlin was ready for another shakedown flight,[6] with the XS-1 in the same configuration as the previous flight and Dick Frost now P-51 chase plane pilot. With the sky over Muroc a bright blue and visibility unlimited, Dow pointed the B-29 down the runway and off for test flight number 12. At 7,000 feet, Goodlin once again climbed down into the XS-1 using the well-tested readiness procedure. With the B-29 circling over Muroc airfield during its climb to 25,000 feet, Goodlin completed the cockpit checkout. As the B-29 reached 230 MPH, Goodlin signaled he was ready and the crew executed the drop procedure. Separation at 12:02 P.M. was again as clean as possible. Unnoticed at the time, Goodlin had activated the NACA recording instruments about eight minutes prior to drop. As a result, only about 30 seconds of data was recovered for study.

Immediately after separation, Goodlin put the XS-1 in a 40-degree dive to about 12,000 feet. Speed as recorded reached Mach 0.71 (420 MPH IAS at 14,000 feet). Coming out of the dive, Goodlin demonstrated a 4g pullout. The pilot again found the XS-1 controls to be highly sensitive. In fact, Goodlin could not remember a plane that exhibited this degree of sensitivity in control responsiveness. The momentary spongy feeling in up-elevator responsiveness remained a mystery, although Goodlin speculated it might simply be low cable tension. During the dive, Goodlin also tested the stabilizer control and found it effective but somewhat slow. He recommended the control motor speed be increased to ensure faster stabilizer response.

After the dive test, Goodlin again lowered the wheels and flaps and checked the spoiler characteristics. The upper wing surface spoilers had caused much debate. Goodlin found that at approach speeds the spoilers were ineffective unless totally extended. At full extension, stall was achieved with mild buffeting, but simultaneously, the nose would drop and the aircraft would roll violently. As a result, Goodlin recommended that the spoilers should not be used at low altitudes until further investigation could be conducted. After a brief flight of approximately eight minutes, Goodlin performed a routine approach and landing. Airspeed during approach was 207 MPH and rate of descent was a very steep 83 feet per second. At touchdown speed had slowed to 134.5 MPH. However, immediately after touchdown, the brakes once again failed. Rollout on the dry lakebed provided the margin of safety.

At the time of Goodlin's second rehearsal flight, an important meeting with significant ramifications for the XS-1 program was taking place hundreds of miles from the desert. It directly concerned the NACA/Bell Aircraft dispute over the structure and pace of the flight program.[7] This meeting was an outgrowth of the October 8 Stanley/Williams clash concerning the pace of the acceptance program. Attending the conference at Wright Field were Clotaire Wood from the NACA Washington HQ, Hartley A. Soule from LMAL, and Trig Blom of the Wright Field NACA Liaison Office. Representing the AMC were Col. George Smith, chief of the Aircraft Projects Section; Col. R. S. Garman and James Voyles of the Fighter Projects Subsection; and Major Shoop from Muroc Field. The conversation focused on a) the Army policy to be followed for the acceptance tests of the first airplane, and b) procedures for handling the research flying of the XS-1.

Stanley's previously stated concern—that the detailed NACA research procedure was inordinately expensive to Bell Aircraft—was the basis for the discussion concerning point (a). The LMAL position, as expressed at Muroc, was that the AMC contract called for the NACA acceptance of the XS-1 only *after* it had exhibited satisfactory flight characteristics. The NACA personnel at the Wright Field meeting again stressed that point. The abbreviated flight program that Stanley implied Bell Aircraft wanted to conduct would be useless from the NACA's point of reference. Not only would there be insufficient data, but in some cases (since not all NACA instruments were ready for the tests) there would be no data. After listening to the discussion, the AMC decided to invoke the clause in the MX-653 contract requiring satisfactory performance up to Mach 0.80 as a lever to force a slower and more orderly program than Bell Aircraft had outlined in its September 20 memorandum to the AAF. As the AMC still believed that the NACA would accept the airplane at the end of the Bell Aircraft contractor tests, it pledged to try to accommodate the NACA concerns on acceptance conditions for the plane. While no definition for "satisfactory performance" was decided, the discussion focused on a requirement for twenty powered test flights to accumulate the information needed to determine airworthiness. Further, the AMC agreed to do anything within reason to ensure that data was recovered from each flight. However, Smith cautioned the NACA that the Bell Aircraft contract was a cost-plus flight contract and thus very expensive for the AAF. The AMC wished to limit the contract's duration to only those flights needed from an engineering standpoint to validate the program.

In regard to (b), the conference agreed that flying the XS-1 re-search program was extra-hazardous and that it could "probably be done most fairly by *contracting for the pilot's services*" (emphasis added). Piloting the research craft was considered an integral part of the flight program and the NACA should negotiate a pilot contract. Since neither the NACA nor Wright Field had funds available for a pilot contract, the AMC recommended that the NACA take up the matter with the Bureau of the Budget, indicating that they believed this issue extremely important and promising to give the NACA full support for the budgetary request. Both parties also recognized that maintenance and servicing of the XS-1 airplanes was a more significant operational problem than originally contemplated and as a result the NACA should receive greater AMC assistance. Thus, the AAF prom-ised to supply the maintenance and servicing for the program and the pilots for the B-29.

With the close of the meeting, an important milestone in the XS-1 test program had been reached. First, the AMC had hedged in defining who was in charge of the acceptance test program by giving each side some room to maneuver. Bell Aircraft would fly the aircraft until twenty powered flights had been reached, which was far more than Stanley wanted to do. By implication, the NACA should adjust its data needs accordingly. The fixed number of flights would thus ensure AMC the program would have an end. The NACA would get its data and set up the flight schedule.

Second, the issue of pilot cost for flying the transonic flights had been put on the agenda. Although the NACA was the reference agency because they would be contracting for the pilot to fly the program, no final decision was reached. From Bob Stanley's point of view, a Bell Aircraft pilot could still fly the program. If Goodlin wanted a personal services contract, here was the way to fund it. Either the NACA could contract with Bell to provide a pilot or Bell Aircraft could, at a later date, recommend some functional alternative (their own program) that would allow them to recoup their Goodlin bonus-program costs. Events were not yet clear as to which alternative might work best for Bell. However, one other financial matter, admittedly a bureaucratic issue, surfaced in the plan to let NACA seek the funds for the pilot program. The AMC had funded the research flying program since its inception. The issue of funding an expensive pilot contract was recognized as integral to the flight program. Yet now the AMC announced that NACA should seek the funds from the Bureau of the Budget. In hindsight, in spite of very real AMC budget problems, it is hard not to believe that this suggestion did not represent an AAF version of the classic Washington ploy, "let's do the program with *your* money." But while the politics of policy continued to twist and turn, events at Muroc continued to move straight ahead.

In anticipation of the next XS-1 test, the Bell technicians worked on loading the aircraft with water ballast to simulate its fuel. Unfortunately, the fuel pressure regulators malfunctioned. Worse, the XS-1's tanks were filled with dirty water, because the fuel trailer's dispensing system filter had not been cleaned before the truck was delivered to Bell, which fouled the lines and tanks. Flushing would require too much time so a decision was made to simply fly the next flight without the simulated ballast.[8]

While the maintenance work proceeded, on October 16–17 Stanley, Williams, and Maj. George Prodanovich (Shoop's deputy) met at Muroc with Voyles and Colonel Garman of Wright Field to again discuss the acceptance testing procedure, following up on the Wright Field meeting. Williams complained about the missing radar and NACA research instruments for the plane. Voyles again promised to try to solve the problem. Of more pressing importance, Garman stated the October 14 AMC policy on the acceptance tests. The AMC wished to accommodate the NACA data collection to the fullest possible extent; therefore, Bell Aircraft should cooperate by allowing the NACA to have its instruments ready for every flight. On another issue, Williams stated the NACA had no problem with Goodlin's enthusiasm (which represented a total change from Williams's initial comments) to perform the NACA flight program. The NACA's biggest complaint was that it was difficult to guess where the XS-1 would be between flights (hangar, pit, on the B-29). Stanley acknowledged the NACA concerns and said that he was sure that once he left Muroc the project pace would slow down. He confirmed his desire to push the project "faster than normal" because his time at the tests was limited. Garman indicated that cooperation was the key, but that if an issue arose that Bell/NACA couldn't reach a consensus on, they should call Wright Field. The group adjourned the proceedings to watch the October 17 flight test. Prior to the flight, Williams indicated to Voyles the NACA interest in the other X-series aircraft. Further, Williams told Voyles that he anticipated good information from the British experiments with the de Havilland Swallow, unaware of the accident that ended that possibility.[9]

On October 17, Bell Aircraft was ready for the third glide flight, the purpose of which was to further examine stall characteristics.[10] The aircraft was unchanged from the previous two flights. The B-29 takeoff and climb to altitude were routine. Separation occurred just after 2:00 P.M. at an IAS of 230 MPH and a pressure altitude of 25,000 feet. The drop separation was clean.

Following separation, Goodlin stabilized the XS-1 at an IAS of 250 MPH in an 80-degree bank to the left. A positive acceleration of 2 g's was experienced and maintained. Goodlin then proceeded to repeat the maneuver, but this time increased the positive acceleration until stall conditions were reached at an observed 2.8 g's (NACA recorded 2.75). As the elevator was applied, slight buffeting developed into moderate buffeting. Goodlin found no tendency for the aircraft to want to roll and recovery from the stall was easily effected by merely releasing the back pressure on the control column.

To finish up the flight, Goodlin descended to 10,000 feet. At 180 MPH, Goodlin extended both the landing gear and flaps. At 5-MPH decrements down to 160 MPH, he tested spoiler effectiveness. No noticeable difference in effectiveness was noted. However, as previously determined, at maximum spoiler deflection the XS-1 experienced moderate buffeting and exhibited a tendency to pitch violently forward while unpredictably rolling to either side. Goodlin theorized the rolling condition might be due to slack tension in the control cables to the spoilers. In any event, Goodlin again warned that use of the spoilers at low altitudes could have disastrous consequences. Finishing up the flight, Goodlin approached Muroc Dry Lake from a southerly direction at a speed of 180 MPH, much slower than on flight number 2. After touchdown at 134.5 MPH and through rollout, the pilot found great difficulty with the brakes. Although they finally worked, Goodlin discovered they required moderate pumping to maintain pressure. More work needed to be done before the brakes would be satisfactory.

Although seemingly successful, the flight was not without the endless problems encountered in the early stages of research flying. Almost immediately after separation, Goodlin noted that the cabin-pressure relief valve was malfunctioning. Cabin pressure was jettisoned at the time of drop, but upon touchdown, the pilot noted the altimeter read 0 feet (Muroc is at 2,300 feet). The remaining cabin pressure of one psi was jettisoned successfully before the XS-1 door was opened, but Goodlin recommended a fix before the next flight.

At the conclusion of the October 17 flight, Bell Aircraft decided to cease immediate testing and incorporate several modifications to the

XS-1. First, the pressure regulators would be changed so that the dome-loading procedure could be executed from the XS-1 rather than strictly from the B-29. Second, the dirty ballast in the tanks had to be flushed before any future powered flights. The fuel mixing and dispensing unit was disassembled and loaded on the B-29 for transportation to Bell Aircraft. Finally, on top of all other problems, 10,000 gallons of the wrong type of alcohol fuel (AN-A-24 instead of AN-A-18) was delivered to Muroc and had to be replaced. This error prevented any ground testing of the engine.

In light of these problems, Stanley ordered most of the Bell team to return to Niagara Falls. Frost and four mechanics were detailed to remain at Muroc with the number 2 airplane. Most of the NACA personnel elected to remain at the test center. On October 22, Dow and Goodlin piloted the B-29 back to Niagara Falls. From there, the B-29 was scheduled to proceed to Oklahoma City for a 100-hour maintenance checkout. However, plans quickly changed once the B-29 was back at Niagara Falls. On October 24, the B-29 journeyed to Rome Air Field, New York, for its maintenance work. At Bell Aircraft, the returning Muroc "veterans" found the number 1 airplane fuselage completely assembled with the new thin 8 percent thickness chord wing and 6 percent tail. The nitrogen spheres and fuel and lox tanks were installed, and the second thin wing/tail units were near completion. It was anticipated that the number 1 aircraft would be ready to travel to Muroc on November 11.[11]

In California, the new "natives" tried to adjust to life in the middle of nowhere. Work on the number 2 airplane consumed some of the time, but the continuing delays in the B-29 repairs meant that the brief downtime stretched into longer than anticipated. Watching other aircraft tests became a part of the new routine, but complaints and complications seemed the real primary occupation of the crew. From November 11 to 14, Muroc witnessed one of the worst storms in its history. The available records kept by the AAF bomber pilots for the test range revealed nothing similar to the storms that occurred over those several days. Terrible rains flooded the area and rendered the lakebed unusable for perhaps two weeks. In addition, temperatures had turned bitterly cold, and for several days in November the NACA offices at Muroc had no heat. Meanwhile, the Bell technicians struggled on with XS-1 work in spite of the lack of spare parts. Everyone still regretted the AAF decision to send the M-2 optical tracker from the East Coast by freight mail. The local betting was that it would not be seen for two months.

Worse than all the weather and supply snafus was the continuing battle over available housing. Early in November, the base housing officer ordered that no additional NACA personnel be put on the housing list. All those on the list could remain and had a chance to receive the "luxurious" base accommodations. In reporting on those accommodations, Williams indicated to Mel Gough that the *good* news was that they had kerosene heaters. The bad news was that they were shacks at best. By mid-November everyone recognized the magnitude of the housing problem. Colonel Gilkey, Muroc's commander, traveled east to obtain authorization to purchase additional land to build 100 new houses for the growing number of personnel. This effort was tempered by the knowledge that the expansion rumors quickly caused the worthless nearby desert land to skyrocket in value.[12]

By November 15, the B-29 had returned to Buffalo for final preparations for the journey to Muroc.[13] On November 20, with Dow as pilot and Slick Goodlin handling copiloting chores, the B-29 began its one-stop return flight to Muroc. The XS-1 number 1 was not with the B-29, since it had been delayed; its new date for completion was December 6. Part of the problem was that no one had told the NACA to have a second set of flight instruments ready for installation. NACA had only one package and one set of spares to *rotate* between the two aircraft. Thus the XS-1 number 1 would go to Muroc without its instrumentation. The bomber arrived on November 21; the rest of the Bell team flew out in the contractor C-47.

In preparation for the first powered flight, Bob Stanley decided to return to Muroc. He flew out on November 27 in a specially prepared Bell P-63 aircraft that had a P-39 engine without the aneriod-controlled automatic coupling of the stages provided in standard P-63s. This modification permitted the relocation of the oil tank forward to allow a second passenger in the airplane. This benefit came at the expense of high-altitude climb performance and ceiling, but did allow for a Bell photographer to fly along to film the next several flights. With Stanley present, the tempo of XS-1 operations gained momentum.

After flight number 13 (Goodlin number 3), Bell's Muroc ground crew altered the tank pressurization system and reworked the instrument panel in the XS-1 number 2 cockpit, changes completed by November 22. Ground tests conducted to ensure the security of the pressure system disclosed several leaks. Additional time during the week of November 25 was required for Frost to finish the repairs. Although the engineers did not "fix" all leaks (indeed one leak, reported as a steady stream out of a one-quarter-inch line, simply drained overboard), Stanley decided to wait no longer. He declared the aircraft ready and ignored Frost's reservations about the security of the last pressure tests.

On December 1, the remainder of the NACA Muroc contingent (Baker, Hinman, and Charles Forsyth) arrived at the test station in the NACA C-47, along with the M-2 optical tracker. The NACA technicians made an effort to use the tracker for the anticipated December 3 ballast test. However, during the attempt to hook up the tracker in conjunction with the SCR radar, Minalga discovered that the latter had been modified. As a result, although the radar was ready on December 3, it was not possible to use the radar in conjunction with the tracker until the radar could be altered, which was accomplished shortly thereafter.

In anticipation of the fourth flight in #6063, Bell engineers filled the XS-1 fuel and lox tanks with water alcohol to about 40 percent of their capacity. This boosted gross takeoff weight to 8,730 pounds. The Bell-initiated design changes to the interior allowed the pilot to now control all regulator and tank pressures from the cockpit. (The previous system required the XS-1 crew chief to climb out into the B-29 bomb bay during flight to dome-load the XS-1 aircraft's first- and second-stage pressure regulators.) On the B-29, the NACA telemeter panel was installed and Heaney was designated as the in-flight operator rather than the B-29 crew chief. A radar beacon had also been installed in the XS-1 to allow automatic tracking since no one was sure the M-2 optical tracker would be ready for the next tests.

A flight was planned for Friday, November 29, but the Bell engineers could not load the XS-1 into the B-29 in time. As a result, the XS-1 was lowered into its pit and the B-29 was placed over it to hasten the operation in anticipation of the next test. The original plan was to ballast-test the XS-1 in three flights.

A powered flight was scheduled for the following Friday, December 6, but Stanley wanted the test as soon as possible. Given that Goodlin felt the ballast tests to be a waste of time, everyone knew it was a good bet that the anticipated schedule would not be the final schedule. This became even more apparent when it was announced that Larry Bell was on his way to Muroc to see the tests. When Stanley found out that

Bell was due to arrive on December 1, he worked even more furiously to get the aircraft ready for the powered flight. Williams noted that this difficult work style was likely to ensure that Stanley would mysteriously end up floating face down in the lake—except the lake had no water. The "Great White Father" (as the Bell team called Stanley) was, Williams believed, "a hard man to understand."

On December 2, the B-29 and the XS-1 were mated and prepared for the drop test.[14] Weather conditions for the day were clear and visibility was unlimited, in continuing stark contrast to the experiences at Pinecastle. (Even though there had been rain on two occasions since early November, the lakebed remained dry and usable.) However, before the actual drop flight, a situation occurred that almost looked comical in hindsight were the problem not so deadly serious. With Dow as the controls and Heaney as copilot, the B-29 took the two aircraft up for a 2-hour, 50-minute flight later listed by the bomber pilot as a ballast conditions checkout flight. Goodlin didn't file a formal flight report. Frost flew chase and observed most of the episode. The flight was much more than routine. During preflight checkout, the Bell ground crew noted that the nosewheel would not lock in the up position, which meant that the B-29 was incapable of landing with the XS-1 mated to it due to extremely close ground clearance. Nevertheless, no one anticipated the possibility of the planes landing still mated and Stanley decided to proceed with the flight. However, after takeoff and during the climb to drop altitude, potential disaster reared its head. Goodlin in the XS-1 could not get the pressure to the lox tank because its Quick-as-a-Wink sleeve valves had frozen open. Now the nosewheel/ground clearance problem became *very* significant.

Frost, flying chase, mentioned that the nosewheel difficulties would prevent the B-29/XS-1 from landing while mated. He recommended that Goodlin retract the XS-1's nosewheel just before B-29 touchdown. Although the wheel would slowly return to the down position, by the time it completed that cycle the B-29 would be on the ground and executing final rollout. Unfortunately, neither Stanley nor Larry Bell in the tower could fathom what Frost was describing. Stanley ordered Frost to return to the ground to explain his solution. Frost was livid, but executed a screaming dive in the P-51 to land and brief the two bosses. Stanley interrogated Frost on the situation (which further annoyed Frost), but finally authorized him to continue with his plan. All these delays produced a plan none too soon. With the B-29's fuel nearly gone, Goodlin finally got the XS-1 gear sufficiently retracted so that Dow could execute a landing. Touchdown was smooth, just as Frost predicted.

On the ground, the Bell engineers discovered that the pair of valves in the lox system had frozen open. Drying and resetting the valves fixed the system. Nothing was done to readjust the nosewheel uplock. After finishing the repairs, the Bell team prepared to repeat the procedure for the anticipated XS-1 drop. By mid-afternoon, with all conditions ready, Dow again guided the B-29 down the runway for the second flight of the day. At 8,000 feet, Goodlin entered the XS-1 and secured himself in the cockpit. He completed the predrop checkout and began the procedure for loading the fuel and lox tanks: first, Goodlin dome-loaded the first-stage regulator, which reduced the XS-1's nitrogen pressure to 1500 psi; then he dome-loaded the two second-stage regulators in sequence to reduce the high inlet pressure. Lox and fuel-tank pressures each built up and stabilized at 315 psi. Just before drop, however, the earlier flight problem recurred; Frost reported that the nose landing gear was slowly creeping down. It was decided to ignore the complication and continue with the test. But even here misfortune dogged the day's work. At the conclusion of the drop count sequence, Heaney activated the release mechanism, but the shackle failed to function. Further attempts also failed to secure a separation. Finally, Mac Hamilton entered the bomb bay and manually operated the release mechanism. The XS-1 began its flight at 4:31 P.M. (another case of Stanley working late since this was after normal Muroc working hours).

Surprisingly, after all the bad luck up to that moment, the separation was very clean and Goodlin noted little difference from previous flights. Shortly after clearing the area of the bomber, Goodlin actuated the fuel and lox jettison valves, allowing the tanks to expel their contents. He recorded that tank pressures dropped gradually, indicating a smooth flow. He noted no center of gravity shifts, further indicating that the two tanks were emptying in equal amounts. After working both valves in isolation, Goodlin jettisoned the remaining propellants and noted that the tanks were empty within 5,000 feet of bomber separation. Due to the nosewheel problem, Goodlin didn't attempt to try for the high Mach number/stall turns that had been the original test requirements for the mission. Rather, at 9,000 feet Goodlin extended the landing gear and flaps and lined up an approach. Landing was satisfactory. After inspection, the nosewheel difficulties were attributed to an improper adjustment of the wheel uplock. This was easily remedied. Goodlin further recommended that based on his observed results the next flight should be a powered flight.

Powered Flight

The possibility of witnessing the first powered flight of the aircraft selected to break the sound barrier proved a magnetic attraction to all interested parties. As a result, during the first week of December 1946, Muroc became the home for many distinguished guests.[15] These included Larry Bell, Bell Corporation's vice-president D. Roy Shoults, and Frederick F. Robinson of Bell's board of directors; James Voyles, the AMC XS-1 project engineer from Wright Field; and Edwin Hartman of the western coordinating office of the NACA. General Craigie was also expected to attend, but did not make it to Muroc for the first flight attempt on December 6.

In the intervening days between the last drop test and the powered flight attempt, Stanley ordered the XS-1 rocket engine put through a series of ground static tests. The first test, on Tuesday, December 3, was not satisfactory as only one chamber started and continued to run, due to low nitrogen pressure at the propellant valves. The propellant valves, which were pressure-operated, were found to be operating at a pressure equal to the fuel pressure. As a result, the engine would not run. The repairs took all day Wednesday. Bell engineers installed a small Grove spring-loaded regulator in the cockpit feed from the first-stage pressure regulator. The line was set to deliver 420 psi to the engine nitrogen manifold, which then distributed the gas to each cylinder head (to keep any propellant leakage from mixing therein prior to the intended start of combustion) and to the respective cylinder solenoid valves. A second test firing on Thursday was judged a success.

One other problem came to light during the static tests on December 3 and 5. The extremely cool temperature of the liquid oxygen revealed an inadvertent design flaw in the XS-1 brake system. The brake lines passed beneath the lox tank through the ventral duct. The extremely cold lox temperature caused the hydraulic fluid in the XS-1 brake lines to freeze, rendering the brakes inoperative. However, after consultation, the Bell engineers decided not to postpone the pow-

ered test since the Muroc runway/lakebed afforded ample rollout space
even in the absence of any XS-1 brakes.

On Friday, December 6, luck regarding the weather at Muroc finally ran out. Although it was clear at sunrise, by 8:30 A.M. it was
raining at Muroc Field. In an effort to be prepared for any possible
weather changes, Bell engineers decided to load the XS-1 with propellants and stand by in case the weather cleared. The loading procedure for powered flight was a variation of that used in the previous
nonpowered flights. First, the B-29 and XS-1 were mated at the loading pit, with the XS-1 tanks empty. The B-29 was then towed to the
nitrogen evaporator/lox storage tanks so that the XS-1 could be filled.
This increased XS-1 weight to 12,012 pounds. It also changed the
center of gravity of the aircraft to 22.2 percent mean aerodynamic
chord. Upon completion of the propellant loading, the B-29 was towed
to the apron for boarding by the crew. After the B-29 engines had been
warmed up, Dow would request clearance from the tower for takeoff.
The newly trained Army radar maintenance personnel had arrived
from Wright Field and all NACA instruments were checked and
loaded. All sat at the ready awaiting the go decision.

By afternoon, the sky over Muroc had cleared and Bell engineers
decided to attempt a flight. The weather to the west of the airfield
remained cloudy, with a stiff breeze of 25–30 MPH on the ground. Nevertheless, a decision was made to attempt a flight. The B-29/XS-1
lumbered down the runway and began the climb to altitude. For this
flight, chase was provided by three aircraft: first, Frost in the P-51
that had been used in all the previous Muroc tests; second, a specially
prepared FP-80 jet aircraft, with fixed nose cameras to record the
XS-1 flight, piloted by a Captain Smith of Wright Field; and third, the
modified Bell P-63 with a corporation photographer in the aft cockpit.
Unfortunately, during the time it took the B-29 to reach drop height,
the weather again closed in on Muroc. Since there seemed little chance
that the weather would soon clear, the drop was aborted. The fuel and
liquid oxygen were jettisoned and the B-29/XS-1 returned to base 1
hour and 20 minutes after takeoff.

After the failure of the December 6 flight, Bell engineers decided
not to schedule another attempt at powered flight until Monday, December 9. The uncertain weather continued to clear over the intervening days, and the lakebed appeared to have suffered no ill effects from
the rain. In preparation for the flight, Bell engineers revised the nitrogen supply system in the XS-1 to obtain higher pressure at the

nitrogen manifold. They also disconnected the attitude gyro and the nitrogen warning light from the pressure system. Finally they installed a provision for the emergency operation of the attitude gyro. This consisted of an on-off manual valve between the nitrogen valve and the gyro. Later, a three-way valve was installed to vent nitrogen directly into the cabin to maintain the 3 psi differential above ambient pressure.

At late morning on December 9, Dow proceeded with B-29 takeoff for another effort at the first powered flight.[16] With weather conditions perfect, the bomber began the climb to drop altitude. At 9,000 feet, Goodlin entered the XS-1 and quickly began the cockpit checkout procedure. During the initial checkout phase, Frost took off in the P-51 to perform chase duties for the flight. Routinely, trouble traveled along with the XS-1 test craft. As the B-29 continued its climb to launch altitude, Goodlin noted an alarming drop in the bleed pressure for the rocket motor. This pressure was normally provided by the nitrogen supply aboard the B-29. Now, Goodlin observed, the source pressure had dropped to 3,700 psi even before the drop. Nor was this the only engine problem. After Goodlin tried to dome-load the lox and fuel pressure regulators he discovered the second-stage lox regulator allowed the lox tank pressure to drop quite rapidly. As a result, he had to keep dome-loading this regulator almost continuously to maintain sufficient lox tank pressure for the later rocket motor operation. Goodlin also noted the cabin pressure differential was only 5,000 feet at the 25,000-feet altitude, clearly indicating that the regulator was not working. As a result, he determined to open the dump valve, letting cabin pressure equalize with outside air pressure. The problems made for a seemingly busy, and certainly not routine, climb to altitude.

Once the B-29 reached 27,000 feet, Dow executed the procedure for the drop sequence. The XS-1 separated cleanly at 11:54 A.M. and began to fall away from the mother ship. The increase in XS-1 weight had one immediate effect—Goodlin noted the rate of drop was seemingly twice the rate it had previously been. The XS-1 now exhibited a slightly tail-heavy condition because of the fuel and lox, which lasted until the two ingredients were exhausted. However, it was not serious enough to warrant trimming out by stabilizer adjustment.

Ten seconds after separation, Goodlin ignited the first chamber of the rocket engine. He immediately realized that without reference to the chamber pressure gauge it was impossible to tell the engine had actually started. Silently and without noticeable effect, the XS-1 began to accelerate. After a brief pause, Goodlin tripped the second chamber ignition and began to feel a slight acceleration. Quickly shutting down the number 2 chamber, Goodlin put the XS-1 in a slow climb that he estimated to be 500 feet per minute. The effortless power and the sense of release caused him to underestimate his continuing rate of climb. (The NACA instruments would later reveal it reached 2,600 feet per minute.) Speed was approximately 330 MPH as Goodlin fired the rocket chambers in sequence, always keeping power confined to one chamber. Goodlin noted the same noiseless phenomenon in the cockpit that Woolams had reported seemingly such a long time ago.

At a pressure altitude of 35,000 feet Goodlin reignited the number 2 chamber and let the airplane acceleration build up. The pilot's gauge rapidly reached an indicated speed of Mach 0.75 and, consequently, Goodlin quickly shut down the number 2 chamber. (The NACA instruments later revealed the actual speed to be Mach 0.793.) During the speed run, Goodlin reported that XS-1 handling was very satisfactory.

Following the speed test, he initiated the second set of experiments. Cutting all power, Goodlin planned to glide down to about 15,000 feet and then reignite all four chambers simultaneously. However, as the XS-1 dropped below 35,000 feet he noted that the aircraft began to snake and weave directionally. The oscillations were of a low-grade nature and were caused by the fuel sloshing around in the XS-1 tanks. This was not the only unusual occurrence during the descent to 15,000 feet; the fuel and lox pressure regulators continued to provide concern as well. Buildup of fuel tank pressure caused Goodlin first to vent some of its regulator dome pressure and then jettison some fuel to keep tank pressure below 350 psi maximum allowable pressure; simultaneously he kept reloading the dome pressure in the other tank's regulator due to falling lox pressure. Goodlin speculated that the causes were a combination of a cracked regulator diaphragm and a leaky valve. In any event, it meant continual work by the pilot to monitor and correct both problems while the plane continued its descent to 15,000 feet.

So far this flight, in spite of all the small problems, had been a field trip. Now came the big test. At 15,000 feet, Goodlin reignited all four chambers and commenced a climb designed to carry the XS-1 to 40,000 feet. With all four chambers running, the XS-1 acceleration was terrific. Goodlin was slammed back into his seat. This time he knew the rockets were firing without reference to any cockpit gauge. The experience was, he later related, somewhat similar to a water injection application on takeoff of a conventional fighter. However, the rocket

The pioneer of rocket-powered aircraft in America, Maj. Ezra Kotcher (*ca.* 1950). (Photo courtesy of E. Kotcher.)

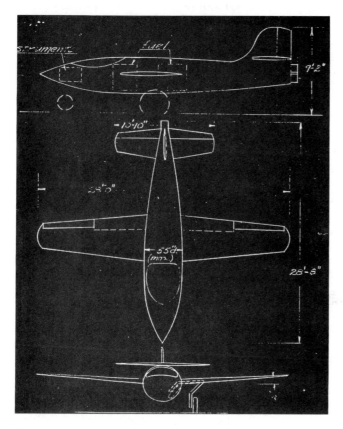

Sketch of the proposed Bell design from the March 15, 1945, meeting at Wright Field.

The XS-1 cockpit layout during construction. (Photo courtesy of Bell Aerospace Textron.)

Harold Dow and the B-29 logo. (Photo courtesy of Irene [Mrs. Harold] Dow.)

Bell employees with the XS-1 on its initial rollout in December 1945. (Photo courtesy of Bell Aerospace Textron.)

BELL X-1
WORLD'S FIRST SUPERSONIC AIRPLANE

6062

KEY TO NOMENCLATURE

NOMENCLATURE

1. PITOT TUBE
2. HIGH PRESSURE NITROGEN SPHERES
3. COCKPIT ENTRANCE R H SIDE
4. CONTROL STICK
5. HEAD REST
6. YAW ANGLE VANE
7. AILERON TRIM TAB ACTUATOR
8. PILOT CONTROLLABLE AILERON TRIM TAB
9. SPOILERS
10. INSTRUMENT COMPARTMENT (RESEARCH EQUIPMENT)
11. WATER ALCOHOL TANK
12. PRESSURE TUBES
13. STABILIZER MOVEMENT UP 5° DN 10° PILOT CONTROLLED
14. RADIO ANTENNA
15. RUDDER—RIGHT OR LEFT 15°
16. RUDDER TRIM TAB
17. BALANCE WEIGHTS
18. ROCKET MOTOR 6000# THRUST
19. FLAPS 60° MOVEMENT
20. AILERONS TRAVEL 12° UP 12° DN
21. TAPERED WING SKIN
22. RETRACTABLE MAIN GEAR
23. LIQUID OXYGEN TANK
24. PILOTS COCKPIT (PRESSURIZED)
25. PILOTS SHOULDER HARNESS & SAFETY BELT
26. RETRACTABLE NOSE GEAR
27. BATTERY
28. SPOILERS CONTROL
29. RUDDER PEDALS

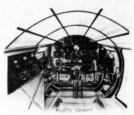

PILOTS COCKPIT

Cutaway of the Bell XS-1. (Photo courtesy of Bell Aerospace Textron.)

Walter Williams, Joel Baker, Jack Woolams, and Capt. David Pearsall. (Photo courtesy of O. N. Hayes.)

The XS-1 secured to the B-29. (Photo courtesy of O. N. Hayes.)

The XS-1 in its pit at Pinecastle. Note the P-47 and B-17 in the background. (Photo courtesy of O. N. Hayes.)

A crash photo from the fourth flight. Note that Jack Woolams does not have on the pressure suit seen in other photos taken at this time. (Photo courtesy of O. N. Hayes.)

The B-17 photographic plane and its unknown AAF flight team from Wright Field. (Photo courtesy of O. N. Hayes.)

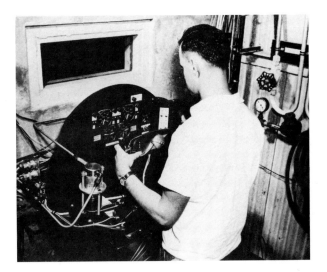

Woolams in the rocket test shack at Bell Aircraft during a pause in the Pinecastle tests. (Photo courtesy of Mary [Mrs. Jack] Woolams.)

The XS-1 just after landing at Pinecastle. (Photo courtesy of Bell Aerospace Textron.)

The XS-1 after the collapse of the nosewheel on the fifth flight. Walt Williams is on the left (*kneeling*). (Photo courtesy of NASA Langley.)

Dick Frost in his official Bell Aircraft photo taken on his first day of work in July 1943. The silk scarf was a Bell studio prop. (Photo courtesy of R. H. Frost.)

Rare photo of Jack Woolams and the completed number 2 aircraft, taken just before his death. (Photo courtesy of Edwards AFFTC-HO.)

The number 1 (#6062) and number 2 (#6063) airplanes at Bell in the summer of 1946. (Photo courtesy of Bell Aerospace Textron.)

Muroc Dry Lake in October 1946. (Photo courtesy of Edwards
AFFTC-HO.)

The Bell flight team at Muroc. (Photo courtesy of Edwards AFFTC-HO.)

The October 9, 1946, aborted flight resulted in damage to the entry ladder from the overpressurization in the cockpit. The resulting pressure blast bent the ladder away from the door. (Photo courtesy of Chalmers Goodlin.)

Harold Dow, Bob Stanley, Larry Bell, unidentified AAF officer, and Chalmers "Slick" Goodlin in early December 1946. (Photo courtesy of Irene Dow.)

Slick Goodlin and Harold Dow. (Photo courtesy of Irene Dow.)

Bell employees George White, Dick Frost, Tex Johnston, Slick Goodlin, Joe Marchese, and Sammy Grey in December 1946. (Photo courtesy of R. H. Frost.)

The B-29/XS-1 in flight at Muroc. (Photo courtesy of Irene Dow.)

Slick Goodlin entering the XS-1 from the B-29 ladder. (Photo courtesy of Edwards AFFTC-HO.)

The first powered XS-1 (#6063) flight, December 9, 1946. (Photo courtesy of Edwards AFFTC-HO.)

The Bell publicity photo of Slick Goodlin and the number 2 XS-1. (Photo courtesy of Chalmers Goodlin.)

Jim Voyles, Bob Stanley, Slick Goodlin, Col. Signa Gilkey, and D. R. Shoults after the first powered flight. (Photo courtesy of Edwards AFFTC-HO.)

Inspecting fire damage after the first powered flight: Slick Goodlin (*wearing flight helmet*), Col. Signa Gilkey (*kneeling and pointing*), and D. R. Shoults (*in suit and hat*). (Photo courtesy of Chalmers Goodlin.)

The XS-1, the loading pit, and its fueling tanks. (Photo courtesy of Edwards AFFTC-HO.)

FLIGHT TEST DATA

PLANE		NO.		DATE		FLIGHT NO.		CARD NO.
PILOT GOODLIN TEST XS-1						TAKEOFF TIME		
GROSS WT.		C.G.		% M.A.C.		GROUND TEMP.		
FUEL		GALS. OIL		GALS.		ARMAMENT		LBS.

RUN	ALTIMETER	INDICATED AIRSPEED	FREE AIR TEMP.	G	MANIFOLD PRESSURE	CARB. AIR TEMP. °C	OIL IN TEMP.	PRESTONE OUT TEMP.	R.A.M.	MIXTURE	OIL SHUTTER	PRESTONE SHUTTER
DATA	ON	AT	DROP									
1	25000	.7		1 to 4.5	1CYL	WIND	UP	TURN				
DATA	OFF				BURN	OUT	FUEL					
DATA	ON											
2	26-24	.7		1 to 4.5	"	WIND	UP	TURN				
DATA	OFF											
3		.65		4.5	PULL	UP		DATA	ON	FOR	EACH	PULL UP
4		.6		4.5	"	"						
5		.5		4.5	"	"						
6		.4		4.5	"	"						
7	STALL	CLEAN		1.0								

Copy of Slick Goodlin's
lap pad for flight, January
31, 1947. (Courtesy of
Chalmers Goodlin.)

Left side of XS-1 cockpit from February 5, 1947, flight. Picture records the accelerometer reading of +8.7 g's during the stall tests, which fulfilled the AMC contract requirements. (Photo courtesy of Chalmers Goodlin.)

Slick Goodlin (*kneeling*) and Joel Baker at Bell Aircraft, March 10/11, 1947, while Baker was being checked out in the aircraft in anticipation of his role as possible NACA pilot. (Photo courtesy of Joel Baker.)

Colonel Albert Boyd, *ca.* 1947, as chief of Flight Test Division. (Photo courtesy of Edwards AFFTC-HO.)

Rare photo of the two XS-1s together, in mid-summer 1947, during the program changeover. (Photo courtesy of Edwards AFFTC-HO.)

The Air Force XS-1 flight team. (*Left to right, front*): Lt. Edward "Ed" Swindell, Maj. Roberto Cardenas, Capt. Charles "Chuck" Yeager, and Capt. Jack Ridley; (*rear*): Lt. Robert Hoover and Dick Frost. (Photo courtesy of Edwards AFFTC-HO.)

The Air Force XS-1 ground team. (*Left to right, top*): Chuck Yeager and Jack Ridley; (*front*): Merle Woods, Jack Russell, and Garth Dill. (Photo courtesy of Edwards AFFTC-HO.)

The XS-1 after landing on the dry lake. Note Chuck Yeager in the middle. (Photo courtesy of Edwards AFFTC-HO.)

The XS-1 and the diamond shock waves. (Photo courtesy of Edwards AFFTC-HO.)

Chuck Yeager and "Glamorous Glennis." (Photo courtesy of Edwards AFFTC-HO.)

The October 20, 1947, private dinner at the Biltmore Hotel in Dayton, Ohio, to celebrate breaking the sound barrier. In a secret ceremony earlier that day at Wright Field, Yeager had received his second Distinguished Flying Cross. (*Left to right*): Maj. Roberto Cardenas, Roy Sandstrom, Capt. James T. Fitzgerald, Jr. (scheduled to soon fly the XS-1), Lt. Bob Hoover, Dick Frost, Capt. Chuck Yeager, Larry Bell, and Capt. Jack Ridley. (Photo courtesy of Edwards AFFTC-HO.)

The Bell team with the Collier trophy. (*Left to right, standing*): Roy Sandstrom (chief design engineer), Jack Strickler (assistant chief engineer), Bill Smith (chief rocket engineer), Bob Woods (vice-president, design), Paul Emmons (aerodynamicist), Dick Frost (project engineer); (*kneeling*): Joe Marchese (deputy project engineer), Harold Dow (drop pilot), Stan Smith (former project engineer), and Benson Hamlin (senior flight research engineer). (Photo courtesy of Edwards AFFTC-HO.)

engine results were much more substantial and certainly more immediate.

With all chambers firing, Goodlin prepared to put the XS-1 through a series of handling maneuvers during the ascent. Suddenly, after only a few seconds of full power, Goodlin discerned a howling noise in the XS-1. He immediately assumed this indicated an improper (lean) fuel mix going to the rocket motor and shut down all four rocket chambers. Simultaneously, lox line pressure decreased about 40 psi.

But this problem was no sooner handled than a far more serious one appeared. Goodlin saw the fire-warning light on his instrument panel come on, indicating the possibility of an engine fire. He radioed for Frost to look over the XS-1 and determine the true state of affairs. But Dick Frost wasn't around. At 15,000 feet Frost had flown chase about 100 feet behind and slightly to the right of the XS-1. However, when Goodlin triggered all four engine cylinders, the P-51 was rapidly left far behind. From his increasing distance, Frost witnessed the startup and quick shutdown of the rocket chambers, and heard Goodlin's call for visual inspection. While Frost could see a vapor trail coming from the experimental craft's horizontal stabilizer fairing, no fire was yet visible and Goodlin did not detect any smoke in the cockpit. On the ground, Bob Stanley ordered Goodlin to jettison all lox and fuel. From his original vantage point far to the rear it was impossible for Frost to determine whether the vapor was smoke or simply Goodlin executing the jettisoning of fuel and lox, but as he drew closer, Frost verified that the vapor was indeed smoke. He also saw that only lox was dumping, confirming Goodlin's surmise that XS-1 fuel was exhausted. After another 15 seconds, the smoky vapor disappeared and Goodlin was able to bring the XS-1 in for a normal landing. Touchdown for the first powered flight was at 12:13 P.M. Rollout was 7,000 feet but even then difficulties continued as the cabin pressure problem recurred. Goodlin noted the cabin-pressure ground reading as 27,000 feet. He operated the emergency dump valve to equalize the pressure, but it was a closing annoyance to what had already been an unsettling flight.

Program Diversions

On December 10, the press animal cranked up its act to publicize the XS-1's first powered flight.[17] Walter Bonney, director of public relations for Bell Aircraft, set up a press conference in Los Angeles attended by Stanley and Goodlin (representing Bell Aircraft), the general manager of RMI, Colonel Gilkey (Muroc), Jim Voyles (Wright Field), and Walt Williams (NACA). The event was mostly a Bell Aircraft show with sufficient embellishment to accent the dangers of the flight. But while supposedly called to discuss the details of the test, the press continued to ask when and who would make the first supersonic flight. The press conference story appeared nationally under an Associated Press byline.

On December 11, the *New York Times* dramatized the story in a front-page article entitled "Hail Rocket Plane in First Test Hop."[18] The coverage was selective to say the least. The flight under power was reported to have lasted only seconds because the team members were not yet ready to see what the craft would do. Goodlin was quoted as saying everything about the flight was "beautiful," with *no* mention of the fire. Bell Aircraft and the AAF received prominent mention, with the NACA getting only passing reference. Reportedly, no one doubted the plane would do 1,700 MPH. In fact, the only jarring note came from an seemingly innocent remark attributed to Goodlin. When asked how the firing of the engine felt, Goodlin commented that when ignition occurred, the XS-1 shot forward with a "momentum comparable to that of a carrier plane leaving a flight deck under propulsion of a catapult," a comparison that must have made the Army men visibly wince. Here was an AAF program stretching the boundaries of flight, and the pilot gives the Navy a favorable comparison in public, as if the Navy had anything comparable to this great adventure. Unfortunately, the analogy as reported was not what Goodlin said. He had never even taken off from a carrier. What he said was the engine power was like being kicked "in the rump with a lead boot." This was just one in a number of regrettable incidents involving Goodlin and the press.

The *Times* article was only the beginning of the coverage. On December 23, *Time* magazine also provided a long story with a photograph of Goodlin.[19] Under an article entitled "Army and Navy: What Comes Naturally" the magazine went over the familiar ground of the flight. But the emphasis was on Goodlin, the "handsome 23-year-old test pilot" who, "like Columbus, Magellan, and the Wright Brothers," was just doing "what came naturally." This man was a true hero, *Time* reported. Facing death in the air, they reported, Goodlin found time to swim, read, and listen to classical records. This pilot was "no daredevil or exhibitionist," but a cool calculating observer. At first, the sound barrier presented such an obstacle that Goodlin retained the privilege of recommending the first supersonic flight be performed pilotless. However, after Goodlin flew the plane and analyzed the data, he stated, "I know what this plane can do. I think it will fly a thousand miles an hour and I think I'll live through it."

Two weeks later, *Life* magazine got into the publicity act with a multipage article featuring very detailed photos of the plane and its pilot.[20] However, unlike the previous stories, Bell and the Army Air Forces were virtually unmentioned. While AAF public relations personnel thought *Life* was going to put the XS-1 or Goodlin on the cover, the magazine wrote a story that took many by surprise—in the *Life* article, NACA was the agency pushing the outer edges of the envelope of science and aviation. *Life* spared no words to dramatize the contest between science and the great unknown that lay out in the supersonic range. In describing the effort to break the sound barrier, *Life* said "before any plane can go through the air at such a speed a strange and fearful zone will have to be passed." This zone would cause the plane to experience "fierce buffeting" which in the past had caused other planes to be "literally torn apart." Those lucky fliers who lived through this terrible experience had been unable to even explain what had happened. To address this problem, the NACA had conducted wind tunnel tests. A plane had been built in conjunction with Bell Aircraft. Soon the NACA would take the lead and would fly the test program with or without Goodlin. But no matter what the preparations, the tests would be dangerous. The pilot couldn't even bail out if things went wrong. In short, *Life* implied, this was a true test of man's courage and an agency's (NACA) scientific skill. Bell Aircraft and the AAF found little enjoyment in this reading material.

As a result, the *Life* article really frazzled some normally even-tempered people. Frost was upset that some of the story implied that Bell Aircraft was doing nothing, but that was minor compared to some other objections raised about the increasing amount of press coverage. On December 12, Walt Williams wrote to Hartley Soule to complain that the Bell/Army publicity was about an aircraft that really didn't even exist. Williams believed the two agencies should discuss the XS-1 airplane that they were currently flying. They "should admit that the airplane was not designed as a supersonic aircraft but rather a high transonic airplane." All the talk of supersonic speeds came only due to being "forced" to use a B-29 to obtain altitude and flight time. Worse, Williams told Soule, public comments about the lack of swept wings on the XS-1 because it was uncertain that such wings should be used on a supersonic airplane were "for the birds." Swept wings were not on the XS-1 because no one (except Germany) knew much about swept-wing aircraft when the XS-1 contract was negotiated. It might be embarrassing to discuss these facts, Williams stated, but at least it would be truthful.[21]

Nor was the AAF fully satisfied with the press handling of the new project. Two issues seemed to bother the military men. First, the continued calls for Goodlin's time finally led to problems in mid-January. When *American* magazine asked for its turn, AAF HQ in Washington had seen enough. Notifying Muroc that it was denying the request, HQ specifically cited the previous agreement that little further coverage would be promoted pending additional scientific results. Second, the Bell Corporation kept intimating that the XS-1 was the forerunner of a production aircraft. General Craigie decided to set the record straight. In a speech he gave in early January, Craigie stressed the XS-1 was not a military plane.[22] It was a research vehicle produced through the "harmonious program of the Army Air Forces, Bell Aircraft Corporation, and the National Advisory Committee for Aeronautics." It was, Craigie stated, a "flying laboratory" with a mission to "investigate and study flight and design research problems" found at speeds greater than contemporary aircraft. The AAF did not expect the research process to be smooth, but rather a series of problems that the combined Army/Bell/NACA team would solve and then move forward to the next phase. Future flights would prove Craigie right.

In addition to dramatic license and inaccurate information, two other issues bothered many people about the press coverage. First, the XS-1 was supposedly a secret program. The fact that it received such extensive coverage managed to raise more than a few eyebrows and led to some detailed discussions about flight information. As a consequence, on January 21, General Craigie sent a teletype to the Commanding General AAF in Washington to complain of the recent treatment of press credentials. Specifically, Craigie reminded Washington that the XS-1 was a classified program at a restricted base. The habit of approving press credentials without adequate checking by security and without proper notification through channels might have unfortunate consequences. Craigie requested assurance that future press relations would be handled with proper care.

Second, the continuing coverage prompted some unflattering comments. Especially severe was the NACA internal treatment of Goodlin's *Time* remark about the handling of the plane. The NACA already knew there was a great deal of slack in the controls of the XS-1. The magazine's characterization that Goodlin "had to handle the controls like a surgeon" certainly was not supported, the NACA team privately said, by the instrument data aboard the aircraft. Anyone who saw the force records, Walt Williams reported, with Goodlin having to tug and pull just to get the controls to move, would chuckle at that remark. But all the talk of press coverage hid one other issue. Privately, Voyles and Williams both agreed that they expected Bell Aircraft to cut loose the power sometime during the tests and exceed the magic Mach 1.0 barrier. After all, had not Bell Aircraft given a supersonic kind of buildup to Woolams during the past summer? The existing AMC contract with Bell only required Mach 0.80. The only way Woolams was going to exceed the Mach 1.0 limit was if Bell Aircraft got the follow-on contract or Woolams simply turned the XS-1 loose, flight program or no flight program. A follow-on contract was still possible for Bell Aircraft. But so was the possibility that Stanley would simply just authorize Goodlin to increase the XS-1 flight profile to include an *accidental* breaking of the sound barrier.[23] Under the current program command structure nobody was sure just what might happen.

The extent of the scrutiny by the press and the motives of Bell Aircraft were seen as problems by almost everyone. The XS-1 was the first purely research airplane constructed in America and, as such, garnered quite a bit of media attention. Throw in the technological danger of a supposedly solid wall such as the sound barrier and the exotic appearance and color of the plane, and the press had the makings of a real story. Bell Aircraft wanted a production contract for the XS-1. Mockups of the plane as a fighter were discussed, but the AMC seemed disinterested. Publicity was a way of keeping the issue on the table.

A secondary problem arose because no one coached the 23-year-old pilot on how to handle the experts of the press. In light of security issues, Goodlin did not always know what to say to correct the erroneous information the press seemed to accept as true. Goodlin's media coverage began to cause friction with the military men back at AMC. For example, press reports in Los Angeles linked Goodlin to a certain Hollywood starlet, although Goodlin didn't even know the woman. Also Goodlin gave interviews that the AMC did not believe were cleared by censors. Although the stories had been cleared, the time lag between initial discovery and final agreement that all was in order only compounded the difficulties with AMC. Rumors had it that Goodlin had a Hollywood agent and was testing for a movie part. Neither was true, but the rumors once again raised eyebrows.[24]

As long as the program moved forward and the initial AMC contract was in force, these annoyances were acceptable. If circumstances changed, however, these grievances were the foundation for other potential courses of action.

Even with the glamour of press conferences and full-page stories, life at Muroc had to continue. A board of inquiry was immediately convened to investigate the damage and causes of the mishap on the powered flight.[25] Dick Frost, Walt Williams, and Frank Iwanowsky, RMI's Muroc Field representative, teamed up to write an analysis of the accident. The initial inspection revealed significant melting of the aircraft wiring as well as scorching of the XS-1 bulkhead concentrated near the centerline of the rocket motor. The NACA instruments had significant visible damage to the control position transmitter for the rudder and stabilizer. Also, both the strain gauge installation and the elevator position transmitter revealed charring and damage. The engine revealed only minor, but very important, fire damage. The causes for the damage were, however, harder to pinpoint.

The December 11 initial report speculated that the engine igniters were triggered in too rapid a sequence, resulting in overheating, as evidenced by the melted spark plug electrodes. Concurrently, when the pilot simultaneously shut down all four rocket cylinders, he produced a water hammer effect. The review team speculated that this action may have caused a cracked fuel line or loose fittings in the fuel/lox system. In either event, fuel/lox sprayed from the lines and was ignited by contact with the red-hot engine igniters; the ensuing flash fire was of short duration since the engine fuel was almost exhausted. The review team speculated that the smoke witnessed by Frost probably came from the melting wiring and was a result, not a cause, of the problem. In conclusion, the team recommended seven changes be incorporated into the XS-1. These included;

1. In starting the rocket cylinders, each one should be allowed to stabilize in operation before another cylinder is ignited.
2. If more that one cylinder is in operation and the pilot wishes to shut them down, a lag time of one-half second between shutdowns is desirable. If human error makes this impractical, the team recom-

mended an automatic delay be built into the rocket motor control system.

3. The auxiliary supply of nitrogen to the engine manifold from the lox pressure regulator should be regulated.

4. Additional fire-warning transmitters should be installed on the plane in areas most likely to ignite.

5. Ram-air scoops should be installed on the XS-1 to create a positive pressure differential between the interior of the fuselage and the ambient air flow.

6. Stainless steel tubing should be used in the rocket instead of aluminum alloy tubing.

7. All wiring in the vicinity of the engine should be fire-resistant.

The damage to the engine necessitated a halt to the flight testing while a new engine (RMI Serial No. 6) was brought out from Buffalo. The damaged engine was returned to Bell Aircraft for testing and analysis. RMI participated in the testing. Thus it was not until March 28, 1947, that a final report on the fire was issued, pinpointing a loose igniter connection caused by the rapid firing of the rocket cylinders as the source of the spilled fuel. Combustion occurred when the fuel sprayed on the red-hot igniter. This intense heat caused several nitrogen pressure lines to burst and these, not Goodlin, shut down the engine. Once the igniters cooled off, there was little chance of additional leakage from the connections. The final conclusion recommended that gaskets for the igniters be heat/expansion–resistant and that the igniter fittings be fastened extra tight before flight. Other internal changes were recommended now that lab testing had pinpointed the exact cause of the damage, including the fabrication of a one-piece woven asbestos pattern to fit around each cylinder in the tail cone of the aircraft. This pattern was painted with water glass to give the cloth rigidity.

Once the immediate damage from the fire had been addressed, the Bell Aircraft and NACA technicians turned to the other problems of the program. Goodlin recommended the cabin pressure regulator be dismantled until a workable model could be secured. He was greatly disturbed that this piece of equipment had been "an annoyance of every flight thus far." However, thorough inspection of the system revealed several leaks that accounted for the variations in pressure. These were all repaired. Additionally, ram-air scoops were installed on either side of the fuselage forward of the tail. It was believed possible that a low-pressure area might exist in this section. Fire damage to the aft section wiring was repaired and the new engine was installed in preparation for a pressure test. The fuel and lox tank pressure regulators were disassembled, cleaned, and repaired by installing stainless steel diaphragms.

Not everyone was sure the XS-1 fire was the only flight problem. After reviewing the NACA instrument data, Walt Williams expressed some concern over the snaking motion Goodlin experienced on the descent to 15,000 feet. The instrument data seemed to indicate the motion might not be attributable to fuel sloshing since the data lines revealed that the snaking oscillation actually was closer to being a dutch-roll oscillation. No evidence of rudder movement was seen in the data. Second, the NACA data also revealed that the airplane had begun to tuck under, that is, the elevator appeared to be moving downward as the speed increased. This situation merited close inspection as it might be symptomatic of a serious problem above the Mach 0.80 boundary.[26]

Back to Business

With the completion of the first test and the conclusion of the press coverage, the Bell Muroc team returned to the routine. Stanley departed for Buffalo on December 14. The presence of so much Bell Corporation leadership at Muroc, especially Robinson, had made him particularly edgy. In addition to being on the board of directors, Robinson was a banker and major Bell stockholder. With Bell's financial problems and potential proxy fight looming just over the horizon, Stanley wanted everything perfect for Larry Bell and Robinson. Their departure over the weekend of December 7/8 allowed the tension level to subside, even if the missed powered flight brought disappointment. The Bell Aircraft team spent most of the intervening time between the fifth and sixth flights of the XS-1 getting the aircraft ready to fly. On December 19, Dow and Heaney checked out the B-29 in a 50-minute flight test and judged it ready to continue the test program.[27]

By the following day all XS-1 preparations were finished, and Bell Aircraft was ready to conduct the sixth flight test for the number 2 plane.[28] During the morning an overcast existed at 32,000 feet, but otherwise weather conditions were acceptable, and the decision was made to launch the aircraft. Shortly thereafter, the B-29/XS-1 cleared the runway and began the climb to altitude. With full tanks, the XS-1 again weighed in at 12,012 pounds, but this time center of gravity was calculated to be at 23.5 percent mean aerodynamic chord, a change Goodlin was instructed to check out during the flight. As rehearsed, Goodlin entered the XS-1 cockpit and began the predrop checkout. Just before separation, the nitrogen pressure in the XS-1 began to fall precipitously. Frost decided that Goodlin should jettison lox and fuel and then land. On the ground, it was discovered that pilot error had been responsible for improper valve operation. Fuel loading of the XS-1 immediately commenced to allow another flight attempt during the afternoon.

As soon as ready, Dow guided the B-29 down the runway for the second launch attempt of the day. Joel Baker followed in the P-51 as NACA observer. Goodlin entered the XS-1 cockpit just below 10,000 feet and readied the aircraft. Although nitrogen pressure revealed a significant improvement over the first flight, mainly due to the correction of several leaks, Goodlin quickly observed that there would be more problems with the regulators. During the dome-loading operations, he noted that the first-stage pressure slowly fell, indicating a leaky diaphragm in the regulator dome. However, this was not considered serious enough to delay the launch.

The drop was normal and the planes separated cleanly at 3:35 P.M. The tail-heavy condition noted in the first flight was still present but not as pronounced, leading Goodlin to doubt the current calculation of center of gravity. Approximately six seconds after drop, Goodlin activated the first chamber ignition. Activation was satisfactory and all pressures were observed to be normal. The original plan had been to follow the first flight profile and fire all chambers individually while the XS-1 descended to 15,000 feet. At that altitude, Goodlin would simultaneously fire all four cylinders and climb to 35,000 feet, where he was supposed to level out and use whatever fuel remained to make a level run at a speed just below Mach 0.80. After that test was conducted, Goodlin was to put the XS-1 through a series of steady turns at increasing Mach numbers up to the limit. Unfortunately, as on all previous flights, technical problems rapidly interfered.

During the XS-1 descent to 15,000 feet, considerable nitrogen pressure was lost due to the gas being cooled by contact with the lox in the half-full tank. Thus, the pressure dropped too low to operate the motor propellant valves. Goodlin could not fire the engine and in fact only operated two cylinders at any one time during the entire flight. Since he was without full engine power, it was decided to jettison lox and fuel and land the airplane. On the way down, Goodlin attempted to operate the stabilizer control handle and found even that effort wasted. The handle moved, but nothing happened. Landing was normal and rollout went 7,000 feet. No NACA data was obtained since Goodlin forgot to turn on the instrument package prior to drop.

Postflight inspection confirmed the leaky diaphragm in the first-stage regulator. The stabilizer motor lacked adequate lubrication, probably due to the altitude, and a special high-altitude lubricant was substituted. Discussion of the low nitrogen pressure issue terminated in a decision to control power-off flying to lessen the lapse time between drop, glide, and rocket utilization. Finally, a visual review of the aft section of the XS-1 revealed that the rudder position string was being scorched by an afterburning flame entering the fuselage. Modification was initiated to prevent a recurrence.

No flights took place from December 30 to January 3. During this time, Bell technicians were waiting on parts for the dome-loading panel and new stainless steel diaphragms from the Grove Company in Oakland. A flight was planned for Friday, January 3, with the new panel and the old regulators. However, the best-laid plans often go astray and this was no exception. The base machine shop failed to fabricate the new panel parts in time for the Friday flight. During the delay, the new regulators arrived from Grove Company and a decision was made to ground-test them before conducting a flight. NACA was notified there would be no flight until Thursday, January 9, unless the weather threatened rain. In spite of the past heavy rains, the lakebed was drying rapidly, thanks to strong winds, and Bell didn't want any more delays.

The holidays were a busy time for the NACA personnel as they tried once again to remedy their housing difficulties. Fortunately, the increasing workload at the various military sites had prompted the need for new quarters for the support personnel. Williams found space at Mojave more convenient than his abode at Palmdale. The government's $28.50 per diem a week didn't go far when one had to rent expensive housing *and* drive long distances.[29]

Back at Buffalo, the new year found the Bell Aircraft engineers engaged in proof tests and strain gauge calibration for the long-delayed number 1 aircraft. The proof tests included checkout of the nosewheel landing gear, the main wheel fairings, the fuselage, and the elevator controls. The strain gauge work included calibration on the horizontal tail, the vertical tail, and approximately 50 percent of the wing. It was anticipated that this shop work would continue until January 17. That was the good news. More serious for Bell were the continued failures involving the General Electric loxygen pump. The experiments conducted in the rocket test facility at Bell had been a total disaster. The GE pump bearing had failed as soon as pumping was initiated. The system was returned to GE for a total redesign of the turbine pump bearing. The setback meant an additional delay in bringing the new system into use. The XS-1 program, however, would not wait.[30]

On Monday, January 6, Bell technicians gave the B-29 a 25-hour maintenance inspection, after which Dow and Heaney took the B-29 aloft for a 1-hour, 45-minute checkout flight.[31] The B-29 was ready. At the same time, new regulators were installed in the XS-1, which featured the new stainless steel diaphragms. The stainless steel replaced the rubber diaphragms, which had a tendency to become brittle at low temperatures. The Grove Company advised Frost that silicon rubber diaphragms would better withstand the cold but would also prove too weak, so stainless steel diaphragms were substituted.

Other changes accompanied the ongoing preparations for the flight. As per the Frost fire report recommendations, new ram-air scoops were installed on the XS-1 to vent the engine compartment in case of fire. Also, a new cockpit camera was installed to record the cabin instruments in flight. Bell technicians planned to use the ground time to engine-test the XS-1 and to subject the new regulators to a failure-rate test. Once these tests were completed, Frost announced Bell's intention to conduct a flight test the following day. Once again the NACA personnel were caught by surprise as the date was sooner than they had expected. But on Tuesday the B-29 suffered a last-minute problem with its elevator control system and Frost scrubbed the mission.[32]

On January 8, the B-29 and XS-1 were ready to conduct another flight test.[33] After liftoff from Muroc Field and at an altitude of 7,000 feet, Goodlin moved to the XS-1 and was secured in the cockpit. The bomber leveled out at 30,000 feet and proceeded with the drop sequence. However, the problem of cabin pressure continued to haunt the project. At the 30,000-feet pressure altitude, the cabin differential was noted to be 20,000 feet. By utilizing the dump valve prior to the XS-1 release from the B-29, Goodlin was able to lower this to 7,000 feet, but it was clearly still not a satisfactory situation. Concurrently,

Goodlin dome-loaded the first- and second-stage regulators and noted that their pressure remained fairly constant. It appeared that progress had been made on that problem.

With all in readiness, Dow executed the drop procedure at 3:35 P.M. XS-1 separation from the B-29 occurred at an indicated airspeed of 230 MPH. As with all previous drops, this one was clean and routine. About 10 seconds after drop, Goodlin sequentially ignited the number 1 and 2 engine chambers. The XS-1 started a shallow climb at Mach 0.65. Shutting down the number 2 chamber as he passed 34,000 feet, Goodlin allowed the XS-1 to coast up to 37,000. This allowed the aircraft to remain below Mach 0.80. After leveling out, Goodlin dropped the nose of the XS-1 and began a dive to 35,000 feet. Speed built up to Mach 0.795 and Goodlin noted a low-amplitude buffeting like that experienced during an approach to an accelerated stall condition. Discounting the possibility of engine problems, Goodlin noted no movement of the control column and surmised the buffeting was the result of the new ram-air scoops. In any event, he shut down the engine and glided down to 28,000 feet. At that altitude, he tried to reignite the number 1 chamber. Although it fired, low chamber pressure and an improper fuel mixture made Goodlin shut down the engine once again and jettison all remaining propellants. Several strong tugs were required to operate the tank pressure blowdown valves lever to vent the tanks.

Gliding back to Muroc Field, Goodlin executed pull-ups at Mach 0.40, 0.45, and 0.50. A partially stabilized turn was made at positive 2g acceleration at Mach 0.55. Slowly Goodlin tightened the turn until a stall condition was reached. Positive acceleration reached 3g and Goodlin noted stick force seemed to lighten as positive acceleration was applied. He believed that at a higher acceleration the XS-1 might experience stick force reversal (rate of change of force reverses). He advised the NACA that this condition merited further analysis. Touchdown and rollout proved satisfactory.

Postflight inspection revealed the customary list of problems. Between the preflight check on the seventh and the test, someone had damaged the tubing running from the airspeed indicator to the instruments. As a result, no airspeed records were obtained. In all other respects, the NACA team was pleased with the data obtained from the flight. The heavy workload, however, continued to tax the capabilities of the NACA personnel. As a consequence, the NACA decided that help was badly needed at Muroc and finally took the steps necessary to increase the personnel stationed there. On January 14, Mel Gough,

Walt Williams, Clyde Bailey, and E. C. Buckley arrived at Muroc. Bailey was scheduled to remain at Muroc for about four weeks to learn the operation and mechanical features of the XS-1. On January 13, De E. Beeler and LeRoy Proctor arrived at Muroc. Beeler was scheduled to replace Aiken and later head the research engineering section at the NACA station. Proctor was scheduled to work on instrumentation. With the increase in ground personnel came other flight duties for the NACA team. In preparation, NACA made additional arrangements for more working space and equipment.

On January 17, the plane was ready for another flight, and the number of observers grew. In addition to Gough and Buckley, and Lewis Rodert of the AERL, five individuals from the NACA Ames Laboratory flew into Muroc early on the morning of the test only to discover that B-29 engine trouble had delayed the flight until the afternoon.

By early afternoon, the Bell technicians had lowered the XS-1 into the pit and the B-29 was pulled over it.[34] Loading quickly commenced and fueling finished shortly thereafter. The only change to the experimental aircraft was the installation of an intervalometer to operate the cockpit camera, which would reduce the 16mm film speed to one frame per second. B-29/XS-1 takeoff from Muroc was in the mid-afternoon. Goodlin entered the cockpit at 5,000 feet and proceeded with XS-1 checkout. Since the full load of fuel had caused a tail-heavy condition on several past drops, Goodlin set the XS-1 stabilizer one degree up (XS-1 nose down) prior to launch. During the dome-loading, Goodlin noted that nitrogen pressure was higher than normal at 4,200 psi. A decision to execute launch was confirmed and Dow initiated the procedure. Heaney executed drop separation at 3:16 P.M. at 16,000 feet.

Approximately six seconds after separation Goodlin sequentially fired chambers 1, 2, and 3. A high-speed level run was performed at 15,000 feet. Speed rapidly increased to an observed Mach 0.77. Slight buffeting was experienced but Goodlin reported it was not serious. Not wishing to overshoot the Mach 0.80 limiting speed mark, Goodlin eased the XS-1 into a climb. With three chambers in operation and the aircraft in a climb of about 20 degrees, the XS-1 speed remained constant around Mach 0.70. During the climb, Goodlin attempted to fire chamber number 4, but the chamber failed to ignite. Goodlin continued the climb on the three operating cylinders. Once he reached 30,000 feet, Goodlin shut down the engine. Frost instructed Goodlin to again try to start the number 4 chamber. It was here that Goodlin discovered, to his embarrassment, the explanation for the chamber failure. He had forgotten to close the panel switch. (The four rocket cylinder panel switches on the XS-1 were series-wired with the master engine wheel switch. Both had to be closed to operate a chamber.) Chamber 4 fired and Goodlin used up the balance of the XS-1's fuel. The residue was jettisoned. The glide back to Muroc afforded Goodlin the opportunity to execute several steady turns at varying Mach numbers. At 10,000 feet, wheels and flaps were extended for landing. Rollout was routine and confined to the Muroc runway.

Data was taken on this flight from all the NACA instruments. Unfortunately, the timers failed, causing some difficulty in correlating the data. One surprise discovered by the NACA technicians was that Goodlin had hit Mach 0.828 on the level speed run. He had definitely proven the aircraft safe up to the limiting factor for AMC/NACA acceptance. Yet in spite of Goodlin's remarks, Walt Williams was less comfortable about the "slight" buffeting that the pilot reported in the test. The NACA data revealed the aircraft experienced pronounced tail buffeting at Mach 0.82. The wing also received buffet, although in all cases no trim changes or elevator reversal was noted. In any event, a decision was made to remove the ram-air scoops to see if that cured the slight tail buffet experienced on both flights 7 and 8. At the same time, Bell technicians decided to repair the lox tank pressure regulator. Inspection uncovered the fact that the valve was not seating, which explained the slight variance in pressure that Goodlin noted during the flight. The lox liquidometer was also reinstalled for the next flight and a successful engine ground test was made to ensure that all systems were still functional.

The Ames lab observers were suitably impressed by the XS-1 test facilities. The NACA instrumentation was impressive and the Bell team quite professional. Goodlin's ability to confine the XS-1 to the Muroc runway was noteworthy and caused a revealing exchange between observer Lawrence Clousing of the Ames lab and Mel Gough. Commenting on the utility of using a loading pit, Gough told Clousing that the NACA intended to build a similar facility at Langley for use with the XS-1. Clousing expressed surprise, but Gough assured him that the XS-1's distance landing was no impediment to operating at Langley. The old idea of tests on the East Coast still seemingly had life at LMAL.

In preparation for the next flight, the Bell technicians removed the ram-air scoops and replaced the inlet doors with flush-mounted doors. On January 20, Bell Aircraft readied the XS-1 for the fifth powered flight in the program. After takeoff, Goodlin entered the XS-1 cockpit at the pressure altitude of 6,000 feet. Upon completion of the predrop checkout, he initiated the process of loading the domes of the first-stage lox and fuel regulators. However, after execution of first-stage loading, Goodlin noted that output pressure was not up to normal. Therefore, he slightly reopened the dome-loading valve to allow an additional flow to the two second-stage regulators, and then proceeded with the loading. During the second operation, Goodlin neglected to close the first-stage valve. As a result, pressure continued to build up in its dome. When the pressure reached approximately 1,800 psi, the first-stage pressure relief valve opened and began venting the nitrogen into the atmosphere. With nitrogen pressure falling rapidly from its preflight level of 4,800 psi, Goodlin radioed ground control to notify Frost of the problem. By the time the open first-stage valve was discovered, approximately 2,000 psi had vented into the atmosphere. As only 2,800 psi pressure remained, all agreed it was too hazardous to initiate the drop. Frost told Goodlin to reset first-stage pressure to 1,500 psi and proceed with jettisoning fuel and lox. The B-29 landed with the XS-1 still attached. No adjustments were made to the XS-1, but the failing light prevented another attempt at launch.[35]

On the morning of January 21, the Bell team again attempted the fifth powered flight.[36] With Dow and Heaney as B-29 flight crew, the mission proceeded. Joel Baker took off in a P-80 (#44-85004) to provide chase. At 5,000 feet, Goodlin made the quick trip down into the XS-1 and executed the drop check. After completion of the checkout, Goodlin proceeded with the dome-loading, but noted that the lox pressure began to rise. Goodlin jettisoned the lox to lower the pressure, but within a short time it had risen back to its previous high level. Dumping the first-stage regulator stopped the rise in lox pressure and seemed to indicate the problem was confined to the lox system. Frost instructed Goodlin to jettison fuel and lox preparatory to landing. The B-29 with the XS-1 attached landed without incident. On the ground, the lox regulator was removed, disassembled, and inspected, where the valve rod slide bearing was found to be gouged. Repairs and re-installation consumed the balance of the day.

With two failures in two days, the Bell technicians approached the next effort with some concern. All preparations were completed by mid-afternoon on January 22, and the B-29 rolled down the runway for the third attempt at this nineteenth flight (fifth powered flight) of the program.[37] Baker again flew chase in the P-80. At 6,000 feet Goodlin entered the XS-1 and executed the checkout. After completion, he began loading the domes. This time, there were no visible problems. Goodlin was instructed to make a speed run up to the edge of the buffet boundary, obtain data, and conduct a full-power high-angle climb. At 16,000 feet pressure altitude, Dow initiated the drop procedure. The XS-1 separated from the B-29 at 3:31 P.M.

Once clear of the B-29, Goodlin ignited engine chambers 1, 2, and 3 in rapid sequence. A high-speed run was conducted at 15,000 feet. At a calibrated Mach number of 0.76 (NACA recorded Mach 0.79), Goodlin again found the buffeting noted in the earlier flights, ending speculation that the buffeting was attributable to the air scoops. As per the flight plan, he discontinued the speed run and put the XS-1 into a 26-degree climb. Slick quickly ignited cylinder number 4 and continued a full-power Mach 0.70 climb to 31,000 feet. Rate of climb was an impressive 15,000 feet per minute due to the operation of the additional cylinders. Goodlin found that although pressure was stable, it was 10 percent below normal.

At 31,000 feet, Goodlin discontinued the climb. Fuel and lox were starting to run low, so the engine was shut down to avoid any possible damage. Power-off maneuvers were conducted to obtain data for the NACA. Turns were executed at Mach 0.70, 0.65, and 0.60. Goodlin reported a decrease in stick force per g on the XS-1 at around 3.5 g's. Unfortunately, by this time the film in the NACA camera was exhausted and the airspeed channel in the telemeter was not operating so useful data was lost. At 9,000 feet, Goodlin lowered the flaps and landing gear in preparation for touchdown, but the cockpit light that signified the landing gear was down and locked in position was not lit, so Baker in the P-80 closed on the XS-1 for visual inspection. He assured Goodlin that the flaps and gear were down and the landing proceeded. During rollout, the left landing brake failed to function, but no difficulties were experienced due to the available open space on the dry lake. After the flight, the brakes were inspected and repaired and the fuel and lox regulators were removed and checked. Overnight repairs ensured that the XS-1 was ready for the next flight.

On January 23, the Bell technicians began the tedious process of loading the XS-1 into the B-29.[38] Propellant loading took even more time. Thus B-29/XS-1 takeoff didn't occur until very late afternoon. During the climb to altitude Goodlin entered the XS-1 cockpit at 7,000 feet. Baker departed Muroc airfield for the chase duties in a P-80A (#44-85309). After XS-1 checkout and fuel/lox dome-loading, the B-29 reached 16,000 feet and Dow initiated the drop sequence. Separation was clean and routine at 4:21 P.M. The lower altitude utilized in the drop procedure quickly appeared in one other aspect of the proceedings. In his flight log, Dow listed the B-29 takeoff to touchdown elapsed time as only 30 minutes.

Approximately six seconds after drop, Goodlin ignited chamber number 1. Shortly thereafter, he fired chamber 2 and commenced a climb to 30,000 feet at an indicated speed of Mach 0.65. Slick attempted to obtain positive acceleration at that altitude, but with 60 percent of the fuel remaining in the plane, the XS-1 experienced some buffeting. No stick lightening was noted. Although he attempted to execute stabilized turns at Mach 0.65 and 0.60 with one chamber ignited, Goodlin reported it impossible to achieve because of the excess speed. A calibrated speed of Mach 0.76 (NACA recorded Mach 0.78) was achieved before fuel exhaustion. A number of constant speed turns from Mach 0.55 to Mach 0.70 were executed on the glide back to Muroc, employing a force of 2 to 3 g's. Landing and rollout were uneventful except that the brakes failed once again.

Postflight review of the test data revealed the normal mix of good and bad news. While the purpose of the flight was to investigate the buffet boundary, the NACA instruments failed to get all of the meager data that was available because the film drum jammed in the telemetering recorder. The good news was that the NACA team got wing- and tail-loads data from the turns. The rest of the flight was rather routine. The brake failure, especially after they had just been inspected, was certainly cause for some alarm. Consequently, the brakes received new discs and biscuits and the lines were bled. A new asbestos seal was also fabricated and installed to provide a closer fit around the outer ends of the rocket cylinders.

Since Goodlin again experienced the moderate buffeting seen on several of the previous flights, it was doubly confirmed that the ram-air scoops, which had been removed prior to the last flight, were not the culprit. Thus the scoops were reinstalled on the aft fuselage section. Also, a decision was made to install tufts on the inboard surface of both wings with the idea of photographing them in flight. However, no camera was installed at that time.

A slight rain fell during the first part of the week of January 27 and the Bell team decided to allow the lakebed to dry thoroughly before attempting any flights.[39] On January 30, all preparations for the twenty-first flight of the XS-1 program had been completed and everything was ready to go. The weather was clear and visibility was unlimited as Dow guided the B-29 down the runway and off for the climb to the day's test altitude. Joel Baker would again fly chase in the Bell P-51. The purpose of the XS-1 flight was accelerated stalls, both fully loaded with fuel and empty. On the climb, Goodlin entered the XS-1 at near the 10,000-feet safety level. No problems were recorded on entry and checkout. The B-29 climbed to a pressure altitude of 27,000 feet where Dow initiated the drop sequence.

Approximately six seconds after clearing the B-29, Goodlin ignited chamber number 1. Once it stabilized, he ignited chamber 2 and executed a constant speed turn at a recorded Mach 0.75. Goodlin tightened up the turn and performed an accelerated stall at approximately 2.65 g. Buffeting was encountered and a stick force of 12 pounds per g was recorded. Goodlin next put the XS-1 into a climb. Upon reaching 32,000 feet, he attempted to ignite chamber number 4. The cylinder started, but the chamber operation was very rough, exhibiting a severe pulsation. Goodlin immediately shut down the number 4 chamber and attempted to restart the number 1 cylinder. Once again Goodlin discovered the same rough operation exhibited by the number 4 cylinder. Simultaneously Goodlin noted the other three cylinders revealed low and unstable chamber pressure. Not wishing to take unnecessary risks, Goodlin shut down the engine and jettisoned the remaining fuel. However, even this precaution failed to curb a new problem. The chamber pressure on the number 1 cylinder was observed to be at 80 psi and *rising*. Goodlin notified Frost who quickly advised him that the emergency engine cutoff switch should be depressed and held for 10 seconds. After an interminable pause, the engine pressure slowly dropped to zero.

Despite all of the distractions over the engine, the XS-1 still could provide useful data. Goodlin put the plane into a dive to about 25,000 feet. At Mach 0.70, he initiated a turn but quickly broke off the maneuver when the XS-1 exhibited severe fish-tailing and a roll. Goodlin had never noted anything even close to this in the previous flights. Not until the plane's speed fell below 200 MPH did the XS-1 regain stabilized flight. At 15,000 feet, Goodlin attempted to stall the aircraft in a clean configuration. Surprisingly, the XS-1 revealed new tricks as it stalled at an indicated 150 MPH and fell off to the left. Heretofore, the XS-1 had always stalled at 141 MPH in a straightforward manner. At a loss to explain these events, Goodlin decided to bring the XS-1 back to the airfield. Landing and rollout were normal.

The Bell and NACA technicians tore into the airplane to discover the reasons for the latest problems. Key to analyzing the rough engine situation was the cockpit camera records. The film revealed that shortly before the attempt to restart cylinder number 1 all chamber bleed pressures were near zero. which meant that engine nitrogen manifold pressure was very low. The technicians surmised a maladjustment of the nitrogen regulator and a resultant slow drop in manifold pressure, which would cause valve chatter and eventually cause the valves to close entirely. Concurrently, the pressure buildup in cylinder 1 was postulated to come from icing on the spark plug, thus providing a false pressure reading. Following the flight, the Bell technicians conducted a seven-minute ground run of the engine. In all, approximately twenty faultless starts were made and the engine was declared ready for further flights.

With the engine problem seemingly resolved, the Bell team began the search for the culprit that caused the fish-tailing and found approximately 60 gallons still in the lox tank. This remaining fuel would have altered the XS-1's center of gravity from its normal 22.4 percent mean aerodynamic chord to a new one of approximately 18.9 percent. The extra lox was doubly surprising because Goodlin specifically reported jettisoning the tank's contents and the lox liquidometer in the cockpit revealed the tank to be empty. But unlike jettisoned fuel, the chase plane pilot could not visually verify lox jettison. As a result, the liquidometer malfunction caused Goodlin to prematurely discontinue the jettison operation. In the future, the pilot was instructed to continue to jettison for 30 seconds after fuel exhaustion.

Having discovered solutions to two of the three abnormalities experienced during the flight, the Bell team looked for the source of the final problem—the higher than usual stall speed. The P-51 chase plane had calibrated the XS-1 stall to be at an indicated 165 MPH. This situation was not readily explainable, but the technicians surmised that something had disrupted the aerodynamic flow over the wings. The chief suspects were the wing tufts installed for the cameras or possibly a slight maladjustment in the spoilers. The Bell technicians decided to remove the wing tufts for the next flight and inspect and readjust the spoilers.

Preparations for the next flight allowed the XS-1 to be ready by the next day, Friday, January 31.[40] After B-29 takeoff, Goodlin entered the XS-1 cockpit at approximately 10,000 feet and quickly went through the aircraft checkout. After Goodlin completed loading the domes of the first stage and fuel/lox regulators, Heaney released the XS-1 airplane at 20,000 feet. As per the flight plan, Goodlin rapidly ignited chambers 1 and 2 and put the XS-1 into a climb. At 25,000 feet he shut down both units and as a test attempted to restart chamber 1. In spite of repeated attempts, Goodlin could not achieve reignition of the chamber. Goodlin could even hear the igniter trying to fire the cylinder, a noise that didn't cease until he put the throttle control lever in the off position. Worse, chamber pressure was slowly rising, much as on the previous day's flight. Fuel and lox were jettisoned and Goodlin prepared to execute several maneuvers so that the NACA could obtain data from the flight.

After reaching 25,000 feet in a dive, Goodlin executed a Mach 0.70 turn, tightening until an accelerated stall condition was reached at 3.9 g's. Pull-ups were then made at various speeds, including Mach 0.65, 0.60. 0.55, 0.50, 0.45, and 0.40. Goodlin noted the maximum acceleration recorded during these pull-ups was 4.7 g's, achieved at Mach 0.60. No stick force reversals were felt during the pull-ups although again there was a lightening effect. No snaking was evident, which convinced Goodlin that the condition had probably been caused by either the wing tufts or the lox remaining in the airplane's tank. Next Goodlin attempted to check the higher stall speeds noted previously. A stall was executed in a clean configuration and, once again, the airplane stalled at about 163 MPH with a noticeable roll to the left. After corrective measures were undertaken, Goodlin stalled the XS-1 twice at an indicated 149 MPH. Upon conclusion, landing and rollout were routine.

When the Bell technicians examined the XS-1 to find the source for the latest problems, they found that the oxygen line to the igniter for the number 1 chamber had ruptured. Most likely, they surmised, the nitrogen bleed line had closed before the restart, which permitted fuel to enter the oxygen line and explode when restart was attempted. Fortunately, the engine had suffered only minor damage. Since Goodlin forgot to turn on the power switch, only the mechanical instruments collected information on airspeed, altitude, and acceleration, which aided NACA technicians in better defining the buffet boundary. The flight tests also eliminated some possible causes of the stall difficulties. Although the same higher speed stall characteristics and roll to the left had reappeared, Goodlin was now inclined to look to the spoilers as the culprit. Before the next flight, the Bell technicians decided to bolt down the spoilers to prevent an asymmetric stall that might occur from one of the spoilers extending improperly.

No new XS-1 flights were attempted until Wednesday, February 5.[41] Earlier in the week, the NACA team had been busy shuttling between the tests at Mojave and Muroc. Now they were ready to proceed. Similarly, the Bell team had used the pause to conduct another ground test of the XS-1 engine, which went perfectly. With the sky clear and visibility unlimited, the Bell XS-1 team decided to attempt the thirteenth flight in the number 2 aircraft series. It would be the ninth powered flight designed to examine the buffet boundary and to allow Machmeter calibration. The only change, in addition to bolting down the spoilers, was that the landing gear and flaps had been altered so that their extension pressure now came from the 420 psi engine nitrogen pressure regulator rather than the lower 325 psi lox regulator. An Air Force officer would fly P-80 chase for the mission. Goodlin's flight plan specified a high-speed turn and level-speed runs at varying Mach numbers to assist in the Machmeter calibration effort.

After takeoff, Goodlin entered the XS-1 cockpit at 9,000 feet. Checkout proceeded smoothly and the drop was executed at 16,000 feet. Shortly after clearing the B-29, Goodlin ignited the number 1, and then the number 2 chambers. The XS-1 initiated a climb to 25,000 feet where Goodlin executed a 2.8g positive accelerated stall at Mach 0.65. With 60 percent of the fuel still remaining in the XS-1, Goodlin continued the climb to 35,000 feet. Short stabilized runs for calibration were performed at Mach 0.75, 0.71, 0.65, 0.55, and 0.50. NACA data reduction revealed that the cockpit Machmeter read 0.05 low at an observed Mach 0.50, and 0.07 low at an observed 0.75.

During the last of the test runs, the XS-1's fuel was exhausted and Goodlin put the aircraft into a dive. Upon reaching an indicated speed of Mach 0.65, Goodlin performed an accelerated 4g stall. Once completed, Goodlin again put the XS-1 into a dive to an altitude of 15,000 feet. Speed was an indicated Mach 0.73 (NACA recorded Mach 0.76). At that point, Goodlin executed an abrupt pull-up attaining a positive 8g (NACA recorded 7.3) acceleration. Next Goodlin performed a second pull-up at indicated Mach 0.70 (NACA recorded 0.79) and achieved an 8.7g positive acceleration. Upon completion of the second test, Goodlin attempted a third and final pull-up at an indicated Mach 0.55. The resultant g force was not reported by the pilot on this run. However, he did note a slight breeze on his left cheek which he believed indicated the glass windshield panel had become dislodged. Goodlin felt no stick reversal, but the same lightening sensation of the controls was again noted. No abnormal tendencies were discovered upon release of the back pressure on the control column. The aircraft recovered nicely and continued its reputation as a grand flying machine.

After conclusion of the tests, Goodlin glided down to 13,000 feet where he attempted to execute a stall in a clean configuration. The Bell technicians sought to discover if the bolted down spoilers would prevent the unusual stall conditions of the previous flights. However, once again, the XS-1 stalled at 163 MPH and demonstrated a tendency to fall off on the left wing. Goodlin noted the rolling tendency could be controlled above 149 MPH, but not below that speed. He remained baffled by the characteristic as stalling speed at landing with wheels and flaps down continued to be normal. Landing at 135 MPH and rollout were routine with the brakes working satisfactorily. Rollout on this occasion was confined to only 6,000 feet of the hard runway.

Back at the hangar, Goodlin was surprised by the ground observers' report of a strange noise they heard during the flight maneuvers. The unusual sound was a double cracking noise that sounded as if the plane had experienced a structural failure. Not only had the radar crew heard the noise, the personnel in the tower had also noted the phenomenon. After a thorough inspection, no source for the strange sound could be discovered. Later reflection led those observers to believe that Goodlin had created the first piloted sonic boom.[42]

For once, everyone seemed relatively satisfied with the results of a flight. Goodlin was pleased to note engine operation on this test was completely satisfactory. The pressures for the engine were normal and the liquidometer functioned perfectly. One other program milestone, the AMC 8-g's requirement, had been surpassed. Concurrently, the NACA personnel were equally pleased with the significant volume of data they received and the results observed in the film. Little between-flight maintenance was required for the plane although the fuel regulator valve was replaced as a precaution since Goodlin had observed a slight rise in pressure during operation.

On Thursday, February 6, Dow and Heaney took the B-29 up to 30,000 feet for a temperature check flight.[43] The 1-hour, 45-minute flight proved satisfactory. On the following day, the same team guided the B-29 down the runway for the tenth powered flight in XS-1 number 2. The only change from the previous flight was that Bell engineers had removed the celluloid tape over the wing pressure pickup orifices and smoothed the wing finish. The aerodynamic fix was an attempt to pinpoint the cause for the new stall characteristics of the XS-1. Goodlin was to repeat the previous flight using slightly different speeds, following with the execution of several stalls and pull-ups.

At approximately 8,000 feet, Goodlin entered the XS-1 cockpit and began the predrop checkout. At 16,000 feet, Dow initiated the drop sequence. After clearing the bomber area, Goodlin ignited the number 1 chamber. When it was stabilized, he ignited the number 2 chamber and commenced a climb. Upon reaching 25,000 feet, Goodlin executed an accelerated stall with approximately 60 percent of fuel still aboard at a constant Mach 0.60. Upon completion of the stall, Goodlin again put the XS-1 into a climb. At 35,000 feet all fuel was exhausted and Goodlin leveled the XS-1. To complete the tests, Goodlin executed a dive at Mach 0.60 to 27,000 feet where he conducted another stabilized turn at minimum weight. After finishing the turn, Goodlin executed a series of abrupt pull-ups at indicated Mach 0.63, 0.65, and 0.68 (recorded Mach 0.68, 0.71, and 0.74), three points suggested by NACA. The Langley technicians had determined that maneuvers at those points could better define the peak of the buffet boundary. Maximum positive acceleration attained on the flight was 7.2 g's.

Upon completion of the third pull-up, Goodlin noted a peculiarity in the handling of the XS-1. To maintain level flight at a constant IAS of 200 MPH, he had to exert an estimated 15 pounds of forward pressure on the control column. Worse, the abnormal force requirement would continue until landing. As a result, Goodlin executed a dive to 15,000, where he performed a stall with the airplane in a clean configuration. The performance results were similar to those experienced on the last three flights. Once again, the speed at stall with landing gear and flaps retracted was about 10 MPH higher than on flights up to number 11, and the XS-1 continued to exhibit a tendency to roll to the left. To add to the mystery, stall speed with flaps and landing gear deployed remained the same as in all the other flights. Rollout was again routine.

After the XS-1 had been returned to the hangar, the technicians quickly discovered the source of the latest problems. The change in stick force was caused by a short section of the right-hand elevator's lower leading edge being caught under the lower trailing edge of the stabilizer. The point of interference was approximately 12 inches inboard of the tip at the spot where the elevator leading edge had been cut away to permit passage of a pressure pickup hose. Additionally, ground control reported that channel "D" had gone out after only one transmission. Inspection revealed that the fiberglass fin cap over the VHF antenna on the vertical fin had split. Repair was impossible and no spares existed at Muroc. A call was placed to Buffalo to expedite shipment of a new cap to California. The delay caused the program to close down temporarily.

The lull allowed time for reflection on the achievements to date. In the flights up until number 10, the XS-1 and its engine had functioned satisfactorily in most respects. The actual testing procedures had settled into an efficient routine. In addition to the data collected for the NACA, accomplishments included a satisfactory demonstration of the engine at 75 percent and 100 percent of thrust, including during a steep (35-degree) climb. The spoilers had been utilized to stabilize the landing patterns, and speeds up to Mach 0.80 had been achieved at 15,000 feet. In reference to previously reported defects, Bell Aircraft had apparently found satisfactory solutions to the problems of excessive cooling of the nitrogen gas, the cracking of the regulator diaphragm, the failure of the cabin pressure regulator, and the engine compartment ventilation. Of the two additional problems Bell recognized, Bell engineers believed that they had a remedy for the tank blowdown valve failures; and the inability of the P-51 to perform in a powered XS-1 flight regime could be handled by replacing the fighter with a P-80.[44]

The Grind Continues

By the end of the first week of February, the Bell contractor team had developed an excellent understanding of the "liquid" costs of the XS-1 program. Between October and February, the program had utilized 1,100,000 cubic feet of nitrogen, 1,870,000 cubic feet of oxygen gas, and 5,800 gallons of fuel. Upon closer inspection, however, "utilized" turned out to be an inexact expression for what was really happening. For example, in reference to nitrogen, 47 percent of the total was lost to tank storage boil-off and only 10 percent to actual flight tests. The balance was consumed in aborted drops, ground runs, and pressure tests. Similarly, 37 percent of the oxygen was lost in boil-off while 60 percent of the total fuel was used in ground tests or jettisoned in aborted flights. The size of the storage tanks for the liquefied gases caused the waste. While smaller tanks could be utilized, thus cutting boil-off losses, the increase in service costs for handling additional fill-ups would cut significantly into any proposed savings. The only good news from this inventory of propellant use to date was that throughout the test period costs per 100 cubic feet of gas had fallen substantially because of the increased volume of purchases.[45]

During the week of February 10 to 14, no flights were attempted.[46] Slick Goodlin returned to Niagara Falls in the Bell C-47 along with several of the Bell team. On February 12, he flew to Patterson Field to be fitted with a pressure suit. At the same time, the remaining Bell engineers continued work on the XS-1 at Muroc. Technicians again filled and smoothed the wing. At the same time, the tops of the wing surface were repainted and polished to eliminate any possibility that XS-1 buffeting was caused by the condition of the wing. Paint was removed from the left rear fuselage as it was impossible to keep the paint finish smooth in this area due to jettisoning of lox in flight. Finally, the Bell technicians removed the single three-position fuel/lox jettison switch and replaced it with two SPST switches. This change was designed to prevent the pilot from accidentally throwing the switch to a jettison position.

When the repairs and modifications were completed, the Bell technicians conducted another ground run for the engine, to find the maximum current number of chambers that could be fired at one time. The test demonstrated that it was possible to fire all four chambers simultaneously if nitrogen, fuel, and lox pressures were at their proper levels. However, engine performance was rather rough.

In evaluating the most recent tests of the XS-1 (flights 11–14), the NACA team expressed satisfaction at the amount and the quality of the data recovered from their instruments. It was anticipated that the data recovered would better define the buffet boundary, buffeting tail-loads, longitudinal stability in accelerated flight, stall data, and longitudinal stability in straight flight up to Mach 0.82. The team, however, was unhappy about the poor quality of the stick force versus g data, which was due to the friction forces in the control system which masked the aerodynamic forces. The good news in the preliminary data was that the maximum tail load reached in the most severe XS-1 maneuvers was less than 3,000 pounds, which was considerably below the aircraft's design load. The sturdy strength quality built into the plane now seemed conservative but still very comforting.

During the time between flights, another issue again appeared that was to influence profoundly the XS-1 program. While the flight tests at

Muroc continued to unfold with increasing regularity, the paperwork half of the program began to develop problems over the question of the ultimate disposition of the two XS-1 aircraft. On February 14, Walt Williams wrote to Hartley Soule at Langley to discuss the most recent developments in the flight test program.[47] The success of the recent flights and the February 13 visit of Col. George Smith from Wright Field led Williams to speculate that five additional flights on XS-1 number 1, after its return to Muroc, would be sufficient to end the Bell contractor program. Additionally, the NACA would want several additional flights on the number 2 aircraft after its return from Niagara Falls. The trip east was expected to occur in the near future to allow the second XS-1 to receive several flight modifications. The NACA would not have all the data below Mach 0.80 they desired but, Williams believed, they would have all they could reasonably expect to get from Bell Aircraft.

On February 18, Williams speculated that Bell would not attempt another test flight until the following week. One day later the XS-1 was fueled and ready to try another flight. Communications between Bell Aircraft and the NACA were still not perfect. The plan for the flight was for Goodlin to execute several pull-ups at approximately 12,660 feet. The NACA data pointed to the probability that the XS-1 could achieve an 8g pull-up at Mach 0.70 at that altitude. Another pull-up would be executed at Mach 0.80 to whatever g force was attainable. Other tests would measure directional stability (sideslips) and dynamic stability (rudder kicks). The NACA thought two flights would be necessary to get all of this data. The Bell team was determined to try it in one test.[48]

By the time all preparations were finished, it was late in the afternoon on the nineteenth.[49] The B-29 rolled down the runway for the fifteenth flight of the Muroc series. Weather was clear and visibility unlimited. Goodlin entered the XS-1 at a pressure altitude of approximately 8,000 feet. Approximately three minutes before drop, he loaded the domes. By the time Goodlin finished the loading operation, Dow had reached the 17,000-feet drop altitude. Dow commenced a shallow dive in the B-29 and released the XS-1 at 16,300 feet. Speed was an indicated 220 MPH. Six seconds after drop the XS-1 had fallen to 15,900 feet. Goodlin ignited number 1 and 2 engine chambers in rapid sequence. After almost one minute of operation, chamber 2 was cut off and the XS-1 stabilized at a speed of Mach 0.65.

In an effort to conduct the maneuvers required by the flight plan,

Goodlin began a series of tests designed to gather data for the NACA instruments. First, he executed sideslips to the left and right while attempting to maintain five degrees of displacement. Next, with speed remaining the same, Goodlin vigorously applied approximately 50 percent right rudder, then completely released it to check the dampening characteristics. Although ailerons and elevators were held neutral, the aircraft developed a slight rolling tendency in the dampening process. Even though the oscillation diminished quickly, a low-amplitude effect remained for some time. The pilot attributed this condition to the fuel sloshing caused by the initial abrupt rudder change. Finally, Goodlin flew the XS-1 around Muroc using only one chamber while he burned fuel down to an indicated 10 percent level to lower the aircraft weight. At that time, Goodlin noted the old problem of rising lox tank pressure. After all the previous problems with line pressure increases, Goodlin knew the source of this malfunction. The second-stage lox regulator valve had obviously been unable to close and pressure would continue to rise. Thus, even though he had not had the opportunity to complete the pull-ups, Goodlin shut down the engine and jettisoned the remaining propellants. Approach and landing were routine, except that once again the XS-1's brakes failed to operate.

On the ground the lox regulator was disassembled and inspected. Foreign matter was found in the regulator valve system, which prevented the valve from seating. The regulator was cleaned and reassembled. The brakes were bled and overhauled. However, the minor mechanical repairs were offset by the discovery that the yaw angle recorder had jammed, which meant that no data from the maneuvers performed were recovered. The NACA felt the flight was "completely useless."

Concurrent with these flights, another planning process was unfolding far from the dusty flats of Muroc. The LMAL timetable for the next several weeks anticipated XS-1 number 1 (#6062) to be ready to travel to Muroc sometime between March 1 and 15. When that trip occurred, the NACA would remove the number 2 plane's instruments and the B-29 would ferry #6063 back to Niagara Falls. After returning to Muroc with the number 1 aircraft, the B-29 would fly to Oklahoma for overhaul and possible engine replacement. Between March 15 and April 1, while the XS-1 number 1 was being readied for research flight, the instruments from the number 2 aircraft would be returned to LMAL. A specially selected NACA pilot and crew would proceed to Niagara Falls for flight and maintenance indoctrination on the XS-1 number 2. When the B-29 returned to Muroc, the NACA team at Niagara Falls would also proceed to California to observe the XS-1 tests. During the month of April, the #6062 would complete a sufficient number of flights to provide NACA with proof of low-speed airworthiness of the thin-wing concept. At that time a major programmatic decision was to take place.[50]

On February 21, preparations for what turned out to be the final flight in the initial series of XS-1 number 2 tests was ready to go.[51] The weather was perfect and the XS-1 had undergone no substantive changes since the previous flight. The stated purpose of the test was to conduct abrupt pull-ups and perform additional stability investigation. Takeoff was routine and Goodlin entered the XS-1 cockpit at 8,000 feet. Checkout and dome-loading proceeded smoothly. Baker in the P-80 (#44-85309) again provided chase. The XS-1 drop occurred at 16,850 feet and an IAS of 225 MPH. It was a slower speed than normal, but Goodlin reported no problems. Within 40 minutes, the B-29 was back on the ground. The drop team seemed to have this launch business down to a science.

After falling 100 feet, Goodlin ignited the first chamber and let it run until stabilized. After igniting chamber 2, Goodlin watched the XS-1 stabilize at Mach 0.65. He had to retard the power to maintain constant speed. Sideslips to the right and left were carried out in the vicinity of 16,000 feet. Once again Goodlin found that these maneuvers created fuel sloshing that the rudder could not cure. Attempts to fly the aircraft with hands off the control column revealed a tail-heavy condition and corrective measures were required. Goodlin applied slight forward pressure on the control column and the XS-1 responded with level flight. Although Goodlin surmised the tail-heavy condition might be caused by unequal lox/fuel burning, he could see no evidence of the condition on the tank pressure gauges. In fact, their observed readings should have meant the XS-1 would become nose-heavy. In any event, Goodlin now noticed a far more serious problem.

While cruising over Muroc using the number 1 cylinder as he burned propellants down to the desired 10 percent level, Goodlin heard the howling sound associated with too lean a fuel mixture, despite tank pressure readings indicative of a somewhat rich mixture. He started the number 2 chamber and shut down the number 1 cylinder. Goodlin then observed that chamber number 2's pressure was at least 10 percent below normal. Shutting down chamber 2, Goodlin ignited the number 3 chamber. With remaining fuel down to 10 percent, Goodlin increased speed to Mach 0.80 (NACA recorded) and executed an abrupt pull-up. No serious difficulties were observed or felt in engine operation during the maneuver. Goodlin then executed a second pull-up at 12,000 feet. The XS-1 experienced a recorded 7.7 g's during the maneuver with no effect on engine operation.

After shutting down the engine, Goodlin performed several power-off maneuvers for the NACA data records and then at 9,000 feet he lowered the landing gear and flaps. Landing and rollout were routine. Postflight investigation revealed the tail-heaviness experienced during the flight was caused once again by the same problem as on flight number 14—the right-hand elevator's lower leading edge became caught under the lower trailing edge of the stabilizer about 12 inches inboard of the tip. The problems with the two cylinders experienced in the flight caused the Bell technicians to ground-test the engine. A decision was made to replace it with a spare engine. The RMI representative concurred since the installed engine had over one hour of running time on it. Unfortunately, the new engine performed very poorly in the ground test, primarily because it had not been run in several months.

During the XS-1 engine overhaul, the B-29 also received overdue maintenance attention.[52] On February 26 and 27, in anticipation of its 50-hour overhaul, Dow and Heaney took the B-29 up for test flights. The setback encountered in the XS-1 engine ground run changed all plans. At the completion of the second B-29 flight, the XS-1 number 2 was secured beneath the bomber. On February 28, with Dow at the controls and Goodlin flying copilot, the B-29 lifted off from Muroc on a 7-hour and 5-minute return flight to Niagara Falls and the Bell factory. Prior to departure, all NACA instrumentation was removed from the XS-1. On March 3, all the Bell technicians except Frost and one mechanic departed in the C-47 for Niagara Falls. Mark Heaney also remained behind to assist with other Bell project work. The same day Baker left for LMAL in the NACA C-45, carrying the recently removed NACA instrumentation. It was the NACA's intention to install the equipment in the number 1 aircraft while it was still at Niagara Falls. That NACA team visit occurred on March 17.

While at the Bell Corporation, the XS-1 number 2 was disassembled and modifications were initiated to provide it with equipment similar to that installed in the number 1 aircraft. The number 1 aircraft had recently passed its engine ground test and was ready for the NACA instrument installation. When modifications were complete, both planes would be identical except for the number 1 aircraft's 8 percent wing, 6 percent tail, and hydraulic surface dampers. With the exception of a new canopy, the number 3 aircraft remained unchanged. On March 10–11, Williams and Baker visited the Bell Aircraft Niagara Falls facility. Baker utilized the opportunity to perform pilot familiarization work in the rocket test shack.

On March 6, Dow piloted the B-29 to Rome Army Air Field for the 50-hour checkout of the mother ship. However, heavy snow accumulations at the airport caused delays in getting the B-29 into the hangar, resulting in the initial overhaul schedule being postponed by a week. Once the B-29 was in the hangar, the AAF workload further delayed the overhaul. Finally, on March 19 eight Bell technicians were dispatched to Rome by C-47 to assist in the B-29 work. Therefore, the original date for the number 1 aircraft to return to Muroc was postponed. The project retained its record of being incapable of maintaining any schedule.

The pause in the program allowed program participants time to objectively view the results to date. In spite of the difficulties caused by differing goals, the AMC/Bell/NACA team had managed to put together a working relationship that had provided a safe and functional basis for the supersonic program. Roughly halfway through the contractor flights, the XS-1 had yet to experience anything like the major problems suffered by many of the other prototype aircraft flying (and many times not flying) at Muroc. To date, the data had revealed the XS-1 was capable of providing a reasonable platform for the transonic flights anticipated after the contractor tests. The rocket engine had weathered some difficult moments, but had performed satisfactorily. The flight duration times had continued to cause problems, but not of the anticipated kind. The engine runs had proven the power of the aircraft, but the myriad problems had limited the ability to perform all of the tests NACA had desired. But events outside the program began to alter its anticipated future. A new actor was about to appear on the scene: one that had been discussed in 1944 and now was actively working to reenter the picture.

5

The Changing of the Guard

Although during World War II the NACA was accused by some of a late entry into the arena of high-speed research, the postwar period of late 1946/early 1947 revealed sweeping gains in the development of the U.S. transonic aviation program.[1] The extensive ongoing efforts in high-speed research since 1944 seemed to be paying rich dividends for the United States as other competitor nations appeared to be having difficulty maintaining the technological pace of advancement. However, severe budget cutbacks in federal spending for fiscal year 1947 devastated any hopes of procuring the necessary funding for the expansive wind tunnel program outlined by the Raymond panel. But, for the NACA, the political muscle acquired through past experience worked to provide an increase in FY 1947 congressional funding from the $24+ million of FY 1946 to over $30.7 million at a time when almost every other federal agency witnessed sharp cuts in appropriated funds. These budgetary cuts devastated the military services' conception of a vast new program in aviation research. The continuing political struggle led both the NACA and the AAF to search for allies to restore the rationale for higher levels of funding.

The Raymond panel's failure to find a congressional consensus for the supersonic wind-tunnel program did not mean the end of the behind-the-scenes political struggle to fund the research activities outlined in the plan. To "sell" the program to Congress, the requested price of the Raymond panel components was cut and efforts were made to strengthen the constituency for such a program. To achieve the latter, the NACA attempted to increase industry participation by appointing people to its consultative committees that would convince their congressional friends that the NACA's priorities were justified. Concurrently, the military augmented the AAF's existing Scientific Advisory Board with the creation of a new Research and Development

Board reporting directly to the new secretary of the National Military Establishment (which became the Department of Defense in 1949).

Throughout 1947, each agency sought to make its case. Yet the logic of a coordinated effort continued to point toward a joint program. During the war, the interlocking agendas of the NACA, the military, and industry had found a congressional constituency. The military provided the sponsorship needed for funding of particular projects and the rationale for pursuit of that objective; the NACA provided the fundamental design criteria and instrumentation; and the industry furnished the development and production facilities. In spite of the desire to create independent efforts or to subdivide the process, each group had benefited by the overlap—the cross-fertilization—of scientific knowledge and equipment. NACA's agenda was research, but the Joint Chiefs of Staff had a larger agenda to follow. In the final analysis, it was harder for the two groups to pursue the same objective with the same intensity, even if in hindsight the AAF, from an apparently weaker position, utilized its political assets more skillfully than the NACA did. In any event, logic directed some continuing effort at coordination, even if only for appearances. As part of that effort, a new $1-billion joint NACA-AAF-Navy Joint Research Development Board (JRDB) program was unveiled to continue the progress demonstrated by the recent transonic tests. But the continuing severe postwar financial difficulties, coupled with the congressional desire to return the budget process to a peacetime pattern, rendered moot any hope of such a modest (by Raymond panel standards) FY 1948 research effort.

Each agency viewed its program as politically justifiable and scientifically necessary. Speaking before the president's Air Policy Commission late in 1947, Dr. Vannevar Bush defined the research needs of the nation in terms of programmatic control. The military effectiveness of the United States, Bush stated,

> rested squarely on the shoulders of the military services.
>
> The military services have the *central* responsibility for seeing to it that the entire procedure operates to produce the needed end result, all the way from fundamental research in universities and governmental laboratories through applied research, military specifications, engineering design, prototype, evaluation, and final production [emphasis added].

Needless to say, the NACA strongly disagreed with that position. Interestingly, the Bush argument really read as a reversal of his comments of June 1946 at Wright Field. Then, diversity was key to the process. Now, centralization—under military control—was the center-point. The NACA rebuttal given to the Air Policy Commission, iron-ically, sounded like the former AAF position. Dr. Hugh Dryden, the new director of aeronautical research (replacing the retired George Lewis) stressed the utility of overlap and decentralization. Indepen-dence, Dryden commented, fostered diversity of views and clarity of reporting to the policymaking group. In retrospect, the bureaucratic in-fighting continued for two more years. However, for FY 1948, Con-gress did boost the NACA research spending by over $12 million to $43,449,000, which almost *doubled* the NACA's 1947 spending level.

Proponents of the increased NACA funding pointed to the data the first transonic research efforts had provided to industry. Congress wanted more such successes and was willing to assist the NACA with its wind tunnel needs, its rocket-powered model tests, and its wing-flow research. The XS-1 airplane tests were to become the centerpiece of this effort as soon as the Bell Aircraft acceptance flights were com-pleted. July 1947 was targeted as the month to acquire the XS-1 for the NACA. At that time, the NACA would initiate a three-phase program focusing first on the XS-1 and the new Douglas D-558-I. The second phase, as outlined to Congress, would involve the new Bell XS-2 and the D-558-II aircraft, both of which would feature sweptback wings and lower aspect ratios than the first-phase models. Phase 3 would involve two new airplanes that were not yet even in construction: the XS-3 and a proposed D-558-III (which was never built). Consistent with the desires of LMAL's John Stack, the NACA plan outlined for phase 1 only involved transonic research. Phase 2 was to explore speeds *up* to supersonic. Research above supersonic speeds was not to be conducted until phase 3. This pattern of caution reinforced the AMC and Bell Aircraft concerns that the NACA was too slow, and was only one of the ongoing problems with the supersonic program.

On December 12, Walt Williams wrote to Langley to suggest that the NACA clarify the ongoing dispute with Wright Field and Bell Aircraft over management of the program. This conflict had continued from the first days at Muroc. On December 20, Hartley Soule outlined in a seven-page memorandum what LMAL felt should be the NACA policy for all XS aircraft research. This memorandum was in response to an AAF request for amplification as to the necessity for further meetings on the issue of XS research jurisdiction. Soule divided his memo into five parts and in so doing highlighted the differences in focus between the NACA program and current Bell Aircraft objectives. Soule indicated that the NACA's goal was to procure aerodynamic, structural, and propulsion data from the research to assist in the development of supersonic military aircraft. With that goal in mind, Soule wanted the NACA and the AMC to reach agreement on the specifics of redesign cost, and data (speed) exaggeration. Soule also touched on the need to clarify the NACA needs with the AAF's contracting methods. The NACA was not a typical AAF contractor, and therefore the AMC should not question its needs and operational procedures as if it were a private company. More significantly, Soule indicated, the NACA must exert more pressure on the AAF to match that which Bell Aircraft utilized every time there was a LMAL/Bell disagreement. Soule hoped that the situation would change once the NACA provided the pilot and took over the program. However, in the interim, important tests and future XS program decisions were being made without complete NACA input.[2]

On February 6, 1947, a conference on the XS-1 flight program was held in Washington, D.C.[3] The high-level attendees included Col. J. S. Holtoner representing AAF HQ, and Col. George Smith, chief of the Aircraft Projects Section of the AMC at Wright Field, accompanied by XS-1 program civilians Charles L. Hall and James Voyles. The NACA personnel present at the meeting included John Crowley, with Russell Robinson, Milton B. Ames, and Clotaire Wood (all from the NACA HQ) and Hartley Soule from LMAL. These major program decision-makers gathered to focus on the ongoing issues of supersonic research.

Colonel Smith framed the discussion at the opening by indicating that lack of funds would preclude the AAF from pursuing construction of the XS-3. Therefore, the meeting would focus on the XS-1, XS-2, and XS-4 programs. In general, the NACA and the AAF reached four broad agreements on the current flight test programs. The AAF would assign a higher priority (1-b) to the research aircraft program than to its tactical aircraft program. The AMC also agreed to review the NACA requests for design changes on each research aircraft on a case-by-case basis consistent with priority and funding. (Design changes would, however, be limited to those that increased stability and control at higher speeds.) Further, the AMC agreed to furnish spare parts for all research aircraft, while the NACA agreed to provide flight crews, fuel, and maintenance for the planes. The AMC stipulated that if additional services were required and available at any AMC test facility, such services would be provided by the AAF in the interest of avoiding duplication. Finally, the AMC issued a gentle reminder that the NACA was responsible for flight test programs and that in the future they should participate in the programs from the beginning at their own initiation.

This not-so-subtle hint was to remind the NACA that the AMC footed the X-craft bills and that design changes were very expensive propositions. However, this AAF admonition must certainly have fallen on shocked ears, given the long history of the NACA's attempts to even find out what was in the AAF/Bell Aircraft contract, much less their ongoing efforts to interject their views into Bob Stanley's closed shop efforts at flight testing. In any event, having finished their general points, the two agencies turned to the specifics of the XS-1 program.

Consistent with past agreements regarding the future of the XS-1 flight program, the AMC set forth the conditions under which the NACA would take over the program. First, the NACA would furnish the flight crew, fuel, and maintenance for the XS-1. Second, the AAF would furnish the B-29 drop plane, fuel, maintenance, and spare parts, as well as the bomber pilot and crew "if possible." Finally, the AAF would continue to supply the SCR-584 radar located at Muroc. The NACA HQ would examine requests to divert the radar to other uses, as long as it did not interfere with the primary mission of the XS-1 program. With those agreements, the conference adjourned.

This meeting was an important milestone in the evolving XS-1 program. Prior to this point, the XS-1 had been a prize dangling just out of NACA's reach. Now it appeared that a fundamental decision had occurred. The NACA would not only get the plane, it would fly the program. What caused this change in attitude at AMC (for Bell still was not of that mind, as will be seen) is not clear from the records. Perhaps it was NACA's success with Congress during the FY 1947 budgetary hearings. Possibly it was the evolution of the plan (as will be discussed shortly) to fly both aircraft simultaneously, with the AAF controlling one and the NACA the other. In any event, the decisions reached clearly suggested the future direction of the flight research program. However, hidden behind the polite discussions were several ongoing battles over control of the XS-1 program: battles that would recur with a fury over the next several months.

Specifically, the NACA concerns on design changes directly related to their continuing fears over the operation of the XS-1 stabilizer control system. Simply stated, the NACA believed that neither the actuating control switch mechanism in the cockpit was satisfactory nor was the limited supply of nitrogen pressure for the pneumatic motor sufficient. Although the NACA acknowledged in private that it might be impossible to fix the XS-1 number 1 before it returned to Muroc, they still thought it possible to alter the number 2 airplane's control system when it returned to Niagara Falls for repairs.

The agenda for the radar was less obvious. The NACA had made no formal requests of the AMC Wright Field to use the SCR-584 on any project other than the XS-1. However, the new Navy-sponsored Douglas D-558-I was being prepared for testing at Mojave. The NACA was considering a "request" by Douglas Aircraft to become more involved in the flight research program. (If they had not, it would certainly have surprised Navy's BuAer, after all the earlier conversations on cooperation.) The NACA would want to use the radar for those tests, but that might require additional NACA personnel, as well as moving the radar from Muroc. Apparently, the AMC was strongly opposed, and indicated that nothing must interfere with the XS-1 tests. As a result of the AMC's firm stand at the February 6 meeting, the NACA personnel chose not to bring up the proposal to relocate the SCR-584 radar for use at Mojave. Instead, as Soule later told Walt Williams, the NACA must accept the fact that the D-558 would probably have to test at Muroc, not Mojave, whenever the radar was needed.

Finally, the AMC complained that the NACA reports from Muroc to Langley took about three weeks to get to Wright Field. This delay caused innumerable command problems (read: we can't intervene in policy decisions). The NACA promised to speed up the process by sending reports directly to Wright Field. However, the Soule internal NACA memorandum to Williams advised the NACA Muroc station to "eliminate any argumentative points that should be handled through Langley." Of course, what constituted argumentative points was left to Williams's discretion.

As a followup to the meeting in Washington, Colonel Smith flew out to Muroc in mid-February to meet with the program participants. Soule had advised Walt Williams on February 11 of the substance of the February 6 meeting and thus the NACA was ready for the AMC visit. On February 13, Smith and Col. Donald Putt met with Walt Williams, Dick Frost, and Major Shoop (the AMC representative at Muroc).[4] Once again, conversation focused on the flight test program. Williams indicated that the NACA did not yet have all the information they desired for the aircraft, but if future flights went as smoothly as the most recent ones, then the NACA would be satisfied with XS-1 number 2. Twenty powered flights would be the desired goal, but not all flights had to be on the same aircraft. In any event, the NACA would still want at least five flights on XS-1 number 1 with the thin wing. The NACA would like to investigate several additional items of interest with this aircraft, including the characteristics of the "new" control systems (the reduced vertical space due to the 8 percent wing/6 percent tail), the stall characteristics with the new wing, and some accelerated flight loadings. In conclusion, Williams indicated that when XS-1 number 2 returned from Buffalo, the NACA would like to conduct several additional flight tests regarding the modifications to that aircraft. Once completed, the NACA would be ready to take over the test program and the two aircraft.

Discussion then turned to the XS-1 engines. The NACA technicians agreed that little additional engine development work was needed since the test program had accomplished this end. It was anticipated that the remaining flights would see satisfactory engine operations. Maintenance, however, was another concern. After Frost reported on the composition of the Bell team at Muroc, Colonel Smith outlined which personnel the AMC would be responsible for supplying in the coming takeover of the program. Basically, the NACA would provide a project engineer, foreman, XS-1 crew chief, electrician, and mechanic. The AMC would supply the B-29 crew including a pilot and copilot, flight engineer, and two mechanics. The NACA would also supply a nitrogen evaporator operator and an instrument technician.

As a result of Williams's letter to LMAL, Soule suggested to Mel Gough, the chief of flight research, that the NACA again reiterate its conditions for acceptance of the XS-1 aircraft,[5] including a sufficient number of demonstration flights, so that a determination could be made of the research craft's airworthiness. Specifically to be examined were the mechanical elements of the plane, the control and stability on drop and landing, the XS-1's basic structural integrity, and the various modifications recommended to fix the observed deficiencies of the program. Soule again stressed that at least twenty powered flights were needed to verify the proper functioning of the rocket motor and pressure fuel system. Any changes to these systems would require additional tests to verify the propriety of the corrections. Soule further added that XS-1 number 3 was not a part of these discussions since the indefinite status of the turbine pump rendered premature any decisions on that aircraft. Soule concluded his report to Gough by indicating that proof of the aerial qualities and structural integrity (obtained from the NACA flight instrument records) would be ready shortly. Additionally, the Muroc test group would soon finalize the list of modifications needed before acceptance. At the current rate of progress, Soule anticipated that the first aircraft would be ready for the NACA acceptance by May 1. Without unforeseen difficulties, the NACA would accept the aircraft and operation of the program by LMAL and the AMC would commence.

Mel Gough was much too careful to rest the NACA's final acceptance decision on the opinion of any one individual. In preparation for the March 5 meeting at Wright Field, Gough asked Joel Baker to prepare a pilot's evaluation of the two aircraft. Although he had not flown either craft, Baker was perhaps the best pilot outside the program to discuss its characteristics since he had been involved with the airplanes since the first flight. Baker informed Gough that many of his observations were made after consultations with Goodlin. The seven-page report Gough received on February 25 stated that the XS-1 could be accepted and used in its present configuration. However, Baker expressed serious flight concerns from an NACA perspective, including poor pilot access/egress, limited visibility, dangerous high-pressure systems for power, and a slow and unreliable longitudinal trim control. All of these would require a major airplane redesign, which was not likely given the AMC's financial problems. Baker also acknowledged several additional issues that could be fixed with minor changes: an improved nitrogen control panel, a new instrument panel, a different longitudinal trim control mechanism location, new brakes, and adjustable rudder control pedals. Baker left the final recommendation for acceptance to higher authorities and confined his conclusion to pilot details. But the lengthy list of problems certainly weighed heavily with the cautious Gough.[6]

While NACA prepared for its important March meeting, the Aircraft Projects Section at Wright Field was also doing some background preparation. In anticipation of the meeting, Colonel Smith asked Charles Hall to prepare an overview document for the XS-1 program. This brief summary of financial and performance issues, submitted to Smith on February 27, would serve as the AAF position for the conference. In the report, Hall discussed the XS-1 project from the perspective of the Fighter Projects Section, taking special care to frame the present complaints about the handling of the project to date. In his remarks, Hall indicated that Bell Aircraft believed that they had *al-*

ready met the acceptance requirements of the program, which was not the Wright Field position. Second, the NACA never provided documents directly to the AMC. Rather, Langley received the documents and later—sometimes critically later—forwarded them to Wright Field. As to the future of the program, Hall identified the problem: *money.* Bell wished to believe they had met all requirements. Thus, any additional tests would receive a new fee. This money would be difficult to find in spite of the past plan to offer a follow-on contract to Bell Aircraft, the NACA, or someone else to fly supersonic. If the Fighter Projects Section said that the initial XS-1 program requirements were not yet complete, then Bell Aircraft could be requested to continue flying the XS-1 as a program overrun that could be paid with previous years' funds to the extent they remained available. In short, good options were slim.[7]

The main problem for Colonel Smith, and indeed for everyone in the Engineering Division, was the continuing congressionally mandated cutback in federal funding. The decline in support for new research projects also hit the existing aircraft programs as dollars were shaved to spare outright cancellations of existing aircraft programs. Fortunately, the XS-1 program was far enough along in its funding cycle to be somewhat immune to the worst effects of the cutbacks. For the entire XS-1 program (three planes), the AMC had committed $230,000 of government-funded equipment. The total program cost (with changes) was $4,278,537. (The entire Aircraft Projects Section budget for *all* research and development programs in FY 1948 was only $29,175,000.) By the end of December 1947 only 2 percent of the XS-1 funds would remain uncommitted. Thus, Smith faced the problem of extending the work on the XS-1 program through the sound barrier with very limited financial flexibility. The shortage of funds would soon have a dramatic impact on the program participants.[8]

One other problem intruded into the XS-1 debate. The XS-1 was a pure research airplane designed to explore the boundaries of transonic flight. When the program began in 1945, transonic flight was beyond the capability of existing aircraft. By 1947, planned *production* aircraft were opening the possibility of transonic and supersonic flight. The new AAF fighter plane prototypes combined swept wings with turbojet engines and rocket boosters to *promise* dramatic increases in performance. The McDonnell XP-85 (650 MPH), the North American XP-86 (650+ MPH), the Curtiss XP-87 (600+ MPH), the McDonnell XP-88 (over 700 MPH), the Northrop XP-89 (550+ MPH), and the Lockheed

XP-90, Republic XP-91, and Convair XP-92 (all supersonic) suggested the breadth and depth of the potential advances in aviation design. These aircraft, supposedly soon to be production planes, threatened to outpace the results of the XS-1 and D-558 programs.[9] The Air Force decision to pursue many new types (including theoretical models, some based merely on wind tunnel calculations) provided ample evidence that the pace of aeronautical advances was accelerating rapidly. However, the margin of doubt remained in place on whether supersonic flight entailed an unacceptable risk or fatal flaw. One of the two research aircraft should remove that doubt. In March 1947, it remained to be seen which aircraft, XS-1 or D-558, that would be.

On March 5, representatives of Bell Aircraft, the NACA, and the AMC met at Wright Field to review the status of the XS-1 program.[10] The primary purpose of the meeting was to discuss the acceptance of the XS-1 aircraft by the NACA, in keeping with the understanding reached by all parties at the beginning of the program and confirmed in correspondence and discussions in May 1945 and reconfirmed in February. Representing NACA were Hartley Soule, Milt Ames, Mel Gough, and Walt Williams. Attending for Bell Aircraft was Bob Stanley. The AMC was represented by colonels George Smith and Osmond J. Ritland (Smith's deputy), plus civilians Charles Hall and James Voyles.

The discussion in Colonel Smith's conference room focused on the NACA's taking over phase 2 of the XS-1 flight test program. The primary question—had Bell Aircraft met the requirements of the March 1945 construction contract—was outlined and a set of concluding objectives identified. In summary, Bell would surrender control of the program to NACA and would only be required to provide an engineering advisor for the duration of the tests to supersonic speeds. The AMC insisted that this advisor be of the professional caliber of Dick Frost. The NACA stated that it believed the airplanes were structurally sound, exhibited no abnormal aerodynamic qualities, and were suitable from a safety point of view considering they were research aircraft. Further discussion centered on when the transfer of responsibility would take place. The NACA insisted that Bell complete twenty powered flights with either one or both aircraft, but at least five must be made with each aircraft. Consistent with the initial construction contract, the issue of Mach 0.80 and capability to handle pull-ups of 8 g's was again stressed as the NACA acceptance requirements. Further, the NACA wanted the junction box on the rocket motor strengthened and demonstrated at 8 g's with all four cylinders operating.

The NACA concluded its part of the conference by indicating that three of its pilots (Joel R. Baker, William E. Gray, and John P. "Jack" Reeder) were under consideration for flight duty with the XS-1 program. Additionally, at the request of the AMC, the NACA indicated its approval for the AAF to test the XS-1 rocket motor up to 60,000–70,000 feet. The NACA informed the group that LMAL would immediately send technical personnel to Niagara Falls to analyze the maintenance and ground requirements for the X-craft program. For the flight tests, the conferees agreed that the AAF would provide the B-29 and its crew. Working and living quarters at Muroc would be acquired from the existing Bell Aircraft facilities.

As to the technical details of the program, the NACA expressed several reservations about the existing aircraft. In a decision based upon the earlier Baker memorandum and consistent with NACA safety procedures, the agency desired that Bell Aircraft make the following changes in the XS-1: first, the rudder pedal, designed especially to accommodate Jack Woolams's height, needed to be corrected to fit an NACA pilot. Second, the emergency shut-off switch on the pilot wheel, if released, could restart the engine (thus, the pilot must make a second action with his other hand to avoid an accidental restart). Since this switch duplicated the action of at least four other controls (the master selector switch on the left-hand side of the pilot wheel, the battery switch, the rocket selector switch, and the Quick-as-a-Wink valve), the NACA recommended it be eliminated. Third, the dash panel had several instruments, switches, and control valves that needed to be adequately marked for identification. Fourth, the wheel brakes required excessive preflight maintenance; either a fluid reservoir should be added or larger master cylinders installed. Fifth, the reliability and rate of the stabilizer change mechanism must be altered; a three degree per second rate of change was recommended. Sixth, the present arrangement of gauges and instruments was not considered functional; the NACA proposed moving several cockpit gauges and adding a bank-and-turn indicator. Seventh, the NACA wanted separate controls for the operation of flaps and landing gear. Actuating pressure for three cycles should be provided in the system. Finally, a manual control switch must be installed for fuel/lox jettisoning in case of electrical failure. Having finished the technical problems, the NACA now listed its other requirements for acceptance of the aircraft. The lengthy list of fixes was about to grow longer.

So that the XS-1 program never again experience a repeat of the windshield glass–fogging problem that had proven so dangerous, the NACA recommended that a clear glass panel be installed on the lower left forward portion of the windshield. The glass would be inwardly removable so that cabin pressure would aid in the sealing process. Further, the NACA still didn't like the high-pressure nitrogen system. However, rather than restate this concern, the NACA requested *all* information on the tests conducted on this system so that they might better understand it. Since Bell Aircraft had submitted all contractor reports directly to the AMC, the NACA requested that all Bell information, reports, and memos on the XS-1 program be made available to them for review. The NACA also asked the AMC to ensure that the Air Force's chase pilot and P-80 aircraft be continued on assignment to the XS-1 program. Finally, the NACA requested that all engine pressure lines (with the exception of the nitrogen pressure line) be changed to stainless steel. The NACA list of changes, clarifications, requests, repairs, and modifications was indeed a lengthy one. The bottom line was that the NACA wanted the airplanes changed. The AMC was short of money but wanted to please the NACA to keep them involved in the effort. Bell Aircraft stood silent—for the moment.

The records that still exist of the March conference seem to be written about separate events. The NACA document is a detailed technical critique of the faults and corrections necessary to provide a minimally acceptable aircraft to the agency for powered flight tests. The AMC document is a summary document of the agreements and procedures to be followed in conducting the remainder of the program. Both seem very detached in tone. Thus, it is the third document, a Bell Aircraft memorandum prepared by Bob Stanley on the following day, that is most revealing.

On March 6, Stanley prepared a confidential internal report for Larry Bell and Roy Shoults. In it Stanley shed additional light on the colorless documents prepared by the NACA and the AMC. First, Stanley repeated the agreed upon requirements for a satisfactory number of flight tests on the two aircraft. But regarding the key question— "Do flight records show compliance by the contractor of the contractual guarantees?"—he outlined the NACA's reluctance to respond. Although Soule actually answered in the affirmative, it was Gough's call as to Bell Aircraft's technical responsiveness in the flights to date. Here, Stanley reported, Gough was as "timorous as an old maid" and "quibbled" about the design specifications of the airplane. Gough could not, given his technical concerns, bring himself to directly answer whether or not Bell complied with the XS-1 contract specifications. In his concerns, one can readily distinguish the results of the Baker memorandum of February 25. As a result, Stanley reported, most of the meeting focused on the issue of the AMC personnel trying to elicit a definitive answer to the contractual question. Finally, after a lengthy debate, Gough affirmed that Bell Aircraft had met the contract specifications. But he could not resist a parting shot by stating that in truth, the original "specification was loosely written."

Stanley went on to say that he was shocked to hear that the NACA had little interest in XS-1 number 3 (with the turbine pump) because they could get all the information they desired out of the present aircraft. (If Stanley understood the NACA position correctly, this would mark a significant departure from the debates of summer/fall 1946.) To Stanley, the NACA attitude smacked of defeatism. To Bell Aircraft, the NACA attitude would end a promising aircraft program in a cash-short postwar era when such contracts were desperately needed by all aircraft companies. Technically, the XS-1 number 3 promised to provide an additional 500 MPH of speed and thus greater information in the supersonic range. Privately, Stanley advised Larry Bell that he had heard that the NACA was merely marking time with the X-craft until the Douglas D-558 was ready for supersonic work, although this attitude, he acknowledged, might simply be an extremely pessimistic view of the NACA actions. In any event, Stanley was fearful, as indicated by his questions, that the NACA plans for the XS-1 were no more aggressive than their previous work with the P-80 and P-84. Apparently, Stanley reported, the NACA intended to fly the XS-1 at low altitudes (20,000–30,000 feet) until they encountered trim changes or buffeting and then stop. Since a P-80 could already do this at low altitude, Stanley saw little gain from testing the XS-1 in that manner.

As to the oft-asked question, "Who will fly the XS-1 airplane for the NACA?" it was clear, Stanley wrote, that the agency still had made no decision. But the Stanley memo threw a interesting monetary twist in the story. Stanley indicated that Baker, because of his longevity with the program, had the inside track "if NACA can come to terms with him." In short, it was a apparent repeat of the Goodlin situation. In any event, Bell Aircraft would do its best to train the NACA pilot and crew. Bell Aircraft's only other involvement would be to supply Frost to the NACA program. The NACA did not object to the thought of tests at 60,000–70,000 feet; they simply were not interested in being a part of that project.

With so much of the program seemingly decided and the course of action clearly delineated, Colonel Smith now threw a curve into the proceedings. Smith asked Stanley directly how he felt about the flight arrangements. When the XS-1 program was originally conceived, Colonel Smith had advised NACA's Dr. George Lewis (on May 31, 1945) that the ATSC (AMC) intended to execute a separate contract with Bell Aircraft for the research flights. Now, that no longer seemed the operative plan, but Bob Stanley didn't accept it. Bell Aircraft wanted that contract, perhaps for the money, certainly for the prestige, but above all because Larry Bell and Bob Stanley were aviation pioneers. The XS-1 was pointed toward making aviation history—they intended to be a part of that ride.

Stanley took matters into his own hands and informed Colonel Smith and the conference that Bell Aircraft would prefer to see *both* aircraft employed simultaneously: one under the jurisdiction of the NACA for detailed data gathering; and the other operated by Bell Aircraft to precede the NACA efforts and qualitatively explore the high-speed regime in advance of the NACA detailed research. In short, Bell would break the sound barrier while the NACA did its technical research. "This suggestion was not well-received by NACA" is how Stanley phrased the agency's response to his concept. Worse, the AMC politely replied that they didn't currently have the funds for two simultaneous programs. The meeting broke up on that note and everyone went to lunch with General Craigie, chief of the Engineering Division at Wright Field.

General Craigie had been busy most of the morning preparing a report for Congress which he was due to give the following day. Thus, he had been unable to attend the detailed meetings that preceded the lunch. Craigie was an old friend of Larry Bell and had been most supportive throughout the XS-1 program. As a result, Stanley took the opportunity, while having the group all together with the head man, to revisit the issue of the dual flight test program. Stanley stated his continuing belief that the program would "probably go faster if both airplanes were employed simultaneously." Craigie then recommended that Stanley work out the details with the NACA.

When the conference adjourned right after lunch, Bob Stanley saw his opportunity.

Stanley waited several hours until Colonel Smith could free himself from previous commitments. Then, in the presence of Colonel Ritland, Stanley detailed his views to the two officers on the undesirability of the flight program that had been worked out at the morning meeting. The issues Stanley discussed went right to the heart of the developing military-industry partnership. In his later report, Stanley stated:

> I explained in some detail my views concerning the public relations situation, the general tempo of the programs under NACA, our company reputation as probably handled by the press when they learned that Goodlin was no longer going to fly the XS-1, and the *probable wrath of General Spaatz* as to the arrangement that had been decided upon [emphasis added].

In short, Stanley stated that Bell Aircraft had a lot of prestige and corporate goodwill riding on this program. Bell had been most cooperative in the past and this was not the only advanced research program then underway. Nor would it be the only program to come down the road. Further, the loss of prestige after the publicity generated on behalf of Bell Aircraft would do no one any good. But it was the last part of the presentation that revealed much about this meeting. Larry Bell and Gen. Carl Spaatz were old friends. Both were dedicated aviation pioneers—men of action, not pencil pushers. Stanley clearly took the measure of the issue when he indicated that Bell Aircraft was not above using that connection if the program seemed to be falling into the hands of other interests. Indeed, given Stanley's strong personality, the quotation from his report appears a masterpiece of understatement. In any event, Bell Aircraft would not allow this issue to drift.

The presentation riveted the military men's attention. Smith was agreeable to the general line of reasoning that Stanley put forward but was obviously troubled by an unstated problem. Ritland readily agreed to Stanley's proposal, or so it seemed to Stanley, and appeared to champion Bell's cause with Smith.[11] Ritland was an old test pilot from Wright Field and had strong reservations about the NACA's ability to execute a flight research program. "There is no doubt that NACA will do a thorough flight test job but they will take forever" is how Ritland summarized his argument. Thus, once again, the issue came down to money. Smith, after listening to Stanley and Ritland, finally conceded that money was not the real AMC problem. (According to Stanley, this is what Smith said, although the author's familiarity with the funding status of the XS-1 project implies that this is not quite what Smith meant.) Rather, Smith indicated, it was his fear that any Bell Aircraft continuance in the flight test program would certainly offend the NACA and have powerful political (that is, Washington and funding) repercussions. Smith stated that the AAF was not in a position (financially or politically) to sponsor a dual flight program unless initiated by the NACA. Unofficially, Smith suggested that Bell Aircraft go directly to Dr. Lewis at NACA HQ to work out the details of such a program. If the NACA agreed to allow Bell to fly a hazardous pioneering program to pave the way for the NACA research, that was the NACA's decision. At that point, with all the cards on the table, the meeting broke up.

In summarizing the meeting for Bell executives, Stanley clearly indicated the position he believed Bell Aircraft should adopt at this time. First, the NACA program agreed to at the morning meeting would not meet the approval of General Spaatz and would certainly harm Bell Aircraft from a public relations point of view. Second, the NACA would string out the program for an exceedingly lengthy period of testing. Third, Bell Aircraft needed to be in Washington as soon as possible to meet with Dr. Lewis and get the NACA approval to provide a continuing test flight program. Unstated about the last proposition but clearly understood by the nature of the flight tests to be conducted was that it would be Bell Aircraft, and not the NACA or some other entity, that would first break the sound barrier. The professional prestige of the record-breaking event would become a corporation victory as well.

Although Stanley believed the power of his remarks had swayed the two military men, in reality Smith and Ritland were already concerned about the NACA's lack of enthusiasm toward the XS-1 project.[12] In recent meetings with the agency Soule had continually expressed doubts about the XS-1's ability to fly supersonic. The AMC was left with the opinion as expressed to Stanley that the NACA would proceed slowly and very cautiously. This was exactly what the AMC did not want from this program. As a result, when Stanley came to see the two AAF officers, he found men willing to explore alternatives to the just agreed upon program. But both men recognized the political dangers in pursuing the alternative options without the concurrence of the NACA. Hence, Smith directed Stanley to reach an agreement with the NACA. However, one other alternative, undiscussed with Stanley, was also available to Colonel Smith, one that involved a close AMC colleague. Although such a choice would represent a significant departure from traditional AAF experimental aircraft research and test procedure, it was one that was to make a surprise appearance within two months.

On March 18, the AMC responded to the Crowley letter of February 19.[13] Specifically, the AMC granted the NACA the right to use the scarce SCR-584 radar at the NACA's discretion on any tests at Muroc or the vicinity as long as it did not interfere with the XS-1 tests. Indeed, the AAF promised increased help if additional tests stretched available NACA manpower. Further, the AMC indicated that since it had already procured 60,000 gallons of fuel (to be delivered at 5,000 gallons per month to end in December 1947) the NACA should notify the AMC if they desired to use this fuel. Additionally, the AMC stressed that the AAF desperately needed the newly acquired flight data on other ongoing aircraft projects. It was imperative that this data collection process be continued and expedited. Finally, the AMC reiterated that, in spite of other remarks made at the March 5 conference, the AMC was ready to turn over office and hangar space at Muroc to the NACA—space currently occupied by Bell. The AMC seemed to imply that the transfer of the program was decided, but not everyone agreed that the removal of Bell Aircraft from the program was a done deal.

On March 21, Larry Bell played his best political card. The inside power game of Washington was one that Bell knew first-hand. At NACA HQ, Bell, Stanley, and General Craigie met with Dr. Lewis to discuss the possibility of Bell Aircraft utilizing one of the XS-1 aircraft for transonic research. Discussion centered on the availability of Army funding and NACA cooperation. The flight plan outlined dealt with attempting to break the sound barrier and conducting a high-altitude (60,000 feet) research program. Larry Bell expressed his company's interest in providing the service to the AAF. General Craigie expressed the AMC's interest in funding *Bell Aircraft* to provide the flights. But he wished to clarify whether NACA would continue to cooperate and provide assistance if Bell Aircraft continued in the program. In hindsight, Lewis was cornered and politically squeezed. In the resulting agreement, Bell Aircraft would fly an experimental research program in one of the XS-1 planes. NACA would assist Bell Aircraft by furnishing instrumentation and engineering help. Details of how the two programs would interface was postponed to a later conference at Wright Field. In the interim, Colonel Smith would contact Bell Aircraft to notify them of the actual agreement.[14] Bell Aircraft had apparently turned the corner and rescued its role in the program.

On April 8, Dick Frost convened a meeting of all key XS-1 personnel at Muroc to discuss the recent letter from Colonel Smith to Bell Aircraft.[15] The details of the letter simply covered the results of the Lewis/Bell/Craigie meeting in Washington on March 21, but added that the new Bell program would now be "a part of the NACA program." Walt Williams was confused as to whether this meant the NACA could finally set the pace of flights to ensure that all NACA instrumentation was working. (After the meeting, he raised this very issue with LMAL to see if they could obtain an AMC clarification.) The Smith letter concluded by requesting the contractor (Bell Aircraft) submit to the AMC a tentative program and cost estimate for the transonic phase of XS-1 research flights and suggested a later meeting to discuss this tentative program. Frost indicated to the assembled Muroc personnel his intention to treat the Smith letter as the signal to proceed with the extended Bell contractor tests. In detail, Frost stated that Bell would continue to operate the B-29 for the contractor tests and for the NACA program. Further, he indicated that Bell Aircraft would "possibly" provide the nitrogen evaporator person and that its crew would be responsible for fueling and loading the XS-1.

Having provided the basic information, Frost now dropped a bombshell. The Bell Aircraft proposal would focus on a high-speed/high-altitude flight program to conquer the sound barrier. In the past, Stanley had verbally indicated in several meetings that the follow-on Bell program would be relatively short, ten to fifteen flights, culminating in a quick attempt to crack the sonic barrier. Now, Frost indicated, the Bell Aircraft proposal to the AMC would focus on a program of some 30 to 60 weeks in duration and include fifty to sixty flights. The thrust of the program would be to carry the XS-1 initially to Mach 1.35 and later, utilizing the efficiency of the turbine pump (XS-1 number 3), all the way to Mach 2.0. The stepped approach would feature an initial flight at Mach 0.80 for 184 seconds. If satisfactory, the XS-1 would undertake a four-degree climb until fuel was exhausted. Subsequent flights would shorten the initial level-flight time and consequently increase the speed and altitude. If all went as planned, the final phase would involve full-power near-vertical flights from the moment of B-29 launch. While some consideration had been given to attempting a transonic program at zero lift by a gradually increasing dive, it had now been rejected in favor of the 45-degree climb procedure. Bell Aircraft believed this would provide the pilot a sufficient margin of safety and maximize results.

Williams was stunned to hear this new schedule. If Frost had announced the intention to pursue the previously discussed short program, that was understandable. However, a flight program of fifty to sixty flights was more than the NACA scheduled work which was to culminate on flight 50. In writing to LMAL, Williams briefly summed up the NACA position: the previously discussed Bell program was reasonable, but this new proposal with a longer "accelerated" flight program was clearly of dubious benefit to the NACA. Seemingly, the NACA didn't understand the purpose of the proposed longer "accelerated" flight plan. In the new program, Bell Aircraft was merely responding to a favorite topic of General Craigie. On numerous occasions over the last several months Craigie had publicly stressed the need for high-altitude research. Bell Aircraft believed the combination of the two efforts, high speed and high altitude, was a natural blend. In reality, their accelerated flight proposal would do more than merely pierce the sound barrier—it was a significant research program using a unique research vehicle. This lengthy program would be expensive for the AMC, but Bell Aircraft believed the results would justify the costs. For Bell, it would also maximize publicity, resolve the number 3 airplane situation, and continue the profitable work on a program that clearly would never become a production vehicle.

By the middle of April, Bell Aircraft had reason to be pleased. Round 2 in its struggle to retain the follow-on XS-1 contract had seemingly gone as planned. Stanley's suggestions to Larry Bell of March 6 had been executed and the public-private pressure had kept Bell Aircraft in the game. The Wright Field notification letter provided the justification for renewed participation by Bell Aircraft in preparations at Muroc. But unknown to Bell Aircraft, the game was not yet over. The new Bell flight program still awaited final approval by the AMC. In that decision, someone else at Wright Field had yet to make his presence felt.

Return of the Original

While all the official maneuvering at the various conferences continued, preparations to move the number 1 aircraft back to Muroc concluded. On April 4, Harold Dow flew to Rome Army Air Field to pick up the B-29. A 1-hour, 20-minute test flight confirmed the bomber was ready for service. Later that day, Dow piloted the B-29 back to Niagara Falls where the XS-1 number 1 was mated to it. The modifications to the number 1 aircraft since its last flight were extensive. In addition to the thinner 8 percent wing/6 percent tail assembly, power-actuated tank blowdown valves were installed, scoops and a curtain to improve motor compartment ventilation and fire prevention were added, and the nitrogen bleed fitting was changed to a quick disconnect type. Fuel and lox jettison valves were replaced for better corrosion control, and a more rigid landing gear control handle, stainless steel second-stage Grove regulators, and a redesigned fuel tank outlet fitting were installed. The other maintenance changes included replacement of the aft fuselage wiring with high-temperature wire, new tires and brakes, and new strain gauge wiring. On Saturday, April 5, Dow piloted the XS-1 number 1 back to Muroc by way of Tinker Field, Oklahoma. Bell Aircraft's chief test pilot, Tex Johnston, flew as copilot. On Monday, April 7, the NACA C-47 arrived carrying the LMAL personnel and spare parts and recording instruments for the tests.[16]

The first three days of the week of April 7 were spent preparing the XS-1 for flight.[17] While Frost conducted his April 8 meeting with the NACA personnel, bad weather and some difficulty with the NACA instruments continued to prevent a test. The last Florida flight of the number 1 aircraft had concluded on March 6, 1946. Now over one year and one month later, a different (due to the aforementioned changes) XS-1 plane would take to the air. The Bell technicians decided that the first flight would be a glide flight, to allow Goodlin a chance to safely examine the new handling characteristics of the plane.

On Thursday, April 10, the Bell team was ready to proceed.[18] The weather finally returned to the perfect conditions typical of Muroc as Goodlin reported clear skies and unlimited visibility. The XS-1 weighed 7,163 pounds on takeoff and its center of gravity was at 28.4 percent mean aerodynamic chord (25.3 percent on the number 2 aircraft). As usual, Dow would handle the pilot duties on the B-29; however, he would have a "new" sidekick. Since Mark Heaney was unavailable for this flight, Dick Frost filled in as copilot on the B-29. Tex Johnston flew chase for the first time in the XS-1 series. Takeoff and climb to altitude were routine. Dow released the XS-1 at a pressure altitude of 25,000 feet.

Because the new handling characteristics on the number 1 aircraft were unknown, a decision was made to place the stabilizer in the neutral position. After B-29 separation, Goodlin quickly discovered the number 1 aircraft exhibited the same tail-heavy tendency as the number 2 aircraft. Slick had to apply considerable forward stick simply to maintain flight speed. The horizontal stabilizer was placed two degrees up (aircraft nose down) to correct the condition. After that diversion, Goodlin proceeded to wring out the XS-1 to determine its feel. At speeds up to an observed 250 MPH, he found it similar to the number 2 aircraft with its thicker wing. However, Goodlin did note the number 1 aircraft seemed to have a larger turning radius. After trimming the aircraft, Goodlin released the control column and noted that it was unnecessary to move any of the controls to maintain level glide flight. Two stalls were executed during the descent to Muroc Field. The first stall was conducted with the plane in a clean configuration. Slight center section buffeting was encountered at 165 MPH, and continued and increased in severity down through 155 MPH. Stall occurred at 151 MPH, accompanied by a slight tendency to roll. The stall with the flaps and landing gear extended occurred at 131 MPH, the same speed at which the number 2 aircraft stalled. Goodlin observed slight buffeting and a tendency for the nose to fall straight forward. All conditions experienced in the stalls were considered minor and correctable.

Descent revealed a slightly steeper approach angle than Goodlin remembered on any of the number 2 flights. While landing and rollout were routine, landing allowed Goodlin to try the number 1 aircraft's spoilers. With just slight use, he noted a positive impact on handling. The effect of the spoilers was somewhat more satisfactory than on the number 2 aircraft with its 10 percent wing. Goodlin rated the number 1 aircraft ready to proceed with powered flight. Once the plane was on the ground, the NACA personnel removed their data records. Unfortunately, the NACA instruments failed to record any useful information on the stalls. As a result, preparations for another flight as soon as possible began.

The following morning, the Bell technicians took the XS-1 to its station. An engine ground test was run and deemed a success, and the plane was loaded in the B-29 for refilling all systems in preparation for flight.[19] As on the previous day, Dow and Frost would pilot the B-29 while Johnston provided chase in the P-80. With fuel in the number 1 aircraft, the plane's weight increased to 12,344 pounds and the center of gravity shifted to 24.4 percent mean aerodynamic chord. The purpose of the flight was to allow Goodlin to determine the plane's flight characteristics under power. The NACA flight plan called for a Mach-meter calibration, a repeat of the previous day's stall tests, and a record of the landing.

With the weather again perfect, Dow guided the B-29 off the high desert airfield and began the climb to drop altitude. Goodlin entered the XS-1 cockpit at 7,000 feet to begin cockpit checkout. At 17,000 feet, Goodlin commenced the dome-loading procedure. By 20,000 feet, the number 1 aircraft was ready and Goodlin noted with satisfaction that all pressure readings were steady. This was a welcome change from practically every flight in the number 2 aircraft. Due to the tail-heaviness experienced in the previous flight, Goodlin set the stabilizer in the two-degree up position prior to release.

Immediately upon separation, Goodlin perceived a new sensation from the aircraft. The XS-1 fell away from the B-29 in a stalled condition. Its fall resembled that of a bomb being released rather than the smooth glide Goodlin had experienced on all previous flights. (Although Goodlin did not list a B-29 drop speed in his official report, it is probable that the bomber released the XS-1 at too slow a B-29 speed, causing a stall. This condition would also later occur in the Air Force flights.) Falling away, the nose fell forward until Goodlin estimated the dive angle was 30 degrees. As the nose dropped, the plane picked up speed significantly, but Goodlin felt no response from the controls. In all his years of flying, he had never experienced such a performance or such sluggish response to the controls. He quickly fired off chamber number 1 and rapidly regained control of the aircraft.

Once in stabilized level flight, Goodlin sequentially ignited all four engine chambers. Only a brief flight was necessary before Goodlin informed ground control the engine seemed to perform much better than the one in the number 2 aircraft. All four chamber pressures were consistent and reached 235 psi--considerably higher than observed previously. Engine power remained strong and smooth as about one-half the fuel was consumed. Speed runs were then executed at Mach 0.65, 0.60, 0.50, 0.45, and 0.40 in accordance with the NACA flight plan. Despite using only one cylinder at a time during the initial tests, Goodlin found it impossible to hold the XS-1 below Mach 0.7 in level flight and, as a result, the varied speed tests were conducted in a climb. After the completion of the airspeed calibration tests, only 20 percent of the fuel remained in the aircraft. Goodlin jettisoned the fuel/lox and prepared for power-off maneuvers. Until this point, the test had gone as smoothly as any conducted in the program. But misfortune had a way of making untimely appearances in the XS-1 program. It now chose to reveal itself in a most unusual way.

As Goodlin continued the jettison procedure, he noticed the breathing oxygen level had dropped quite low. He reached over to reset the regulator control to the auto-mix position. Instantly, a burning sensation filled his lungs. The isopropyl alcohol used to de-ice the windshield was leaking fumes into the cockpit. The sensation became so uncomfortable that Goodlin switched back to 100 percent oxygen from the breathing system. He notified ground control of the problem and decided to land as quickly as possible. At 13,000 feet, he lowered the landing gear and flaps. Nitrogen source pressure was also observed to be low, and as a result the flaps hardly extended at all. This would ordinarily be no problem, given the length of the runway and the miles of smooth dry lakebed beyond its eastern end, but the increase in landing speed meant that touchdown on the east-west runway would be approximately three-quarters of the way down the strip.

During the final approach, Goodlin exhausted the last of the breathing oxygen. Back to breathing the cockpit air, Goodlin discovered the cabin was now literally full of the isopropyl alcohol fumes. The burning sensation returned with a vengeance. Goodlin almost gasped for air as the cockpit seemed to close in around him. Touchdown at 155 MPH was the last Goodlin remembered until he heard a faraway voice in the cockpit. Tex Johnston, flying chase alongside the number 1 aircraft, recalled Goodlin from the land of dreams with a question that was enough to snap anyone back to consciousness. Johnston asked if the nosewheel had collapsed! Goodlin finally realized he was rolling on the main wheels but skidding on the nosewheel door. After sliding 1,500 feet, the XS-1 finally came to a stop.

Ground observers later filled in the missing pieces for Goodlin. The XS-1 landed approximately 1,000 feet from the east end on the concrete runway. At touchdown, the plane began rollout in a normal fashion. However, it quickly reached the east end of the runway while still at a substantial rate of speed. Frost reported the landing as routine until this point. However, at the east end, the runway dipped onto a slightly downsloped macadam apron (with a "washboard" surface) to the lakebed floor. In his groggy condition, Goodlin never even felt the plane's arrival at the dropoff. In going from the concrete to the lake surface, the XS-1 jumped into the air and came down hard on its fragile landing gear, a leap some observers estimated at 100 feet. Other observers detected no leap at all, but saw the landing strut subjected to strenuous bouncing as the plane connected with the washboard-like apron. In any event, the harsh impact sheared the nosewheel locking structure from the plane and, as at Pinecastle, the nosewheel collapsed. The accelerometer in the cockpit registered loads exceeding those in normal landing conditions. (Specifically, the instrument registered 8.4 positive g's and 5.2 negative g's.) Although Goodlin walked away from the landing, he suffered a severe headache the remainder of the day.

Since the NACA recording switch had been turned off prior to the landing, no recorded data was available to analyze the mishap. In his official report, Walt Williams only mentioned the "jump" as the cause of the wheel problem. He did not mention the noxious leak. For a meticulous recorder of events such as Williams to neglect to mention this fact was something quite unusual. In fact, Williams included a puzzling sentence in his official report: "This explanation is the one furnished by Bell." It appears that either Bell Aircraft decided not to fully discuss the matter with the NACA, or Walt Williams had a very rare momentary lapse in his reporting.

In his pilot report, Goodlin noted that breathing oxygen was insufficient for the longer flights now being contemplated, a problem demanding immediate action. Likewise, the nitrogen supply was marginal at best. Goodlin recommended correcting these condition if his next flight confirmed the need. (Further flights would confirm these suppositions.) The damage to the XS-1 was confined to the nosewheel door, the wheel, and the gear-retracting mechanism in the ventral bulge. It was anticipated that, due to parts needed, it would require 10 days to repair the problem. The NACA used the delay to complete the hookup and calibration of their test instrumentation in the number 1 plane. Bell technicians proceeded with implementing the technical orders for modifications and repairs to the B-29 and the P-51 chase plane. On Friday, April 18, Dow and Frost took the B-29 aloft for a 35-minute check flight. With the successful completion of the flight, the Bell team believed the program was ready to expedite the wrap-up of the contractor portion of the project.

On Tuesday, April 29, the Bell engineers were ready to try another flight.[20] As part of the test preparations, the Bell technicians added 85 pounds of ballast in the XS-1 nosecone to change the plane's empty-weight center of gravity from 28 percent to 25 percent mean aerodynamic chord. (Total weight was now 12,255 pounds.) The lead ballast, which was shaped at the base machine shop, would remedy the tail-heaviness exhibited during the plane's first two flights. The only other changes included the nosewheel repair and installation of the NACA equipment. The plan for the flight was for the XS-1 to be dropped at 20,000 feet at 260 MPH IAS. Goodlin was to execute a climb to 35,000 feet with speed increasing to Mach 0.8. This was to be followed by sideslips at Mach 0.75 and then, after fuel exhaustion, pull-ups at Mach 0.76, 0.73, 0.70, and 0.65. Finally, Goodlin was to perform an accelerated turn until stall was obtained at Mach 0.70.

With all in readiness, the B-29 lumbered down the runway with Dow and Heaney at the controls. After a routine climb to 7,000 feet, Goodlin entered the XS-1 and began the cockpit checkout. At a pressure altitude of 19,000 feet, he initiated the dome-loading procedure. During the climb, Goodlin again detected the existence of irritating fumes emanating from the pressure sealant. As a result, he set the oxygen regulator at 100 percent. Once checkout and loading were complete, Goodlin notified Dow that he was ready. However, a momentary failure of the interphone system led Dow to delay the drop. As a result of the delay, nitrogen source pressure at drop was below normal levels. Separation occurred at 20,000 feet and 260 MPH IAS.

Goodlin immediately noted the improved flight characteristics of the XS-1. Although the higher B-29 speed may have helped, Goodlin attributed the better XS-1 performance to the effects of the ballast on the center of gravity. Approximately five seconds after separation, Goodlin ignited chamber number 1 and put the XS-1 into a climb. During the ascent, Goodlin ignited chamber 2 and shut down number 1. Both performed normally. During the climb, Goodlin detected an increased sensation of high-frequency nibbling through the rudder pedals at an indicated Mach 0.73. Although it never reached dangerous proportions, it seemed to increase in intensity with the rise in speed. Highest speed attained during the climb was an indicated Mach 0.77. Handling throughout the range seemed normal except for the rudder nibbling and a slight tendency toward tail-heaviness, which caused Goodlin to change the stabilizer setting 1.25 degrees up in the vicinity of an indicated Mach 0.65. Throughout the remainder of the flight, as speed increased so did tail-heaviness. As speed decreased, the tendency seemed to diminish until at an indicated Mach 0.53 it seemed to disappear. At that speed, Goodlin could remove his hands from the control column with little effect.

At 35,000 feet, he leveled out and began the sideslip tests. Sideslips to the left and right were executed at an indicated Mach 0.75. Fuel was quickly exhausted and Goodlin began the glide back to Muroc. During the descent, he executed a pull-up at an indicated Mach 0.73, resulting in a positive acceleration of 4.2 g's. Goodlin immediately noted the difference between the number 2 aircraft and the pull-up in the number 1 airplane. Whereas the number 2 aircraft in a pull-up tended to buffet and produce a slight rolling motion, the number 1 aircraft tended to snap violently. In appearance, the plane seemed to perform an outside snap roll. Once again the number 1 airplane exhibited the previous tail-heaviness immediately before the pull-up. Goodlin could correct this problem only with heavy forward pressure on the control column.

A second pull-up was performed around 23,000 feet at a speed of indicated Mach 0.65. The result was a 4.8g acceleration. The aircraft exhibited less severity in performing the test than on the previous pull-up. However, upon recovery of the aircraft, Goodlin felt the same tail-heaviness as before. He suspected that the airplane had an elevator problem similar to that which occurred on the number 2 airplane during its pull-up tests. As a result, he discontinued the pull-ups and continued the descent to the airfield. At 12,000 feet and 220 MPH, he lowered the flaps and landing gear. In spite of a stiff crosswind, touchdown on the lakebed was routine and rollout was uneventful.

Postflight inspection revealed no visible damage to the elevators. As a result, the technicians believed the increased tail-heaviness might be associated with compressibility effects over the tail surface; the condition would be watched on the next several flights. After Goodlin used oxygen for 19 minutes, the system pressure was down to 100 psi. Because longer flights were anticipated in the future, Goodlin again recommended an increase in the oxygen supply. The only surprise in the inspection was the split fin cap for the antenna, probably as a result of the snap roll during the pullout. Bell technicians immediately set out to replace or reinforce it. Aside from the relative absence of problems, the other surprise was that the NACA data seemed to be complete.

At Muroc, preparations began immediately for another test flight. At Bell's Niagara Falls plant, the pace of operations also quickened. By the end of the month, the modifications to the number 2 aircraft were nearly complete. Bell engineers were determined to ground-test the XS-1 engine before allowing the XS-1 number 2 aircraft to return to Muroc, estimated to take two to three days. As a result, Bell technicians believed the XS-1 number 2 could return to California around May 9. The newly increased pace would improve that schedule.

On Wednesday, April 30, the XS-1 was ready for another test flight.[21] It would be the fourth flight for the number 1 aircraft with the new wing/tail combination, this time to investigate static longitudinal stability and to again examine the buffet boundary. Dick Frost would fly chase in the P-51 and an AAF officer was recruited to fly second chase in an FP-80 photo plane. The XS-1 would be dropped from the B-29 at 20,000 feet at an IAS of 260 MPH. A climb would be made to 35,000 feet where Goodlin would investigate hands-off stability at various Mach numbers. After fuel exhaustion, Goodlin was scheduled to execute an abrupt pull-up at 32,000 feet and an accelerated stall at 26,000 feet. During the descent, Goodlin was to perform three additional pull-ups at varying Mach numbers. In total, it was an ambitious flight plan, given recent XS-1 experience.

Shortly after takeoff, at a pressure altitude of 7,000 feet, Goodlin entered the XS-1 and worked through cockpit checkout. At 20,000 feet he initiated the dome-loading procedure. Once again, he noted the irritating cockpit fumes, although they were much less obvious. He again decided to use oxygen to avoid any potential problems. The B-29 dropped the XS-1 at 20,000 feet and an indicated speed of 260 MPH. As on the previous flight, the XS-1 separation was markedly improved over past attempts. Since the pause between dome-loading and separation was kept quite short, the nitrogen source pressure remained very high. The repetitive number of flights was beginning to pay off in the smoothness of operations.

About five seconds after separation, Goodlin ignited the number 1 chamber. After engine stabilization, he put the XS-1 into a climb and increased speed from an indicated Mach 0.55 to 0.75 in 0.05 increments. The aircraft was trimmed to fly in the hands-off position to check stability. However, this required Goodlin to significantly adjust the stabilizer trim setting. At indicated Mach 0.55, he set the stabilizer at 1 degree up. By the time the plane had increased speed to indicated Mach 0.65, it was necessary to change the setting to 2½ degrees up. From Mach 0.65 to 0.75, no further changes were noted, but Goodlin recognized that this might be fuel use shifting the center of gravity of the airplane or it might simply represent friction loss in the control system. On the climb, Goodlin detected the rudder nibbling that he noted on the previous flight. After two experiences with the condition, Goodlin believed, the high-frequency, low-amplitude nibbling was caused by compressibility effects from the tail surfaces being transmitted through the control system. The condition again appeared at an indicated Mach 0.73 and continued up to indicated Mach 0.77.

At 35,000 feet, the XS-1 exhausted its remaining fuel and Goodlin proceeded with the glide tests scheduled for the flight. About one minute into the glide flight, at an altitude of 34,000 feet, Goodlin heard a loud noise that seemed to come from the aft section of the aircraft. He compared the sound to the muffled blast of a 20mm cannon shot. Weather conditions were ideal and Goodlin detected no disturbance in the controls of the aircraft. Since the FP-80 had lost radio contact, it was impossible to discern whether an explosion had occurred or if the plane had suffered a structure failure. Goodlin decided to get to the ground without conducting the pull-up tests. Frost in the P-51 hastened to catch up to the XS-1 and quickly looked over the aircraft for any signs of problems. At 15,000 feet, he reached the XS-1 and performed a visual inspection. No damage or fire was visible, but Goodlin proceeded with the landing procedure. Landing occurred to the east on the north-east runway at 150 MPH. Postflight inspection revealed no damage to the airplane. Everyone was baffled by the sharp noise and NACA data revealed no cause.

On May 1, Bell technicians again attempted to test fly the XS-1.[22] Frost flew chase in the FP-80 (#45-89406). Since it was his initial flight in the new aircraft, he took off early for familiarization before the drop. During the B-29 climb to altitude Goodlin entered the XS-1 cockpit and began the checkout procedure. However, the string of good luck accompanying the speed-up in flight operations finally came to an abrupt end when Goodlin attempted to load the pressure domes. A premature opening of the pressure regulator relief valve made it impossible for the aircraft to obtain sufficient pressurization. The flight was scrubbed and the XS-1 jettisoned fuel/lox. After 55 minutes in the air, the mated aircraft were back at Muroc. Repairs were initiated and the bomber made ready for another try to drop. On Friday, May 2, the B-29 cleared the Muroc airstrip for another attempt at the fifth flight of the number 1 aircraft series. It was again a very short mission. Once more, the pressure regulator relief valve opened prematurely and ruined any possible opportunity to pressurize the system. The mission was aborted and both aircraft returned to Muroc after Goodlin jettisoned the XS-1's fuel/lox.

By Monday, May 5, the Bell technicians were ready to try again.[23] It appeared that repairs had corrected the relief valve opening prematurely. Joel Baker would make his final appearance in the XS-1 program, joining Frost in the FP-80, to handle chase duties in the Bell P-51. Climb to altitude and dome-loading went smoothly. Separation from the B-29 was routine and rocket engine ignition was satisfactory. Goodlin used the flight to perform abrupt rudder kicks with various amounts of fuel. Goodlin then executed several turns in the XS-1 and performed stalls, first with the plane clean and then with the flaps and landing gear lowered. Landing and rollout were routine. Examination of their records revealed that all NACA instruments, except the altimeter and airspeed indicator, were working during the flight. The next flight was scheduled for Wednesday, but another flight interrupted that plan.

On May 6, Frost received a call from the Bell factory at Niagara Falls informing him that the XS-1 number 2 was ready to return to Muroc. The B-29 was dispatched to Buffalo on May 7 to pick up the airplane. Dow flew pilot on the return flight and Goodlin sat as copilot. Their two-day return flight arrived at Bell early on the morning of the eighth. The number 2 aircraft was loaded on the B-29 and the bomber started the return journey to Muroc. Frost notified AMC that four additional flights were anticipated and the Bell program could be concluded by May 23.[24] But while the test flights were moving forward, the behind-the-scenes maneuvering to define and control the XS-1 program continued unabated. At the same time, visible evidence of the changing of the program came from an unexpected source.

Shortly after the conclusion of the May 5 flight, Joel Baker participated in a three-way phone call from LMAL. While Mel Gough and Herb Hoover were on the line, the three men discussed the future role for an NACA pilot. The Bell decision weighed heavily in the discussion although Baker was not familiar with the details. A disagreement ensued over Baker's role in the program. As a result, one week later Baker resigned to take a job with Chance-Vought. For the NACA the setback was to have profound consequences in the course of the program. At the March 5 conference, Baker had been one of the three likely individuals mentioned as potential pilots. Stanley certainly believed Baker would be the NACA XS-1 pilot as he stated in his March 6 memo to Bell. Further, Baker had traveled to Bell on March 10–11 for rocket engine instruction. (However, on March 17–19, the NACA clouded that view by sending Gray and Reeder to Buffalo to receive indoctrination in the rocket motor operations.) Now, in mid-May, Baker was gone and, worse, the other two pilots would soon no longer be in the picture, due to family considerations. The delay would cause the first NACA XS-1 flight to be without an agency pilot.[25]

The Dance Continues

With the Bell proposal to retain control over XS-1 number 1 moving forward, NACA prepared to take possession of aircraft number 2. In late April, LMAL had notified the AMC that they desired to operate the XS-1 number 2 only *after* the previously discussed changes had been made to the aircraft. To date, no decision had been made on the AMC's ability to pay for any changes. Indeed, the progression of events quickly cast serious doubts on the financial stability of the entire XS-1 program, given the expanded scope of the Bell program as outlined to AMC.

Throughout April the give-and-take over the new Bell proposal provided the AMC with plenty to occupy its attention. However, given the scope, size, and especially the cost of the Bell contract proposal, not everyone at Wright Field was sure the AMC could—*or should*—afford it. Apparently, the original flight/payout graph that Bob Stanley had discussed with Goodlin in their September 1946 meeting over contract terms had made its way to Wright Field. When Colonel Smith of the Aircraft Projects Section discovered the graduated payment system on the graph (increasing sums per Mach number flown), he became quite angry. In a conversation with Charles Hall, Smith stated that AMC would never fund any multiyear pilot bonus similar to that outlined on the graph. Nor was that situation the only reason AMC was upset with Bell Aircraft. Bell was still attempting to use political pressure to get the AAF to buy the XS-1 as a production fighter.

Bell Aircraft had very little ongoing contract business in 1947 and survived on cost-plus-fixed-fee contracts. For the XS-1 research program, the AMC wanted a more active partner: one that risked some of its own assets. However, Bell Aircraft wasn't inclined to risk its money in that manner. In an effort to pursue all avenues to concluding the XS-1 contract negotiations, Larry Bell reopened the old issue of whether the XS-1 could become the prototype for a production contract. During the spring, Bell made several trips to Washington to discuss this matter with key political and military men while Bob Woods approached Wright Field about the issue. Colonel Smith was opposed to purchasing the airplane as a production model, but he did have a compromise in mind. The AAF would not buy the XS-1 as a fighter, but it would award a new contract for four additional improved-type XS-1 airplanes to Bell Aircraft.[26]

Although it seemed the production issue was resolved, the result of these continuing controversies led Smith to ask Col. Albert Boyd, the head of the AAF Flight Test Division at Wright Field, to review the XS-1 project and provide an alternative to the Bell Aircraft proposal. By mid-April the Flight Test Division had sketched out a preliminary program.[27] To a large degree, it closely resembled the accelerated Bell Aircraft proposal. Consistent with previous agreements, the tentative new program acknowledged the NACA airplane would have priority over the AMC research effort. Due to the lack of telemetering equipment and the desire to reach Mach 1.0 as quickly as possible, the AAF decided to promote a very short test program. Instrumentation would be kept to a minimum. The test program might include a series of powered ground takeoffs for the XS-1. Also, if fuel capacity existed to reach 60,000 feet, the Flight Test Division proposed attempting to set a new world's altitude record. Close coordination with the NACA tests was stressed. The AAF believed their version of the XS-1 flight test program would last approximately thirty flights. Like the Bell Aircraft proposal, the Flight Test Division program was a mix-and-match proposal, with something for everyone. The advantage it had was the lower cost. But it was not yet clear that cost would be the overriding criteria.

In detail, the Flight Test Division speculated that five glide and power familiarization flights up to Mach 0.80 would be conducted by its pilot. Next, the program would explore the critical region of speed up to Mach 0.87. Finally, the XS-1 would undertake a series of high-speed/high-altitude flights (perhaps as high as 100,000 feet) that would include an effort to achieve speeds up to Mach 1.10. These tests would be conducted with the airplane in a steep climb configuration. As with the Bell proposal, the AAF believed this method would afford the safest and most practical manner to reach and exceed the transonic region. The AAF high-speed tests would feature a stepped approach to faster speeds. The high-altitude flights would include tests at 70,000 feet and a progression of speed runs at 60,000 feet, 50,000 feet, 40,000 feet, and 30,000 feet. The speed and altitude research would, of course, depend on the results of previous flight research, especially those tests beyond Mach 0.87.

While the Flight Test Division hammered out the design of its proposed program, the debate over the Bell Aircraft proposal continued to occupy center stage. The proposal that Bell Aircraft sent to Wright Field called for the high-speed/high-altitude research program to be conducted on the basis of a cost-plus-fixed-fee contract with no guarantee of results. The AMC countered with a proposal for Bell to conduct the supersonic tests on the basis of a fixed price contract with a stipulated guarantee of what the contractor would do in the tests. The Bell contract price seemed quite high to the AMC. Concurrently, Colonel Boyd had already assured Colonel Smith that the Flight Test Division at Wright Field could conduct the research program flights. Boyd indicated three advantages to letting his pilots fly the program. They were skilled pilots familiar with testing production aircraft. They were already available, and the price to the government—free since they were already on the payroll—was right. Finally, the necessary security clearances and support logistics would be much easier to arrange with the military taking responsibility for its own program.

Throughout late April the XS-1 standoff continued while the AMC searched for money for the follow-on supersonic contract. In the interim, program cost-cutting became a necessity. On April 29, the AMC notified Bell Aircraft by wire to suspend all work on the XS-1 number 3, including Bell's subcontractor work performed by GE. No reply to the Bell Aircraft proposal for the high-speed/high-altitude XS-1 number 1 flight test program was forthcoming from Wright Field; in fact, Colonel Smith put the project in abeyance for lack of funds until Washington could decide among several flight programs which project would surrender money to pay for the new Bell contractor work. Some speculated that the delay might last a month or more. But events defeated that timetable. On May 1, Colonel Smith formally notified the commanding general of the AAF that *Bell Aircraft* had decided not to accept the Wright Field contract proposal for a fixed-fee XS-1 research program. The fact that Colonel Smith notified Washington revealed his prescience in anticipating Bell's response once the finality of Wright Field's acceptance took effect. Smith aimed to head off what would obviously be a behind-the-scenes effort in Washington by Bell Aircraft to restructure the contract. Smith informed General Spaatz that Wright Field intended to pursue the Flight Test Division taking over the program. The exact interplay of reasons for the Wright Field decision continues to be shrouded in the mists of rumor and contro-versy to this day but, as will be seen, the initial "final" AMC decision was not yet final.[28]

On May 6, the AMC forwarded a letter to Dr. Lewis at NACA HQ formally notifying the NACA that Bell Aircraft would not be given a contract for the proposed transonic flight test program they had suggested to Wright Field.[29] Instead, the AMC stated its inten-tion to allow Colonel Boyd's Flight Test Division to provide the pilot for the test program. The AMC stated there was no question the Bell Aircraft proposal for accelerated flights through the transonic zone was of "great value" and to achieve this in the shortest time in conjunction with the NACA research flights would be of "mutual benefit." However, Colonel Boyd had already "indicated his desire to furnish pilots and limited maintenance personnel for the program." This decision, AMC informed Dr. Lewis, "was to a great extent ne-cessitated by the lack of funds within the Command." Since there would be little chance of additional funds, AMC requested NACA concurrence in the new Flight Test Division proposal. Smith recom-mended a meeting in the near future to work out the details of the AAF/NACA program. With this letter, the door seemed to close on the Bell contractor program.

On May 8, the AMC wrote to Hartley Soule in response to his recent phone call requesting clarification of the status of the two XS-1 air-planes.[30] Colonel Smith informed LMAL that the AMC would formally accept the XS-1 airplanes as soon as Bell Aircraft completed the flight test demonstrations as specified in the March 18 AMC letter to the NACA. The XS-1 number 2 would be turned over to the NACA as requested. The XS-1 number 1 would also be made available to the NACA for tests, but only until such time as the AMC Flight Test Division could begin the AMC accelerated transonic flight program as outlined in the May 6 letter to Dr. Lewis. Smith further informed Soule that the AMC would not have sufficient funds to make the re-quested NACA modifications. The AMC concurred in the Soule sug-gestion that NACA accept the planes for the test work below Mach 0.80 and then revisit the issue of design modifications after the decision was reached to fly faster than the Mach 0.80 acceptance speed. The AMC promised at that time to review each modification request on an individual basis. All other provisions of the March 18 agreement were still in force and Smith recommended a conference at Wright Field as soon as practical to work out the final details.

The Flights Continue

On May 8, the B-29 with the XS-1 number 2 attached began the long journey from Niagara Falls back to Muroc. Dow flew as pilot with Dick Frost as copilot, with Tex Johnston and Slick Goodlin along. Johnston was going to Muroc to be checked out in the XS-1 and to perform a test flight. Sometime late that day, Frost turned over his flight duties to Johnston, and Frost and Slick decided to take a nap. When Frost awoke, Dow and Johnston were trying to dodge an area of thunderstorms in the growing darkness. It was clear that the plane was lost as the static caused by lightning had knocked out the low-frequency radio reception. After weaving endlessly to dodge the storms, Bill Miller, the B-29's flight engineer, advised the pilots that the bomber was low on fuel. Frost and Goodlin found broadcast station manuals in the bomber and finally triangulated a fix on Albuquerque, where they landed some 9 hours and 15 minutes after takeoff. They discovered that the bomber was actually *not* low on fuel, causing some hard feelings over the temporary scare. After overnight rest and refueling, Dow and Johnston guided the coupled aircraft on a routine concluding flight into Muroc.[31]

On May 13, Frost went up in the P-51 to perform a weather check, and the 20-minute flight confirmed there could be no XS-1 test that day.[32] On Thursday, May 15, the XS-1 number 1 was ready for flight number 32 in the series stretching back to Pinecastle. The weather was clear and visibility was unlimited. The purpose of the flight was to again analyze the buffet boundary; the only change to the aircraft since the previous flight was the remedy for the flutter damper problem. Bell technicians listed the gross weight of the XS-1 as 12,255 pounds on takeoff and the center of gravity as 25 percent mean aerodynamic chord. Dow and Heaney would handle the B-29 duties while Frost performed the chase function in the P-51. The B-29 liftoff and climb to altitude were routine. Goodlin entered the XS-1 cockpit at approximately 7,000 feet and proceeded with predrop checkout. At 21,000 feet, he initiated the dome-loading procedure. At a pressure altitude of 21,000 feet and an IAS of 260 MPH, Heaney activated the release mechanism to free the XS-1 from the bomber.

Approximately five seconds after drop, Goodlin ignited chamber number 1. Although fuel and lox line pressure both exhibited a sharp drop, Goodlin noted that motor operation seemed normal. However, a problem appeared almost immediately, which virtually ended the flight before it began. Shortly after application of engine power, Goodlin noted a slight rolling tendency in the airplane. To correct the problem, Goodlin applied opposite aileron. Unfortunately, the ailerons exhibited considerable resistance to control wheel rotation, an unusual reaction. The condition existed whether Goodlin moved the control wheel right or left. After exerting additional pressure, Goodlin managed to free the ailerons enough to provide sufficient deflection for control. However, tugging was required for each correction. Goodlin spent the next anxious moments fighting to keep the airplane straight and level. While the XS-1 retained normal rudder and elevator motion, the scheduled tests for this flight required the use of unrestricted ailerons. Not being sure of the real problem, Goodlin judged the risk too great to attempt the tests outlined in the flight plan. As a result, Goodlin put the XS-1 in a climb to 32,000 feet. At Mach 0.70, Goodlin shut down engine power and jettisoned fuel. But, as usual, his problems didn't end here.

During the ensuing descent, Goodlin noted nitrogen source pressure dropping rapidly. He suspected a leak in the system. After actuating the tank blowdown switch, the nitrogen pressure stabilized. To test for a leak, Goodlin returned the blowdown valve switch to the "pressure" position. The nitrogen source pressure once again began to drop. Returning the blowdown switch to the jettison position again stabilized the source pressure. Giving up, Goodlin lowered the XS-1 flaps and landing gear and lined up the plane for a landing to the east. Touchdown occurred at 155 MPH and rollout was uneventful.

Postflight investigation revealed the ailerons to be functioning normally. Thus, Goodlin suspected the problem might have been due to moisture in the flutter dampers, which could have frozen at the cold temperatures of the higher altitudes. If so, ice crystals could have restricted the orifices in the dampers. A second possibility was that the orifices were set too small and the increased viscosity of the silicone oil at the decreased temperature tended to restrict aileron motion. Bell technicians also tested the tank blowdown valves and found they moved from the blowdown position to the pressure position of their own accord. This clearly indicated a gas leak through the solenoid valve controlling the operation of the valves or a leak from the tank pressure system itself into the valves.

On May 16, the Bell technicians completed work on overhauling the right aileron damper. The technicians enlarged the orifice for better oil lubricant flow and made preparations for a check flight that afternoon. With Dow at the B-29 controls, the XS-1 was taken aloft for a 1-hour and 20-minute aileron investigation flight. Frost occupied the XS-1 for the test. The NACA instruments were utilized to check the aileron force readings. The results confirmed the supposition that moisture was freezing in the orifice since the aileron friction became worse with altitude but freed up after the control wheel was rocked vigorously. No drop was attempted and the B-29/XS-1 returned safely to Muroc.[33]

On Monday, May 19, the Bell engineers were ready to conduct another test flight.[34] The weather at Muroc was clear with unlimited skies. The only changes to the XS-1 were that the right aileron flutter damper was repaired and the tank-type blowdown valve reworked. The blowdown control switch was also changed. The flight plan called for Goodlin to climb to 35,000 feet while speed increased to Mach 0.80. After completion of the climb, Goodlin would execute a constant speed turn until buffeting was experienced and then do accelerated stalls at different speeds after propellants were exhausted. Before concluding the flight, Goodlin was expected to perform several pull-ups at speeds between Mach 0.45 and 0.70 (in 0.05 intervals).

With all in readiness, the B-29 lifted off from Muroc for the seventeenth flight of the number 1 aircraft (and the thirty-third flight of the XS-1 series). Frost followed the B-29 as chase pilot in the FP-80 (#45-89406). At 7,000 feet, Goodlin left the B-29 cockpit and proceeded to the XS-1. He began the dome-loading operation as the B-29 approached 21,000 feet. Just as the loading procedure was completed, Dow notified Goodlin that the B-29 was experiencing trouble with its number 1 engine. As a result, a decision was made to drop the XS-1 early. Although the altitude was different, Dow brought the bomber up to its drop speed of 260 MPH. Separation presented no problems and Goodlin allowed the XS-1 to glide for approximately five seconds before igniting the first chamber.

Keeping in mind the past difficulties with the engine, Goodlin closely observed the rocket motor's starting characteristics. It seemed to him that the number 1 chamber labored after ignition and full power only came up after several seconds of operation. Lox line pressure appeared to drop even more than on its previous starting attempts, although Goodlin noted no such results in the fuel pressure system. Frost, flying alongside the XS-1, also reported the sluggish start and observed a gush of liquid for perhaps one-half second, then a pause, and then another gush of liquid before normal engine start was confirmed. Frost speculated that either the lox or fuel propellant valve was opening too slowly, thus allowing too rich or too lean a mixture until the opening was complete.

After the number 1 chamber stabilized, Goodlin put the XS-1 into a climb. During the ascent, the number 1 chamber was shut down and the number 2 chamber ignited. At 35,000 feet, Goodlin put the XS-1 into level flight and proceeded to make a speed run to an observed Mach 0.78. At that point, the airplane experienced slight buffeting, but a far more worrisome situation developed. Goodlin observed the number 2 chamber pressure begin to fall quite rapidly although fuel and lox pressures remained normal. After reaching a reading of 160 psi, Goodlin decided to shut down the engine to prevent possible damage. All remaining fuels were jettisoned.

Upon completion of the jettisoning procedure, Goodlin proceeded with the scheduled maneuvers. As the XS-1 descended to 25,000 feet, Goodlin executed accelerated stalls at speeds of Mach 0.70 and 0.75. At 23,000 feet, he performed an abrupt pull-up at an observed speed of Mach 0.60. Continuing the descent, he performed additional pull-ups at observed Mach 0.65 and 0.70 between 20,000 feet and 15,000 feet. The maximum g load during the pull-ups was an observed 7.7 g's obtained at Mach 0.65. Descending through 13,000 feet, Goodlin lowered the flaps and landing gear, and glided in to execute a routine runway touchdown at Muroc.

Postflight inspection revealed the observed dropoff of pressure in the number 2 chamber at 35,000 feet was due to a malfunctioning cockpit gauge. Goodlin had no theory on the drop in lox line pressure experienced at the time of separation. Bell technicians undertook analysis of the problem to determine why it had appeared on the last three flights. It was especially disturbing since engine operation did not seem to be affected by the change in pressure. Goodlin reported the aileron damper problem appeared cured by the preflight fix. The Bell technicians also disassembled the flutter damper and repaired a piston seizure in the cylinder. It was anticipated that Bell Aircraft would attempt two more flights in the number 1 aircraft and then two flights in the number 2 aircraft. In the former flights, Mach 0.8 and an 8g pull-up would again be the targets. The number 2 flights would be shared by Goodlin and Tex Johnston. Johnston's flight would be the first by anyone other than Goodlin since the death of Jack Woolams.

On Wednesday, May 21, with perfect weather conditions, the B-29 took the XS-1 number 1 aloft.[35] Since this would be a glide flight, no fuel was loaded into the aircraft, and as a result, XS-1 weight at takeoff was only 7,074 pounds. The plan was to release the XS-1 at 30,000 feet to allow time for various maneuvers. Constant-speed turns to accelerated stalls, and abrupt pull-ups were on the flight worksheet. At 7,000 feet, Goodlin entered the XS-1 and began the cockpit checkout. The B-29 released the XS-1 slightly below 30,000 feet at an airspeed of 260 MPH.

Goodlin immediately put the XS-1 into a dive to pick up speed. Only two accelerated stalls in a constant speed turn were executed before the XS-1 was below 18,000 feet. Because of the rapid loss of altitude, Goodlin decided to discontinue these maneuvers and perform the required pull-ups. The first pull-up was executed at an observed speed of Mach 0.71 around 12,000 feet. This resulted in an observed positive acceleration of 8 g's. The second pull-up was made around 10,000 feet at an indicated Mach number of 0.68. This effort resulted in an observed positive acceleration of 8.2 g's. A third pull-up was canceled since Goodlin felt it unwise to attempt further maneuvers below 10,000 feet.

Descending to the landing strip Goodlin now found his gremlin for this flight. Upon completion of the second pull-up, the inside of the plexiglass canopy began to fog over. (Postflight analysis attributed this condition to a defective exhalation dehydrating canister.) As he lined up the landing, the frosting on the glass was almost complete, which substantially blurred his vision. As a consequence, Goodlin took no chances and touched down on the runway approximately halfway down the strip. Rollout was normal. Once the plane was back at the hangar, the Bell team decided that the number 1 plane had finished the AMC contract requirements. Preparations began immediately to ready the number 2 airplane for its first test flight since its return to Muroc. Tex Johnston would handle the pilot duties on this flight, the seventeenth flight of the number 2 aircraft and the thirty-fifth in the XS-1 series.

On the outside Johnston's flight appeared to be a routine checkout flight. However, the continuing difficulties with the AMC over the follow-on contract had provided additional reasons to have Bell's chief test pilot see for himself the characteristics of the plane. On Thursday, May 22, the usual spring weather presented a problem for the Bell team.[36] Strong winds were whipping across the Muroc area from the west-southwest, with gusts at times exceeding 35 MPH. In spite of these conditions, the Bell engineers decided to proceed with the drop. The flight was to be a combined familiarization and check flight on the number 2 airplane. The flight team for Johnston would be the same one as on Goodlin's flights with Frost providing chase in the FP-80. Takeoff and ascent were normal with Johnston entering the XS-1 cockpit at a pressure altitude of 8,000 feet. By the time he secured the XS-1 door, the B-29 was passing through 10,000 feet. During the climb, Johnston noted the restricted visibility from the XS-1 cockpit while attached to the B-29. At 20,000 feet, Johnston loaded the first- and second-stage regulators. At 21,000 feet, the B-29 started its drop run. Diving toward 20,00 feet, Dow pulled the B-29 up slightly and Heaney released the XS-1 at an observed speed of 270 MPH. The drop was satisfactory with the XS-1 falling away quite rapidly from the B-29. Johnston observed the separation was comparable to a −1.5g pushover in level flight. After a brief pause, Johnston proceeded with the attempt to fire the engine.

Flipping the number 4 chamber switch, Johnston learned just how temperamental the XS-1 could be. The igniter on the number 4 chamber failed to fire. Johnston turned off the switch, and tried again. This time the cylinder ignited and operated in a normal manner. Johnston flew the XS-1 through a series of turns, climbs, and shallow dives at Mach numbers between 0.55 to 0.72. He discovered the ratio of forces between elevator, aileron, and rudder was excellent. Control forces were light and the force increase with speed was negligible. However, Johnston noticed that when displaced and released the ailerons did not center, a condition he attributed to friction in the system, which he had already noticed in the ground tests he had performed. Due to this situation, Johnston decided to forego a qualitative lateral stability investigation since the XS-1 obviously flew straight and level without excessive effort. Johnston believed the aileron forces were light and very effective, producing a rate of roll similar to the P-80's, absent the "hunting" qualities of that plane's aileron boost control system. Johnston found directional stability to be very positive, although he noted that fuel sloshing seemed to occur in the aircraft after some of the fuel had been expended which resulted in some Dutch roll.

When Johnston tested the longitudinal stabilizer control, he discovered what he considered to be a major drawback of the airplane. He believed that rapid, precise longitudinal trim changes were "impossible" with the current control system. He pinpointed lost motion in the control linkage as the villain, exactly the same complaint expressed by Baker in his February 25 XS-1 acceptance memo to Mel Gough. (In fact, it was the same complaint that Frost and the AMC had discussed on numerous occasions.) However, Johnston considered the rate of stabilizer actuation to be satisfactory because a more rapid rate of longitudinal trim changes applied during high-velocity flight might result in more pilot/airplane damage than the theoretical demons of the transonic zone.

As Johnston attempted to check out the horizontal stabilizer, he discovered the excessive slack in the control. He recommended that a spring-loaded toggle switch operating the control mechanism of the actuator, or a wheel control of considerable radius, be substituted for the present system. While applying nose-down trim, Johnston discovered just how sloppy the control was when overtravel of the control caused a $-2.6g$ acceleration, followed by immediate failure of the engine. Johnston quickly flipped the ignition switch to the off position, and Frost in the FP-80 immediately observed a wisp of smoke or vapor—he could not be sure which—from the XS-1. Johnston checked for fire but no warning lights were visible. Since Frost saw no more smoke or vapor, the matter was dropped. Johnston reignited the number 4 cylinder and then the number 2 cylinder. The XS-1 began a climb at approximately Mach 0.70. The fuel liquidometer remained in the "full" position throughout the flight, once again upholding the XS-1's record for something electrical failing to perform. Therefore, Johnston could only guess at the true state of fuel availability. He did not wish to lean burn the engine so when the lox liquidometer read one-quarter full, he initiated shutdown procedures and jettisoned all remaining fuel and lox. The first try at using the blowdown valve failed and Johnston reset the switch. This time the system performed in a satisfactory manner.

Johnston dropped the XS-1 toward Muroc and lowered the landing gear and flaps at an observed 210 MPH. The number 2 aircraft deployed the landing gear rapidly. Cockpit noise increased to a noticeable degree and he observed only slight trim change. The seat parachute now provided Johnston with somewhat better vision from the cockpit than he had anticipated. In level powered flight, Johnston noted that he could see the horizon and to the left and right but seeing the ground in the immediate vicinity required the pilot to bank the airplane (the same situation noted by Goodlin and Woolams). However, forward visibility was very good when his glide descent was executed with the landing gear and flaps deployed. The downwind and base legs of the approach were flown at 200 MPH. After the end of the turn the approach was flown at 160 MPH. Touchdown was at 140 MPH on a west heading. Rollout on the lakebed was routine and the brakes seldom applied.

Postflight inspection revealed the fire screen between the aft end of the combustion chambers was completely burned away, apparently the source of the smoke Frost saw. Bell engineers recommended that it be replaced with a less flammable material or an asbestos curtain similar to the one on the number 1 aircraft. The propellant valves on the number 4 chamber were found to be leaking although this discovery was not made until preparations were undertaken for the next flight. The NACA team recovered data from the recording instruments and the cockpit camera. Johnston summarized his feelings about the plane in his pilot report: He considered it a "pilot's aircraft," exhibiting "ideal handling characteristics and satisfactory visibility" for a research aircraft. He found the rocket engine was simple to operate if the pilot was properly briefed and the power, even on only two cylinders, was "remarkable." Johnston stated his interest in flying the aircraft once again at a higher power level. Events were to overtake that desire.

Closing the Door

During the first week of May, the Flight Test Division at Wright Field continued to work up a plan for operating the XS-1 at Muroc. Bell Aircraft was notified that their refusal to continue the XS-1 program was accepted, which was not what Bell Aircraft had in mind. The question of the final disposition of the XS-1 program had boiled down to the simple issue of whether AAF Commanding General Carl Spaatz in Washington would overrule the Wright Field decision of May 1. Absent that reversal, Bell Aircraft was out of the XS-1 flight test program. Flight Test anticipated a final decision by May 20. However, that date passed with no "final" decision, but Colonel Boyd felt positive enough about the Smith decision that his personnel continued with their preparations. On May 7, Flight Test took the step of requesting the use of a faster chase jet than the P-80 currently being employed in that role. The request was accompanied by a plea for haste in filling the request since it would take *three* months to obtain and outfit the airplane with the necessary equipment—a lengthy period of time for an accelerated program.[37]

As May unfolded, the continuing tests brought the contractor program closer to termination. The AMC proceeded with its plans to divide the project between the NACA and the AAF. On May 29, James Voyles of the Wright Field XS-1 project office visited Muroc to discuss the NACA takeover of the program. During his visit, Voyles again told the NACA the AMC's principal problem was a lack of funds for the XS-1 program, which would not allow the expensive Bell contract. No money would also not allow the NACA-desired changes (except the stabilizer fix). Voyles privately advised the NACA that Wright Field already considered the decision final. As a result, the AMC was moving ahead with letting the AAF perform the flights. Within the past week, the new AAF B-29 pilot, a Captain Walker, had initiated the screening of his flight crew and was preparing to take over the bomber duties. The Flight Test Division had already screened several potential XS-1 pilots, but no announcement would be made for some time, Voyles explained, so as to allow for a big press release. However, in his meeting with Bell Aircraft, Voyles's message was not so straightforward. Voyles advised Frost only the official AMC policy line that transfer and acceptance of the XS-1 program to the NACA would be held in abeyance until a "final" decision on the Bell Aircraft proposal was made in Washington.[38]

Although NACA may have known the reality of the XS-1 program, the Bell team at Muroc continued to prepare for one more test as a conclusion to the contractor program. On Monday, May 26, Bell technicians tried to conduct that flight, but nature intervened in the form of extremely bad weather. The weather break in the smooth schedule of events continued through Wednesday May 28. NACA technicians used the downtime to remove all instrumentation from the number 1 aircraft and install them in the number 2 aircraft. The enforced delay was compounded by the discovery of the leaky propellant valve on chamber number 4 from Johnston's flight. The leak necessitated removing the engine, where the rusty valve was replaced by one from a spare engine and the rocket motor quickly replaced in the number 2 airplane. By the afternoon of May 29 Bell technicians decided that the existing weather conditions and flight preparations would allow the XS-1 to fly.[39]

The B-29/XS-1 number 2 lifted off from Muroc ostensibly to conduct the final acceptance flight of the XS-1 program. Goodlin would again pilot the XS-1 and Frost would provide chase in the FP-80. The purpose of the flight was airspeed calibration. The flight plan called for Goodlin to be dropped at 21,000 feet and to conduct a speed run at

20,000 feet altitude and then an increasing g turn in a climb at 25,000 feet. Further tests would be to obtain a negative g reading on engine output and abrupt pull-ups at Mach 0.65.

Goodlin entered the XS-1 at 7,000 feet and began the cockpit checkout procedure. Dome-loading proceeded normally and Dow began the drop routine at 21,000 feet. Release from the B-29 occurred at a speed of 250 MPH. Chamber number 2 was ignited shortly after drop separation. The constant speed run was initiated, but Goodlin noted that acceleration of the number 2 aircraft seemed slower than his experiences on recent flights in the number 1 aircraft. As a result, Goodlin ignited the number 3 chamber and speed rapidly increased to an observed Mach 0.75. After finishing the speed run, Goodlin shut down the number 3 cylinder and started the increased g climb. When it was completed, he was at 27,000 feet. At that point, Goodlin executed a − 1g pushover. The engine immediately experienced a loss of chamber pressure and a reduction in fuel/lox line pressures. Once level flight was regained, the pressures returned to normal within four or five seconds. It appeared from the relative changes in pressure that the engine did not actually shut down during the negative g maneuver. Rather, the motor suffered from a rich mixture caused by the lox being forced away from the tank bottom pickup. Under continued negative g conditions, this would shut down the engine.

As per the stated purpose of the flight, Goodlin also experimented with the rate of stabilizer movement by operating the stabilizer control in the cockpit. It continued to be as bad as everyone had indicated for both planes. When completed, Goodlin jettisoned all remaining fuel/lox and proceeded to glide test the aircraft through the speed range. An observed top speed of Mach 0.73 was achieved during this moderate dive to about 12,000 feet, where he then made an abrupt pull-up at an indicated speed of Mach 0.65. The result was a positive accelerometer reading of 7.2 g's. No restriction of elevator movement was noted, which seemed to indicate that past corrections had finally fixed that problem. Landing and rollout were routine.

Postflight NACA review indicated mixed results. Although the NACA team labeled the flight successful, data results were more difficult to measure. The airspeed calibration went smoothly and most other NACA instruments recorded data. However, the available data on the abrupt pull-up only covered the recovery from the maneuver. Also, no loads data was recovered since the recording equipment was mistakenly connected to the wrong circuit. In any event, the last XS-1 contractor flight had been completed. The flight had certainly proceeded more smoothly than recent tests. Final acceptance by NACA of either one or both planes now rested with Wright Field—or so it seemed to the Bell employees.

Although Voyles had indicated to NACA that Wright Field considered the decision to deny Bell Aircraft a follow-on contract as final, Bell Aircraft did not give up the fight so easily. On June 3, Walt Williams was advised that General Spaatz was still listening to arguments about the "final" decision. The practical result was that in the first week of June the B-29 was still sitting at Muroc (when it should have been at Oklahoma City undergoing the Voyles-mandated 200-hour check) because Bell Aircraft continued to hold the bailment contract on the airplane. No one would authorize the maintenance work until Washington made a final decision on the accelerated flight contract. During the pause, the NACA proceeded with plans to rework the XS-1's cockpit and instrument panels. With both planes in the hangar and little work being done, the NACA team secured unlimited access, allowing them to complete the new instrument panels. Bell assisted in this work to the extent that they could without AMC clearance. Modifications included a snap-action emergency engine cutoff switch, a near total rearrangement of the instruments to suit the preferences of the newly designated NACA XS-1 prime pilot Herb Hoover, and the marking of their identification and operating limits. Additionally, flameproof wiring was installed in the aft section of the number 1 aircraft (number 2 already had this precaution). Time seemed irrelevant while the decisions for acceptance were played out. But, while the modifications and political infighting continued, Bell Aircraft received a surprise request for another XS-1 flight.[40]

Publicity for the XS-1 program made good copy in Washington. In tight budget times, a little air show helped make the case for continued funding for research programs. As a result, on June 5, the Muroc Army Air Forces Command held an "open house" for more than 100 members of the Aviation Writers Association,[41] to present the writers with the latest developments in military hardware. Two special research aircraft were an added bonus. Through Wright Field, Muroc arranged to allow the writers a double dose of the supersonic program. Bell Aircraft would demonstrate the characteristics of the two XS-1 airplanes with and without power. To accomplish this task, Frost would work a ground demonstration of the number 2 aircraft while Goodlin performed a power-on flight demonstration of the number 1 airplane. Since the decision and clearances came literally at the last minute, no NACA instrumentation was in the number 1 aircraft.

The ground run of the number 2 airplane was scheduled as the first XS-1 demonstration for the writers. Goodlin would work the cockpit controls while Frost described the plane's unique features and engine starting procedures. The XS-1 show started in the afternoon. Approximately 2½ hours before the demonstration, the Bell ground crew began the fueling process for the airplane and finished well ahead of schedule. However, the technicians discovered that the fuel pressure regulator's diaphragm was leaking. To replace the diaphragm required dumping the dome pressure of the first-stage regulator, which relieved the pressure on the malfunctioning regulator. Repairing the first-stage regulator initiated a sequence of events that led to a 1½-hour delay. During this diaphragm replacement work, no bleed pressure could be furnished to the engine. After completion of the repairs, the first-stage regulator was again dome-loaded, so the engine bleed system became operational again.

After finishing the repairs and filling the lox tank, the technicians towed the XS-1 into position and secured it at the ground-run demonstration area in front of the writers and air-show spectators. By the time Bell Aircraft was ready for engine ignition, it was past noon. Goodlin was still detained with the press, but Frost decided to proceed with the demonstration. Almost 15 minutes had elapsed since completion of the explanation and preparation for engine startup. Unable to wait any longer, Frost climbed into the cockpit and checked the various gauges. All appeared normal. Frost then ignited the number 4 chamber. The chamber pressure rise appeared sluggish, so Frost shut down the cylinder and tripped the switch to fire the number 3 chamber. It ignited, but also appeared not to be operating satisfactorily. Frost shut down chamber 3 and tried chamber 2, where he found success. The number 2 chamber ignited and ran at full power. Frost quickly shut it down and tried the number 1 chamber. However, the number 1 cylinder never ignited as the engine lost electrical power and the lox tank pressure descended rapidly. Almost simultaneously the fire warning light came on in the cockpit. Jumping down from the cockpit, Frost ran along the plane looking for problems. Literally right before his eyes, the paint began blistering beneath the horizontal stabilizer. No smoke or flame was visible, but the ground crew immediately began taking off the inspection plates as the fire crew raced into action. The quick reaction by Frost, the ground crew, and the local fire team saved the aircraft. Many of the ground observers approximately 200 feet away didn't even know what was transpiring.

The writers had been suitably impressed with the engine's howling noise and the visible plume of fire from the XS-1's tail. With the number 2 airplane's exciting ground performance, it was no surprise that they eagerly anticipated the aerial test flight. With the B-29/XS-1 number 1 already loaded and waiting, the Bell team prepared for the aerial show. As soon as Goodlin was ready, Dow and Heaney took the B-29 aloft on the one-hour mission. Frost flew chase in the FP-80. Goodlin entered the XS-1 cockpit during the climb and proceeded through the checkout and dome-loading procedures. At 21,000 feet, the B-29 positioned itself directly over Muroc Field. Dow proceeded through the countdown and released the XS-1 on the demonstration flight. Goodlin immediately ignited engine chambers 1 and 2 in rapid succession. Both appeared to operate normally. Goodlin allowed them to stabilize and then executed a 30-second engine run. When finished, he jettisoned approximately 60 percent of the fuel and performed a dive toward the ground demonstration area. As he pulled out of the dive at 7,000 feet, Goodlin turned the aircraft to fly over the waiting crowd. Speeding past the observers, Goodlin sequentially ignited chambers 1 and 2. The surge of power, exhaust flame, and ear-splitting sound was sensational. The XS-1 entered a steep climb. At 15,000 feet, Goodlin nosed the XS-1 over into level flight. As fuel was nearly exhausted, Goodlin initiated engine shutdown, then jettisoned the residual propellants as he again dived to the field in front of the reviewing personnel. Goodlin concluded the demonstration with a slow roll at 5,000 feet, followed by a 180-degree turn toward the landing strip. With wheels and flaps lowered, he greased the landing in a westerly direction. Rollout was routine. After 18 months of work and a reported 3 hours and 47 minutes of total flight time, the XS-1 program took a rest.

The next day the Bell technicians began work on the number 2 aircraft. The XS-1 engine was removed to determine the origin and extent of the damage, and a visual inspection immediately revealed the source of the fire. The weld joint at the head of the number 3 cylinder appeared to have been broken by the force of an explosion within the lox side (the front end) of the cylinder head. The break had emitted a blowtorch-like flame at the forward end of the engine within the empennage, which ignited the substantial amount of pressurized and heated fuel spewing from that cylinder's circumferential cooling jacket through the ruptured head. Frost's shutdown of the engine switches in the cockpit caused the cylinder's two propellant valves to close, thereby instantly cutting off the fire's lox supply. However, even though no more fuel could enter the cylinder through the closed fuel valve, all the superheated fuel within the cylinder jacket and the aft-end fuel tube could feed the fire. The explosion in the lox head was due to the inability to maintain bleed pressure to the engine during the morning repairs, which allowed residual fuel in the chamber to seep into the lox. The result was an explosion during the chamber 3 ignition attempt. In addition, the number 4 bulkhead was almost completely burned away, and a hole was burnt through the fuselage and the trailing edge of the ventral duct. The NACA strain gauge equipment and control position transmitters were also damaged. The extent of the damage meant that repairs at Muroc were impossible.

After calling LMAL to tell Mel Gough what had happened to the NACA airplane, Williams attended a meeting with Joe Vensel, Dow, Frost, and Colonel Ritland from Wright Field. Williams reiterated that NACA wanted the number 2 airplane with the 10 percent wing. They were not interested in the number 1 airplane with the thinner 8 percent wing, or in having Bell Aircraft simply repair the damaged fuselage. The NACA wanted the fix done correctly. To that end, Williams offered four options: 1) Bell Aircraft could take the wing/tail assembly off the number 2 airplane and put it on the number 1 airplane; 2) Bell could replace the entire aft section of the number 2 plane with the same from the number 1 airplane; 3) Bell could replace the aft section with the aft section from the number 3 airplane; or 4) Bell Aircraft could simply fix the aft section of the number 2 plane. Williams finished his options by again remarking that NACA was not interested in option 4. Frost would not commit to anything without talking with Stanley back at Buffalo. Ritland tried to steer a middle course by stating that he agreed with the NACA that Bell Aircraft needed to replace the aft section.

The decision to repair the number 2 aircraft was only one of several major program issues on the XS-1 project that now loomed before the AMC. The continued foot-dragging over the final disposition of the accelerated flight contract put Bell Aircraft in an awkward position. Its flight station at Muroc was an extremely expensive operation to continue funding while there were no flights in progress. Further, the B-29 was badly in need of a 200-hour checkout, which would take several weeks at a maintenance base to finish. As a result, Frost notified the AMC that Bell Aircraft intended to shut down the flight station and send its personnel back to Buffalo, pending the disposition of the XS-1 follow-on contract. Few at Bell Aircraft doubted that the hiatus would be very short.

In all the give-and-take over the new Bell Aircraft contract proposal for the high-speed/high-altitude tests, one person was seemingly overlooked in the proceedings: Slick Goodlin. Since Goodlin's departure for Muroc, his attorney in New York, Edmund Burke, had been corresponding with Bell Aircraft to work out the details of the Goodlin handshake flight bonus agreement with Bob Stanley.[42] A contract had been submitted by Burke to Stanley late in 1946. Little progress on the details had been achieved by the spring. In April, Burke phoned Goodlin to advise him that Bell Aircraft seemed to be stalling on resolving the terms of the contract. Goodlin advised Burke that the acceptance flights would soon be over and he would personally handle the matter with Stanley. Goodlin was unaware of the AMC/Bell contract difficulties. But Bob Stanley was well aware of the impact a $150,000 services contract could have on Bell Aircraft—a Bell Aircraft that in May 1947 did not have an XS-1 program.

On June 4, 1947, Joseph E. Conners, Bell's corporate attorney for the proposed Goodlin contract, replied to Bob Stanley that it was his opinion the proposed contract was not legal due to the retroactive clause contained in the document. Further, Conners indicated such a contract would not be in the best interests of Bell Aircraft because of the possible repercussions in dealing with its other test pilots. This type of deal could cause others to request similar treatment. Finally, Conners argued such a contract would be subject to an AMC veto since it was a change in the terms of the existing contract. Conners speculated the AMC would not allow the change since there were other pilots who would be happy to take this work. On June 6, Stanley notified Burke that it "would be impractical for us to enter into any contract with Mr. C. H. Goodlin for his personal services. This applies either to a retroactive contract or to any contract in the future." A somewhat longer letter was sent to Goodlin at Muroc with a notice that Stanley was prepared to pay an additional $3,500 as per the initial agreement with Goodlin. A copy of the letter was put on Goodlin's desk at Niagara Falls.[43]

At Muroc, all preparations were quickly concluded for the return to Bell Aircraft. At 6:55 A.M. on June 9, Dow guided the mated B-29/XS-1 number 2 off the desert airfield for the trip to Niagara Falls. Slick

Goodlin flew copilot with Frost aboard as a passenger. Most of the rest of the Bell team also returned to Niagara Falls the same day in the C-47. The B-29 arrived later the same afternoon following a nonstop 6-hour, 40-minute flight. The next morning, Goodlin found the Stanley letter on his desk. (Apparently the Muroc copy crossed in the mail.) Goodlin immediately went to meet with Stanley. Upon learning that Stanley would not honor the terms of the handshake contract agreement, Goodlin submitted his resignation effective immediately. By the time the resignation was signed, Bell Aircraft had been officially out of the XS-1 program 41 days and nobody had even bothered to tell Slick Goodlin.[44]

With the passage of time, rumors grew as to the cause of the transfer of the XS-1 program to the AAF. In many of these stories, Goodlin and the $150,000 bonus became the rumored prime villain. As can be seen, the issue of total program cost was indeed the *final* cause of the transfer. However, as the detailed discussion provided here reveals, other issues weighed just as heavily in the decision. Nonetheless, the subject of Slick Goodlin, his flight record, and the bonus money became fair game for myth.

To dispose of the myths, the rumors may be divided into three groups. First, the rumor that Goodlin flew too slowly and the program was not making progress: in fact, Goodlin's progress in the XS-1 was handicapped by several factors, none of which he had any control over. The XS-1 was an experimental plane. As seen in the detailed record of the flights, few were ever trouble-free. Many flights were repeated because of the failure of the rocket engine. Some flights were repeated because the NACA instruments malfunctioned. A few were redone simply because the aircraft experienced unusual problems. All of this was to be expected in the initial flights of a unique aircraft. The record of many of the other experimental production aircraft at Muroc were no better and, for several, were quite a bit worse. Indeed, Goodlin's "progress" was really constricted by the AMC/NACA agreement regarding the twenty powered flights required for NACA acceptance of the two aircraft. Bob Stanley certainly didn't want to fly that many nor did the performance of the aircraft require that number. Goodlin flew the XS-1 past the AMC acceptance criteria early in the powered portion of the program. The flights in April/May merely served to repeat the earlier results and satisfy the NACA requirements for data and flights.

Second, Goodlin was only 23 years old when the program was offered to him. Within three months, he was given national publicity of a continuing nature. It was heady stuff for a young man with little formal training in the art of the press. Additionally, the ability of the news media to find a good story in a program unable to offer details obviously led to unfortunate misquotes and erroneous misstatements. Goodlin had rarely been to Wright Field. He was unknown to many of the personnel at the AMC. Worse, his personality was not of the "good ol' boy" nature. He didn't like to hang out at Pancho Barnes's legendary bar on the edge of the base. To some, he seemed aloof; to others he had a devil-may-care attitude. He was a good pilot, as his contemporaries readily noted, but away from the plane his actions were harder to judge. In any event, all of his behavior seemed unusual to some associates accustomed to the totally absorbed, flying and drinking fighter-jock persona. It left Goodlin exposed to the damage of rumors when others might not have been so vulnerable. For example, one story was that Slick had difficulties with the AMC because of his Hollywood press agent. Goodlin did not have a Hollywood press agent, only a book agent.[45] Indeed, the civilian boss of the XS-1 program at Wright Field never even heard that Goodlin supposedly had an agent. But Goodlin did have Hollywood friends. Supposedly, he was spending more time in Hollywood with "those" people than at Muroc. Bell Aircraft and the NACA observers on the scene agree that Goodlin was absent much of the time. However, several add an important caveat. Goodlin had no involvement with the plane other than flying it. It was not his job to be present at Muroc every day or to perform the XS-1 maintenance and data analysis. He did check in by telephone with Frost every day, but was simply not needed to do the routine shopkeeping tasks. This situation was in marked contrast to Jack Woolams's situation at Pinecastle. Woolams had been the test site manager as well as Bell's chief test pilot. His involvement with the day-to-day program decisions was more total: Goodlin's involvement was not as complete. Whether Goodlin should have been more involved is another issue. But the fact is that he was there for the flights and that is what Bell Aircraft paid him to do.

Finally, the issue comes again to the money required to fly the plane. Bell Aircraft (not Slick Goodlin) wanted an expensive contract from AMC to do the job. That contract certainly reflected the Goodlin bonus, but not through any action of Slick Goodlin. Bell Aircraft made the decision to submit a pilot bonus contract of multiyear duration. It was a contract with terms that were unacceptable to Colonel Smith.

This attitude was compounded by the financial condition of the AMC since that agency had little funds available to offer negotiating room. Regardless of the *remembered* facts, the surviving records of that time clearly indicate that the AMC was suffering financial problems on its limited postwar budget. Bell Aircraft did not help its cause with its XS-1 proposal. Its expensive contract offer and lengthy number of potential flights strengthened the case by some for a lower-cost AAF alternative. It is hard not to believe that Bell Corporation's ongoing proxy fight and continuing financial woes didn't contribute to the contract request. However, the documentary record clearly reveals that *Bell Aircraft* turned down the opportunity to continue the program by refusing to accept the AMC fixed-price offer. (To what extent AMC begged this Bell Aircraft refusal by the nature of the fixed-price constraint is difficult to determine from the surviving memoranda.) Bell Aircraft had legitimate business reasons for this decision. It was hard for the corporation to guarantee a successful performance in the uncharted waters of supersonic flight. Further, an undamaged ready-to-use airplane for such an unpredictable, experimental program was also no sure thing. The June 5 runway incident at Muroc demonstrated how little was routine on this project. Yet, the AAF felt justified in requiring some guarantee regarding its limited financial resources.

The expensive Bell Aircraft contract proposal and the lack of AMC funds dovetailed with another factor—the negative publicity within the AMC over Goodlin and the Bell contract difficulties, which gradually brought about a Wright Field attitude that perhaps Slick Goodlin was not the pilot for the program. This opened the door to another possibility. The XS-1 prestige factor was something that several people in the AAF were only too keenly aware of. The legislative creation by Congress of the new independent U.S. Air Force over the spring-summer of 1947 and the continuing budget battles with the Navy provided fertile ground for increased interservice competition. A successful conclusion of the supersonic program by the AAF could provide valuable prestige in the race for congressional support. Colonel Boyd's stated belief that the Flight Test Division could do a better job of flying the XS-1 may simply have been a case of turf battle. But Boyd believed his pilots had something to prove. There had been no strictly military test pilots since the 1920s. This was the opportunity for the AAF to step back into the flight research (instead of just development testing) program. His conviction that the sound barrier could be broken and his remarks that an Air Force pilot should be the one to do it clearly provide the intellectual basis for the conviction that perhaps more was being played out in the Bell Aircraft contract decision than merely a shortage of money. Coincidentally, the Air Force's XS-1 flight test program looked substantially like the Bell program. The Flight Test Division proposal certainly alluded to other flight records that would surely bring prestige to the new Air Force—flights that would add badly needed budget dollars to the emerging service.

All of these issues lay on the table as the program direction was finally decided. But publicity could not be the only issue in the final decision. Flying and the pure challenge of flight made Colonel Boyd want the program. For a test pilot, this was a golden opportunity to do what they paid you for. Thus, in fairness to the participants, the author's conversation with the two prime AMC representatives, Ritland and Smith, reveals little contemporary discussion of the prestige issue.[46] As a result, absent the internal AMC memos and a copy of the Bell continuation contract proposal, it is hard to pinpoint a definitive answer to the story of an Air Force power play. In conclusion, it is easier to argue that the skilled cadre of Air Force pilots then serving in the Flight Test Division were equally as capable as anyone at Bell Aircraft or any other civilian to perform the tests. Perhaps it is best to say that the Bell proposal and the AMC cost requirements were convenient supporting reasons that allowed the AAF to exercise its already determined desire to acquire the biggest public relations opportunity then underway. In the end, the interplay between the three factors of funding, performance demands, and prestige was perhaps predictable: after two years and three months, Bell Aircraft was out of the XS-1 transonic program. The seemingly "final" AMC final decision did not rest easy in Buffalo as might be expected. But it also produced grumbling in another location—one that was certain to cause surprise after the record of uncertain cooperation over the life of the program.

Closure

On June 13, LMAL's engineer-in-charge Henry Reid finally got around to replying to NACA HQ regarding the AMC decision on Bell Aircraft.[47] LMAL made quite clear in their written reply to HQ that it did not like the expedited program as outlined by the AMC. Langley's rationale for this stand acknowledged that they had never liked the Bell Aircraft XS-1 flight program. LMAL had merely gone along with it since a good working relationship with Bell might be necessary to get the aircraft modifications that NACA desired before final XS-1 acceptance. Further, the Bell Aircraft flight tests ensured that NACA pilots would be in a safe airplane at the time of agency acceptance of the planes. Now neither point was clear any longer. As a result, Reid stated, "The elimination of Bell from the project . . . is considered unfortunate by the Laboratory." While the NACA had no objections to the AAF taking over the program, it did want the AMC to acknowledge that LMAL did not agree with the expedited flight program. Langley still preferred its own proposed flight program since it would offer the largest amount of research information.

In spite of the AMC decision in favor of the expedited program, LMAL promised (through NACA HQ) its commitment to the AMC to provide full cooperation in the new joint venture. For now, the NACA Muroc team would maintain the XS-1 number 1 currently at the test site. However, LMAL advised, the AMC should anticipate a requirement for four additional maintenance men to properly maintain their aircraft. Reid also reminded NACA HQ that Langley was in no position to fund oxygen/nitrogen costs for the AMC and new arrangements must be now made. He estimated that a contract with Linde Company for the gases for only one aircraft would be $5,000 per month. A speedy decision was of the essence, Reid reported, as Linde had always expected to dismantle its handling facilities at Muroc as soon as the Bell acceptance tests were finished. Now that time had come.

On June 25, 1947, Walt Williams and NACA test pilot Stefan A. Cavallo made one more journey to Buffalo to visit the Bell plant.[48] The formal purpose was to discuss the glass canopy for the long-delayed XS-1 number 3 and to review the status of the XS-1 number 2. Cavallo was present to be checked out in the operations of the rocket powerplant. Although Herb Hoover had agreed to make the first NACA XS-1 flight(s), LMAL was aware that he did not intend to stay at Muroc. As a result, NACA planned to use Cavallo as the backup XS-1 pilot at Muroc.

A face from the XS-1 program's past, Roy Sandstrom, the chief of preliminary design engineering for Bell, served as their guide. Sandstrom outlined the wrap-up of Bell Aircraft's involvement with the program. First, Sandstrom reminded the NACA personnel that the B-29 was due to be ready by the second week of July. Second, the P-51 had proven unsatisfactory in terms of speed for the XS-1 program. Bell had declined to renew its bailment contract when it expired on July 8. Frost had recommended to AMC that the FP-80 continue to be utilized as the chase plane. In anticipation of turning over the XS-1 number 2 to NACA, Bell had initiated several significant fixes of the airplane. Specifically, Bell had removed the entire aft section of the number 3 plane and used it to replace the fire-damaged section of the number 2 plane. (The lack of AMC funds for completion of the number 3 plane made this the most economical solution.) Also, Bell technicians had installed the NACA instruments and changed the fuel lines in the aft fuselage to stainless steel. The stabilizer control was moved to a switch on the control wheel. Bell Aircraft proposed to AMC that the troublesome stabilizer gears be changed to provide a rate of motion of three degrees per second with 3,000 pounds of force. Bell engineers believed this would be sufficient to correct past lag problems.

Sandstrom indicated the XS-1 number 2 repairs were complete and that Frost would shortly return from RMI with a new motor. Once the new motor was installed, the plane would be ready for pickup by the AMC. Bell would not deliver the aircraft to Muroc since their B-29 contract would expire on July 1 and the engine tests would not be ready by that time. This necessitated some changes to the NACA timetable since they had assumed Bell Aircraft would make an additional checkout flight in the number 2 aircraft after its return to Muroc. As a conclusion to the visit, the NACA personnel looked over the stripped-down XS-1 number 3. The new glass canopy was a reminder that technology continued to move forward even as the number 3 program seemed mired in both bureaucratic confusion and budgetary shortages.

The passage of time certainly changed many things in the XS-1 program. When Bell Aircraft was awarded the XS-1 project from the ATSC in March 1945, it was with the assumption that Jack Woolams would fly the airplane to supersonic speeds. Ironically, two-plus years after the contract award, Bell Aircraft was still in the supersonic business for the AMC, but not with the XS-1. Rather, it was with the XS-2 that Bell Aircraft intended to continue the quest for high-speed flight. Nor was it Jack Woolams, or even Slick Goodlin or Tex Johnston, who would perform the first sound-barrier flight. But, if seemingly much had changed, more remained constant. After thirty-seven flights the AMC contract conditions had been fulfilled. The XS-1 had proven airworthy and the rocket engine reliable in performance without major design modifications. The new research partnership among industry, science, and government had endured, albeit somewhat shakily at times. The transition from contractor acceptance flights to NACA/ AMC ownership had concluded the chain of events opened by the conferences in 1944.

Now the divergent pattern of research only hinted at in 1945 in all the grumbling regarding the NACA performance during World War II would receive its first test. The AMC and the NACA would each develop new procedures and methods for what appeared to be a new course for U.S. aviation research. For the Army Air Forces, it was a new direction. The new *research* flight team, charged with a significant flight mission, was ready to extend the boundaries of knowledge about the speed of sound. The new mission stipulated rapid progress in breaking the supposed sound barrier: just how rapid the accelerated flight program could be would come as a surprise to many.

6

The Air Force Takes Over the Program

Wright Field

The Flight Test Division at Wright Field was a test pilot's heaven. The division had responsibility for developmental testing of all new aircraft. The large hangars at the field were loaded with new and exotic aircraft from America's factories and overseas (many formerly belonging to our World War II enemies). The flight test crew of veteran pilots brought together many of America's brightest engineering types with some of America's best wartime aviators. It seemed to the outsider that being an ace from World War II or obtaining a college degree was a basic requirement.

Colonel Albert G. Boyd was the commander of the Flight Test Division. Boyd, an Army flier since 1927, was a fighter pilot's pilot. His responsibilities for the Air Force included all phases of aircraft development, flight-testing, maintenance, and operations. During his 30-year military career, he managed to fly over 23,000 total hours in 723 different types of airplanes. In June 1947 he would fly a specially prepared P-80R (nicknamed "Racey") over a three-kilometer course at Muroc to a new world's speed record of 623.738 MPH. He was unofficially known as the "Father of Modern Flight Testing" and the "World's Number One Test Pilot." When he spoke about test flying, Boyd commanded respect. However, just in case someone missed the point, the commanding officer in Boyd would take over and provide a clear, distinct order. A tough disciplinarian, with little tolerance for easy familiarity, Boyd was every inch the officer in charge. If there was a mistake, Boyd quickly called the error to the attention of the guilty party. He retained enormous respect from everyone around him and possessed the good judgment and sense of purpose to deserve that respect. When Boyd recruited his XS-1 team he told the group that

Col. George Smith of the Aircraft Projects Section approached the Flight Test Division to do the job. Boyd did not say that he had been involved in securing that request. Boyd believed in the possibility of supersonic flight. He also believed that it was "not only proper . . . but absolutely necessary that an Air Force pilot should make this flight." He picked his pilot based on the certainty of that belief.[1]

In March 1947, Boyd first informed Colonel Smith that his division could handle the XS-1 program. As discussed earlier, in April Boyd's group outlined a tentative flight plan if the XS-1 project should be turned over to them. It involved breaking the sound barrier and much more. The program would achieve a notable first for the soon to be independent Air Force. It would also allow Boyd's test pilots to do what they enjoyed most: stretching the outside of the envelope—discovering the unknown.

In May, the pace of events picked up speed at Flight Test. Colonel Boyd asked for a list of volunteers from among his test pilots interested in flying the XS-1. Of the 125 test pilots in the division, a group of 25 was selected for additional review. One of that select group was Capt. Charles ("Chuck") Yeager. Again and again, Yeager's name kept coming up in the screening process. Finally, his was the only one left.[2]

Yeager had enlisted in the Army in September 1941. After a brief time as a mechanic, he transferred to pilot training. Graduating in March 1943, Yeager was assigned to the European Theater of Operations in January 1944. Shot down two months later over France, Yeager demonstrated the aplomb of an escape artist as well as a pilot. Upon returning to England, Yeager managed to score twelve aerial victories (including five in one day) in sixty-one missions over Nazi-occupied Europe. He received the Distinguished Flying Cross, the Bronze Star, the Silver Star with oak leaf cluster, the Air Medal with six oak leaf clusters, and the Purple Heart. This distinguished combat record opened the door to an even greater opportunity.

Yeager started his career at Wright Field as an assistant maintenance officer, responsible for routine test flights for aircraft after overhaul. His flying abilities and personality quickly brought him to the attention of Colonel Boyd. In November 1945, Boyd assigned him to the Flight Test Division as a test pilot. While Yeager was by no means the senior officer or pilot in Flight Test, he did have two outstanding characteristics. First, he loved to fly *anything,* and his job as maintenance officer afforded him the opportunity to pile up more hours in the air than other officers. Second, his outstanding war record, his cool demeanor, and his acknowledged skills in all types of airplanes promoted him in the eyes of Al Boyd. Cutting against Yeager was his lack of a college education and his junior rank in the division. But Boyd had been carefully watching this young officer and thought he knew his man quite well. Boyd played no favorites. Either you cut the mustard in Flight Test or you were gone. Yeager had passed the character test—the gut check—on numerous occasions. After much soul-searching, Boyd selected Yeager to be the prime pilot on the XS-1 program.

The other two members of the new Air Force team were close friends of Yeager's.[3] Since Yeager lacked a formal college education or background in engineering, Boyd decided to assign a technical individual, Capt. Jack L. Ridley, to assist in guiding and analyzing the program. Ridley was born in June 1915 in Oklahoma and attended the University of Oklahoma. He was a country boy who used his laid-back manner to hide a superb engineering mind. Ridley graduated from Oklahoma University with a bachelor's degree in engineering in 1939 and was assigned to the U.S. Army in the artillery branch. In 1941, he transferred to the Air Corps and received his wings in May 1942. His first duties were at the Consolidated-Vultee Aircraft Corporation as assistant plant representative. In 1944, he was reassigned to Wright Field to attend the Air Corps School of Engineering. Shortly thereafter, he was sent to Cal Tech to obtain a master's degree. He graduated in 1945 and was reassigned to Wright Field in the Flight Test Division. From January to June 1946, Ridley continued his education in the AMC's Experimental Test Pilots School. Ridley's sharp mind, easy demeanor, and obvious personal rapport with Yeager made him a natural for the program. Boyd initially selected Ridley as the engineer-in-charge. In the absence of a program manager, Ridley also filled that role. The other member of the Air Force team was a former combat pilot and prisoner of war from World War II. Robert A. "Bob" Hoover had enlisted in the AAF in 1941. He received flight training in both twin-engine and fighter aircraft. He was assigned to England and eventually wound up in the 52nd Fighter Group. He flew fifty-eight combat missions before being shot down over Nazi-occupied Europe. Hoover spent the remainder of the war as a prisoner. After returning to the United States, Lieutenant Hoover was assigned to Wright Field as a test pilot based on his superb flight skills. While stationed there, he had the opportunity to participate in a rocket program of his own.

The German Me-163 "Komet" was scheduled to be tested at Wright Field and Hoover was assigned as test pilot. Although the program was eventually canceled (after being glide-tested), it provided Hoover with a healthy regard for the perils of rocket flight. During his later career, Hoover would test many other jet aircraft and be acknowledged as one of America's greatest aerobatics pilots.

Flight Test was an old hand at wringing out unfamiliar aircraft, but an experimental supersonic aircraft was something new. Colonel Boyd's charge to take over the XS-1 program put in motion the pre-arranged procedure to be followed for this unusual aircraft. Division estimated that after XS-1 acceptance, 69 days would be required for the project. Wright Field expected that a definite flight plan would be drawn up for each flight. The division anticipated the first XS-1 flight would occur around August 15. Progress toward breaking the sound barrier would be as rapid as safely possible. The XS-1 project officer would furnish information to Muroc for a daily report to be sent to Wright Field. After the experiences with Bell Aircraft, all public relations would be handled only through HQ at Wright Field.

Boyd believed there was no time to waste if the Air Force was to meet this ambitious schedule for the XS-1 flights. As a result, in the third week of June, Yeager, Ridley, and Hoover were sent to Buffalo for a briefing by the old Bell team. At Buffalo, they met with Larry Bell, Bob Stanley, Dick Frost, Paul Emmons, Bob Woods, and the other Bell engineers. They received a complete history of the program. While the XS-1 program's progress was clearly delineated, the dangers that remained were not glossed over. Operation of the rocket motor continued to give everyone reason to pause.[4] But upon conclusion of the visit no one involved doubted the Air Force would not try to keep its accelerated schedule. Back at Wright Field, AMC believed the program was finally moving again after the long delays necessitated by the program changes.

Not all decisions are as straightforward as they seem in hindsight. This was equally true of the change in status for the Flight Test Division. After all Colonel Boyd's efforts to position his group to *fly* the XS-1 through the sound barrier, Air Force HQ decided to get into the act. A memorandum appeared that indicated the true word on the Air Force piloting the XS-1 through Mach 1.0 had not reached all corners of the establishment. On June 24, Brig. Gen. Alden Crawford, chief of the research and engineering division in the Office of the Assistant Chief of the Air Staff in Washington, wrote to General Craigie to instruct the AMC to *finally* cease all negotiations with Bell Aircraft and proceed with a transonic test program using in-house material and personnel. But his next instruction could only come as a shock to those ready to see an Air Force pilot in the high-profile mission. Crawford instructed Craigie to proceed with the flight program using "either piloted or pilotless means, the choice of which is left to the discretion of your command."[5] Choice? There might be some question in Washington, but among the true believers in flying, there was no decision to be made. An Air Force officer—a pilot—would break the sound barrier. Talk of "pilotless" flight missed the entire point (and, by that time, would cost a lot more money).

On June 25, a closed meeting was held at Wright Field.[6] Only Aircraft Projects Section and Flight Test Division personnel were present. From Flight Test these included; Boyd, Lt. Col. Frederick J. Ascani (Boyd's deputy), Yeager, Lt. Col. R. G. Ruegg, Hoover, Capt. K. M. Garrison, and civilian L. H. Sibilsky. From AMC those present included colonels P. B. Klein (head of Fighter Projects Section) and Ritland, and XS-1 civilians Hall and Voyles. The purpose of the meeting was to find internal agreement on how the XS-1 program would be handled by the Air Force. The group reached the following agreements:

1. The Air Force flight test effort would be an accelerated program; it would not duplicate the NACA program.
2. The Air Force would use the XS-1 number 1 as currently configured with the 8 percent wing and 6 percent tail.
3. Flight Test desired NACA cooperation on data and instrumentation.
4. Despite previous procedure, Flight Test wanted their own team to crew the B-29. The current Muroc crew could be used on the NACA flights. No crew had currently been picked for the B-29. Captain Ridley would serve as engineer-in-charge (or Major Butman, if Ridley retired), and Yeager would fly the XS-1 with Hoover as backup. Bell Aircraft would be asked to supply Dick Frost as technical backup.
5. Flight Test would furnish four men to assist with XS-1 maintenance. Fuel and supply would be handled by the supply division using FY 1948 funds.
6. The Air Force would only pay for modifications to the fuel jettisoning system and to the stabilizer actuator. Herculite canopies would

be installed for the transonic flights. Any other XS-1 changes would be safety-related.

7. The B-29 and FP-80 would be assigned to the Flight Test Division. The SCR-584 radar would be assigned to the NACA.

8. Data would be given first to the NACA and then to all other government agencies.

Having reached agreement on the major items of the program, the Flight Test Division was ready to meet with the NACA to work out the details of the cooperative project. On June 27, in anticipation of the NACA meeting, Colonel Smith formally requested that Bell Aircraft loan Frost to the Air Force for a period of six months. The Smith request provided high praise for the abilities of Frost. "The Air Materiel Command . . . believes that the success of the two flight programs depends to a considerable extent upon the proper organization of technical personnel to handle the airplane." The AMC specifically wanted Frost to assist in that setup.[7]

On June 30–July 1, the follow-up conference between the NACA and the AMC was held at Wright Field to discuss the coordination of the XS-1 programs.[8] Attending from the NACA were Walt Williams, Hartley Soule, Clotaire Wood, and Herb Hoover. The AMC was represented by James Voyles and Col. P. B. Klein. Those present for the Air Force Flight Test Division again included Lieutenant Colonel Ruegg, captains Yeager and Ridley, and Lieutenant Hoover. Colonel Klein opened the meeting by announcing the Air Force intention to allow the Flight Test Division to conduct an accelerated flight program on the XS-1. The purpose of the new program would be to achieve speeds of Mach 1.1 in the shortest possible time. The flights would occur at an altitude somewhere between 50,000 and 60,000 feet.

Soule presented the NACA summary of milestones for the project, and briefed the participants on possible technical problems and program procedures. Unfortunately, he spoke very rapidly, making notetaking virtually impossible. Worse, the copy he was reading from was the *only* copy available at the meeting and it was not "presentable." Soule promised to correct and forward a copy to Flight Test, but it was certainly an inauspicious beginning for a combined effort in a complex and dangerous project. In detail, Soule recommended the Air Force personnel read the several NACA-prepared preliminary flight test data reports and become familiar with the wind tunnel data and other test literature that NACA would provide to them. After reviewing the tests to date, Soule briefed the new participants on the potential dangers they might yet expect to encounter. These included fire, high-altitude pressure requirements, unusual stall characteristics, rate of roll abnormalities, and most serious, changes in trim and stability above Mach 0.85. During the discussion, the NACA provided Flight Test with a copy of the NACA flight program and asked the Air Force personnel to see their specific flight program. They were advised that none had been formulated in regard to specific altitudes or Mach numbers. These would be decided upon during the tests and at the recommendation of the pilot after analysis of the data. The Air Force statement raised some concern with the NACA. They could not help but think this sounded a great deal like the Bell Aircraft program all over again. But this time it seemingly mattered less since NACA would have its own airplane.

Having defined the parameters of the project, the group proceeded to discuss the merits and drawbacks of the Air Force proposal. NACA immediately expressed concern that tests at the altitude proposed by the Air Force could lead to serious problems. NACA again emphasized their desire for a phased approach on a slower basis. Although the higher altitude would reduce the risks of structural problems (due to the lower dynamic pressure), it would increase the dangers from high-speed stalls. The danger of explosions and fire remained the same at any altitude, but the safety of the pilot in case of pressurization failure or the necessity to bail out would be lessened at a lower altitude. Even though the Bell tests had demonstrated these dangers were low, as long as the approved flight procedures were followed, they nonetheless remained a potential liability. (However, everyone agreed the risk to the pilot regarding pressure loss could be reduced by the use of the pressure suit along with the pressurized cabin.)

The discussion regarding the proposed organization, crew selection, and facilities was essentially the same as the one that occurred between Fighter Projects and the Flight Test Division during the June 25 meeting. The AMC indicated its desire for NACA guidance in the XS-1 program and that they would cooperate in every way to ensure that cooperation. The AMC again stated they would provide supplies and maintenance for their aircraft, but until the Air Force personnel were trained, it would be the responsibility of NACA to maintain both aircraft. The conference also agreed to confirm the request directed to Bell Aircraft to provide Dick Frost to the program for a period of six months. Frost would give technical assistance on the new flights. The

AMC expressed their pleasure at the professional manner in which Frost had supervised the program and emphasized to all parties their desire that he be accepted as a full participant in the flight tests.

The one new point that was discussed concerned the testing of XS-1 number 2. The B-29 was expected out of overhaul on July 10. At that time it was to proceed to Bell Aircraft to pick up the now refurbished number 2 aircraft. Given the contract termination with Bell, neither the AMC nor the NACA wanted the Buffalo group to conduct any further flights in the XS-1. Thus, both parties agreed that after Captain Yeager had acquired some experience in the number 1 aircraft, he would make one flight in the number 2 airplane to prove the structural integrity of the plane by executing an 8g pull-up. After that test, NACA would accept the number 2 airplane and begin its flight program. It was agreed that the NACA would always have flight priority over the Air Force for their program and that a new NACA prime pilot would be named in the near future.

NACA flights would be dependent on the delivery of the XS-1 number 2 to Muroc and the acceptance of the aircraft after a flight to test its post-fire structural integrity. Yeager would pilot the XS-1 number 1 with Hoover as his backup. Walt Williams would continue in charge of the NACA personnel assisted by Joe Vensel from the Cleveland Aircraft Engine Research Laboratory. Ridley would be the engineer-in-charge for the Flight Test Division. All public relations would be handled through AMC Wright Field and they would decide what information would be confidential. Boyd closed the meeting by stressing that the Air Force program would be progressive and brief. Common sense, engineering experience, and safety would be the key points of the effort. The AMC desired the full cooperation of NACA and was ready to extend the same.

While AMC and NACA continued their conversations about the XS-1, a far larger discussion was occurring in the nation's capital. The continuing congressional debate in Washington over the spring and early summer of 1947 clearly revealed that the United States was about to create a new military service. Thanks to its excellent war record and the intensive lobbying by its political allies, the new U.S. Air Force was authorized by Congress and signed into law by the president on July 26, 1947. The new Air Force was to officially take its place alongside the older Army and Navy on September 18, 1947. In the difficult scramble for postwar military funds, this new Air Force independence did not rest well with its older associates. Both the

Army and Navy feared loss of funding and constituency as the new service acquired its own programs. Yet, in many respects they expected business to continue as usual. Thus, perhaps the biggest surprise of the FY 1948 debate was the political strength of the Air Force lobby. By the end of 1947, the new service would be asking for an end strength of 12,441 airplanes. The need for new aircraft to join the jet age meant that many of the Air Force's tested warplanes were now obsolete. The desire to continue to provide new aircraft in numbers sufficient to meet U.S. responsibilities at the end of World War II required that Congress invest in new technology and production. The call for more "Buck Rogers" airplanes was a pivotal issue in the quest for higher Air Force funding, especially research funding. The Army and the Navy desired similar program capabilities. The net result was a search for programmatic one-upmanship that would serve to rivet attention on competing priorities (especially aviation) and continue each service's strong case for taxpayer dollars.

While the creation of the Air Force had been assisted by excellent publicity, the fathers of the new service were astute enough to realize that success in the political/budgetary arena would require more than press clippings. Speaking to the Air Force Aircraft and Weapons Board in August, new Air Force Secretary Stuart Symington indicated that the old days of caution and publicity were past. If the new Air Force was to grow and continue to lead in the defense of the nation the achievement must be based upon a solid foundation of accomplishment.[9] The attempt to break the sound barrier provided such a program opportunity. Both the Air Force and the Navy were determined to achieve that goal.

But more than fear of Navy one-upmanship drove the Air Force in its supersonic quest. A glance across the water to the Soviet Union produced concerns over the continuing Russian efforts to upgrade the former German Me-262 jet fighter. Throughout 1947, rumors held that the Soviets had pioneered an upgraded version of the high-speed fighter. Perhaps they, too, might have a hand in the race. In any event, the need for haste, mirrored by the XS-1 accelerated flight test, was brought home to Air Force brass simply by monitoring the Navy's D-558 progress. On April 15, Douglas Aircraft conducted the initial flight of the number 1 airplane. During July, the Douglas/Navy team made furious efforts to keep the D-558 in the air. Testing proceeded at an increasing tempo. On August 5, Douglas test pilot Eugene F. "Gene" May reached a speed of Mach 0.85—faster than anything the

XS-1 had achieved. A tentative date of August 14 was set for a try at a new world's speed record. To the uninitiated, the sound barrier seemed to be the next objective. Obviously, the Air Force would not be ready for a supersonic try in the near future. But they recognized if past events were any indication of future trends, the D-558 program was bound to slip timetables. The Air Force wanted to be ready for that delay. But even if it were not, the Air Force decided to provide a little publicity that, while certainly true, sounds more like hedging one's bets than evaluating actual Air Force intentions. At the semi-annual meeting of the American Society of Mechanical Engineers, Colonel Klein of Fighter Branch, Aircraft Projects Section, finally provided the public confirmation of the reservations expressed by Walter Williams back in January.[10] After providing the audience with detailed information on the XS-1's history and tests to date, Klein proceeded to argue that the XS-1 was not a true supersonic aircraft. Rather, Klein stated, the XS-1 was designed to fly the transonic range. While the plane possessed supersonic potential, its real mission was only to provide data on the problems found in the *transonic* range. This data would be used in *later* aircraft designed for supersonic performance. As a result, if the XS-1 failed to reach supersonic speeds, the Air Force "will not be too disheartened because it was never really expected that it would." Taken in isolation, the Klein statement could have been simply a case of military secrecy, but it was certainly a strange comment after all the ballyhoo earlier in the program. More probably, it was a cushion in case of programmatic failure, given the recent strictures by the Air Force leadership on the need for solid accomplishments to be the foundation for the new service. But certainly the statement must have come as a shock to the old Bell Aircraft team now busy on the new swept-wing XS-2 back at Buffalo.

The Quest Continues

On July 3, the Bell contractor team returned to Muroc via the corporation's C-47 and began shutting down the Bell flight test facility. All equipment necessary for future flight testing was segregated, inventoried, and offered to the NACA/Air Force as spares. Bell Aircraft personnel continued to indoctrinate the on-site NACA and Air Force personnel in the maintenance and testing of the XS-1 program. On July

14, the majority of the crew departed Muroc. Frost, together with four other personnel (Mac Hamilton, Jack Russell, James George, the nitrogen evaporator operator, and his wife Elle, Frost's secretary) remained behind to complete minor modifications to the XS-1 number 1 and to the nitrogen evaporator. However, Russell's stay turned out to be longer for a slightly unexpected reason. The Air Force's Muroc Flight Test hangar boss, Harry Clayton, had worked with Russell during the Bell XP-59 jet tests at Muroc during the early 1940s. Now he asked Russell to stay on to help the new Air Force team as the civilian crew chief for the remainder of the accelerated program. Russell accepted the new position, and along with Frost retained a unique perspective on the eventual outcome of the program. (Hamilton also remained at Muroc. However, his duties took him over to NACA as the crew chief on the number 2 airplane.)[11]

Although July turned out to be a busy month the wheels of bureaucracy turn slowly even in an accelerated program. The Flight Test Division was notified on May 1 that they would conduct the Air Force program. This was confirmed in late May, again in early June, and finalized on June 24. However, the actual paperwork transferring the B-29 and XS-1 to Flight Test for the program did not materialize until July 3. It was reviewed until July 8. The accelerated program finally got underway on July 10 when the formal sign-off occurred at Wright Field. Also, during the second week of July, the last principal Air Force participant appeared in the program. Major Roberto L. "Bob" Cardenas was selected to pilot the B-29 and serve as the administrative officer in charge of the project. Cardenas was born in March 1920 in Yucatan, Mexico. He attended secondary schools in San Diego. He entered the California National Guard in 1939 and commenced aviation training in 1940. He received his wings in July 1941. Cardenas served as a flight instructor, glider pilot instructor, engineering officer, executive officer (at Twenty-nine Palms) and flight test officer, operations officer, and finally director of the flight test unit, at the Experimental Engineering Laboratory at Wright Field until January 1944. While serving in the European Theater of Operations, Cardenas flew B-24 missions over Germany earning an Air Medal and two oak leaf clusters before being shot down and captured in March 1944. Escaping from occupied Europe, Cardenas was returned to the United States and eventually assigned as a test pilot at Wright Field. It would be Cardenas's job to pick up the B-29 from the repair facility at Oklahoma City and proceed with the Air Force team to Bell Aircraft to pick up the number 2 aircraft. Cardenas would be accompanied by Lt. Edward L. Swindell, the B-29's new flight engineer. Finally, on July 15 AMC shared with Muroc the necessary information on the details of the Flight Test Division program. The teletype instructed Colonel Gilkey to provide all necessary cooperation to complete the project expeditiously. But just as all the details of the program seemed to be coming together, a dark cloud appeared on the horizon, a cloud over the very heart of the program.[12]

After returning from Bell Aircraft, the Air Force team busied itself with the mundane business of preparing for the trip to California. All was ready in anticipation of the completion of the B-29 overhaul. Suddenly, in mid-July, Colonel Boyd summoned Yeager to his office for an unexpected conversation. In the presence of Lieutenant Colonel Ascani, Boyd began a circuitous series of questions on the XS-1 program. Yeager was confused as to the purpose for the questions and the goal of the conversation. Finally, after testing Yeager with a series of issues, Boyd asked the real question: was Yeager married? Yeager was not expecting that question, but it was clear from Boyd's remarks that Yeager's answer could determine his continued participation in the XS-1 program. While acknowledging that he was married, Yeager framed the answer to show how it would make him more cautious in the air. He would fly by the proscribed program. This seemed to satisfy Boyd and the moment of danger passed. Yeager would continue as the prime pilot. With that issue aside, Boyd added some final counsel and the program was set.[13] While the personnel were ready, however, the equipment continued to cause delays.

As late as July 18, six days after the maintenance work at Oklahoma City was supposed to be finished, neither AMC, Muroc, or NACA knew the status of the B-29. When the B-29 was finally ready, Cardenas and the new Air Force team picked up the bomber and left on July 23 for the flight to Bell Aircraft. Three days later, the newly refurbished B-29 and its Air Force crew left Niagara Falls with the XS-1 number 2 for the trip to the California coast. The bomber proceeded as far as Denver on that day and arrived at Muroc the next day.[14]

Glide flights were anticipated to begin within the week and powered flights were scheduled about two weeks later. The Air Force used the time for routine maintenance and housekeeping duties. As usual, the B-29 needed maintenance work after its recent long haul from the East Coast and the Air Force team began replacing the number 3 engine mounts. Not only were parts unavailable, but the simple things, such as a spanner wrench (for tightening the propeller bearing retainer nut), had to be obtained from another base. The resulting delays halted the B-29 crew initiation into the mysteries of the XS-1 drop procedure. Worse, maintenance work revealed an unanticipated setback.

On July 31, Cardenas notified Colonel Boyd that it was his opinion, based upon the unsuitable condition of the B-29, that Oklahoma City had failed to perform the proper maintenance during its 200-hour overhaul. Since the pickup of the B-29, the Air Force crew had found a leaking fuel injection pump gasket on the number 4 engine, experienced a failure of the number 2 engine rear oil pressure transmitter, and discovered defective engine mounts and spark plugs on the number 3 engine as well as a loose shaft thrust bearing nut on the number 3 engine. If Oklahoma City had done its job, Cardenas stated, then at least three of the items would have been corrected. In any event, four days were lost in delays for parts and equipment. Even after repairs, the B-29 could not begin test exercises until after a checkout flight. On August 1, Cardenas took the B-29 aloft for a routine engineering flight. Once that flight was completed, the Air Force crew believed they were now ready to move to the next stage. The tentative plan was to begin glide familiarization flights for Yeager on Monday, August 4.

Frost used the downtime to conduct a project familiarization school for new personnel beginning on July 29 and scheduled to last a week. The workshops included rehearsals of the detailed duties for everyone and review of the work scripts that Frost had written. Frost also requested that the AMC expedite the XS-1 program acceptance paperwork before the initiation of the flight tests. Procedurally, all three agencies—NACA, XS-1 project office (Cardenas), and Bell representative (Frost)—notified AMC Wright Field that in-flight photography was no longer necessary. Wisely, Wright Field overruled them and instructed Muroc Field to continue to ensure that the FP-80 (#45-89406) be used as chase/photographic plane on the flights (thus preserving the valuable photographic record of this epic adventure). By August 1, Frost had finished the familiarization school. He believed the group ready to attempt the first Air Force flight in XS-1 number 1. Unfortunately, one small problem continued to linger, one no one at Muroc seemed able to fix.

When the Bell crew left Muroc in mid-July, the crew chief and nitrogen evaporator operator remained behind to install new steel tubing in the engine compartment of the XS-1. The equipment would come from the number 2 aircraft if NACA agreed to this procedure. NACA did not agree and so Frost was forced to order the parts from Bell. Unfortunately, Bell Aircraft was closed for vacation and so by August 4 it was beginning to look like it was better to proceed without the parts rather than risk further delay for a simple glide familiarization flight. Yet even then, bad luck followed the program. On the first flight attempt, a brake system leak on the XS-1 was discovered. The flight test was rescheduled for August 5, but delays in fixing the brake system and difficulties in the installation of different springs in the stabilizer control valve postponed a flight. Again Colonel Gilkey contacted Voyles at Wright Field to ask that XS-1 acceptance be expedited due to the imminence of the flight tests. Voyles was requested to contact the Muroc procurement inspector to ensure the smooth transition of this process. However, acceptance or not, Gilkey notified Voyles, the tests would proceed as rapidly as possible.[15]

The necessity for progress did not come simply from the Air Force team's restless anticipation. Other events at Muroc also put the word "hurry" into the Air Force vocabulary. On August 4, *Aviation Week* magazine advised its readers that the Navy would shortly attempt to smash the previous Air Force world speed record set in June by Colonel Boyd in the P-80R.[16] Although the Navy had originally thought the McDonnell F2D Banshee would be the likely candidate to break the Boyd record, they were now touting the new D-558 Douglas Skystreak as the logical recordbreaker. On August 16, the D-558 flew a total of four times while gradually increasing speed. Finally, on August 20, Navy commander Turner Caldwell piloted the number 1 D-558 airplane to a new world speed record of 640.663 MPH. Five days later, Marine Corps major Marion E. Carl raced to 650.791 MPH.[17] This new record, plus the continued appearance at Muroc and on the drawing boards of faster and faster prototype aircraft, made everyone at Muroc doubly anxious to get on with the XS-1 flights. Time was not a spare commodity in Air Force eyes.

On August 6, Cardenas filed a proposed flight plan for the first glide flight.[18] In that document, Cardenas listed the principal purpose of the flight as pilot familiarization. Any procedures to check out XS-1 flight characteristics would be at the discretion of the pilot. The NACA strain gauge, internal airspeed, and altitude recorders were functioning for the glide flights, but the telemeter was not installed so little usable data was obtained.

Takeoff and climb to altitude were routine with Cardenas handling the B-29 pilot duties and a Lieutenant Horn along as copilot; Yeager and Ridley were passengers. As in past drops, at 7,000 feet Yeager descended into the B-29 bomb bay and entered the experimental craft. Swiftly, Yeager completed securing the restraining straps and making the various cockpit hookups. Once finished with the interior work, he motioned to Ridley by tapping on the canopy. At that signal Ridley lowered the cabin door into place. As the B-29 continued its climb in the rapidly thinning air, Yeager and Ridley worked to button up the XS-1. As the plane neared 10,000 feet, the decreasing oxygen made it increasingly difficult for Ridley to perform even simple tasks. Finally, everything was ready and Ridley retreated into the B-29. As the B-29 reached 16,000 feet, Hoover and Frost took off to follow the two aircraft in their P-80s. The B-29's long climb to drop altitude continued and Yeager used the time to check the cockpit instrumentation. After XS-1 checkout, Yeager notified Ridley that he was ready. At 26,000 feet, Cardenas initiated the drop sequence. Approximately three miles west of Muroc Field, the B-29 dove to 25,000 feet at an IAS of 250 MPH. Leveling off, the B-29 released its cargo and Yeager took his first step on a journey into history.

As part of the drop routine, the B-29 pulled away from the XS-1 as soon as separation was completed. Frost, flying low chase outboard of the B-29 and to the right, but level with the XS-1, was stationed to watch the drop. Just before countdown, Frost moved above and behind the B-29, but still to the right. Hoover, flying high chase in the P-80 above and forward of the B-29, followed Yeager during the descent and provided geographic reference information. As soon as the XS-1 cleared the B-29, Yeager executed two slow rolls. As Woolams and Goodlin had discovered previously, the XS-1 glided like a bird. With no sound in the cockpit except his own breathing, Yeager put the XS-1 through its paces. He found the plane very responsive and graceful in motion. Its handling characteristics were a delight, but without power the journey proved to be short. Within eight minutes, Yeager found his altitude below 5,000 feet and lined up a smooth landing on the dry lakebed. The first Air Force step toward the sound barrier proved effortless. Yeager summarized his feelings about the plane to Frost: "Best damn plane I ever flew."[19]

On the morning of August 7, Cardenas took the B-29 aloft for the second glide flight of the Air Force program. Frost and Hoover again flew chase. As on the first flight, Yeager was authorized to wring out the plane at his discretion. Once again, Yeager was impressed with the flying characteristics of the XS-1. During the flight, he surprised his accompanying chase plane by taking his hands off the control wheel and raising them for Frost to see, as if to say, "Look, Ma, no hands."[20] The flight was a routine checkout and landing was without incident. Preparations immediately commenced to proceed with the third and final glide test scheduled for the next day.

On Friday, August 8, the Air Force team took the XS-1 up for its final familiarization flight.[21] With Cardenas once again handling the B-29 flight chores, Yeager entered the XS-1 and prepared to follow the routine drop sequence. Frost and Hoover flew chase. After a clean separation from the B-29, Yeager again put the XS-1 through simple procedures to examine the plane's flight characteristics. However, as with the past tests, Yeager couldn't let this test opportunity go by without trying something novel. Not only did he execute a two-turn spin, but he also spent the descent portion of the flight in a mock dogfight with Hoover. Landing and rollout were once again reported as satisfactory. Yeager's enjoyment was evident. Likewise, the NACA were very pleased with the Yeager reports. But the signs were already visible that Yeager was not going to simply fly the past test cards.

The end of the glide flights marked a pause in the operational proceedings at Muroc. Yeager, Hoover, and Cardenas returned to Wright Field where Yeager received indoctrination in the use of his pressure suit. At Muroc, the oft-repaired B-29 was once again found to be in such poor mechanical condition that a 100-hour (major) inspection was initiated, only 23.5 flight hours after the 200-hour overhaul at Oklahoma City. The maintenance review took over one week to complete and confirmed that the previous maintenance at Oklahoma City had failed to adequately repair the engine mounts. As a result, the bracket for engine number 1 had begun to wear through the exhaust collector ring. Concurrently, the fuel filters were extremely dirty and local speculation was that they had not been changed at the last inspection. While the Air Force team reworked the B-29, the NACA team continued work on the two XS-1 aircraft. NACA instrumentation was checked out, and the missing telemeter was found and installed. On August 15, NACA pilots Herbert H. Hoover and Howard C. "Tick" Lilly arrived at Muroc to begin their XS-1 flight training. Both programs now appeared to be moving. But appearances can sometimes be deceiving, as was soon discovered.[22]

While the NACA team continued work on both aircraft, the Air Force team prepared to return to Muroc. On August 19, Colonel Gilkey received official AMC notice clearing Major Cardenas, captains Yeager and R. D. Smith (the new B-29 copilot), and Lieutenant Hoover to work at Muroc for a period of 30 days. On August 21, they arrived back at Muroc and set to work to prepare for the first powered flight. On August 22, Cardenas took the B-29 up for another engineering test flight, but the anticipated XS-1 engine ground run could not be conducted that day because of a lack of pressure fittings for the pressure lines. Once again, maintenance delays proved troublesome to the program. The needed parts had been ordered by Muroc supply, but some were not standard Army issue and thus were unavailable. Of the ones that were standard, none were available anywhere on the West Coast. Consequently, no fittings were expected to be available before Monday, August 25. Under that timetable, the first flight could not occur until August 28.[23]

While this was not the first time a delay was caused by unavailable parts, the situation seemed to keep getting worse. On August 15, Williams informed Soule that the problem with parts supply was becoming intolerable.[24] Muroc's requests would go to the Sacramento depot and no one would hear anything for perhaps two weeks to sixty days. Williams advised Soule that Charles Hall of Wright Field, currently visiting Muroc, was trying to set up a telephonic inquiry/rapid response system. If Sacramento had the parts, a plane would be sent that day. If they did not, they would advise Muroc where to find the part. If that procedure could receive clearance (and Wright Field would have to provide that clearance), Williams stated, then it might be possible to keep the project on some semblance of a schedule. Without that process, Williams advised Soule, "we are sunk." But what seemed so straightforward at Muroc appeared different half a continent away.

The latest delays in finding parts and equipment was not a problem Wright Field had anticipated. Wright Field had many experimental aircraft programs under development of which the XS-1 research project was only one. The routine criticism contained in Muroc report 8-378 to AMC of the manner in which this problem was being handled by Wright Field prompted a strong reply from AMC to Colonel

Gilkey.[25] In the AMC teletype, Gilkey was requested to have the critic, Walt Williams, catalog the problems and how Wright Field had "hindered, impeded or delayed progress on the XS-1 project." After all, Flight Test personnel operating at Muroc weren't complaining. If nominal planning was executed, Wright Field advised Gilkey, then there would be no problems.

The message could not have been any clearer and the target of the criticism was obviously Walt Williams. It was the last straw for Williams, who fired off an angry memorandum to Hartley Soule at Langley later that day. He noted that the fact that the Air Force group at Muroc was not complaining was not surprising since NACA personnel had practically ceased work on their plane to help maintain the #6062 aircraft. In addition, the parts Williams had complained about had been ordered in *June*. The handling situation at the Sacramento depot slowed the process to a crawl. When Williams had advised Hall of the problem upon his arrival at Muroc, Hall had been sympathetic and forwarded the teletype to Colonel Smith about providing the "Blue Streak" expediting process. Apparently, Williams told Soule, this teletype from AMC was in response to the earlier Hall note. The newest teletype said to notify Colonel Smith—the same Colonel Smith who received the Hall message. Williams calmed down after further conversation with Muroc-based Wright Field personnel. However, as he summarized to Soule, he was forwarding the original memorandum in any case because he felt it was a "very unjust criticism and reflects a poor attitude and absolutely no understanding of the set-up on the part of someone at Wright Field."[26] In reality—in spite of money woes, great distances, and interminable bureaucracy—the XS-1 program was moving forward with increasing speed. Results were just around the corner.

On August 25, the missing parts arrived at Muroc, and preparations for the engine ground tests proceeded.[27] It was anticipated that the Air Force would perform three ground tests, with the first one scheduled for August 26. However, on Tuesday a leak in the fuel line was discovered during a pressure check of the system. As a result, repairs forced the postponement of the engine test until Wednesday, when Yeager made the first ground test for the engine and everything performed normally. Wright Field was notified that additional tests would be conducted on Thursday, August 28; the first powered flight was anticipated for August 29.

Powered Flight

On Friday morning, August 29, the Air Force team arrived at Muroc as dawn came over the horizon.[28] The preflight briefing was terse and businesslike. The first Air Force rocket flight would allow Yeager to familiarize himself with the XS-1 under power and to explore the handling characteristics near Mach 0.80. During the briefing, Frost reminded Yeager of the continuing trouble with the stabilizer actuator. Yeager was instructed to fire only one rocket cylinder at a time to keep speed to a moderate level near Mach 0.82. As a conclusion to the flight, Yeager was instructed to execute a fly-by of the Muroc tower, where observers from Wright Field would be watching the flight. The operative words for the first flight would be *slow* and *by the book*. Frost and Hoover provided some final thoughts to Yeager and the team prepared for the first Air Force powered flight. Cardenas received clearance from Muroc Tower and the B-29 rolled down the runway and off into the high desert air.

At 7,000 feet, as practiced so many times by the three previous XS-1 pilots, Yeager moved back into the bomb bay and climbed onto the ladder. As the ladder dropped into position beside the XS-1 door, Yeager slowly inserted his legs, then his waist, and finally his shoulders and head. There was never any getting used to this contortion act in the arcticlike wind blast. Once Yeager was in the cockpit, Ridley helped put the door in place. Yeager quickly began the process of working through the predrop checklist. Upon conclusion he notified Ridley. who responded with the cautionary "Take it easy, son." Hoover and Frost quickly took to the air and moved into position. At 20,000 feet, with five minutes to go before drop, Cardenas notified Yeager, and the XS-1 pilot started dome-loading the first-stage regulator. Next, Yeager pressurized the fuel tank and the lox tank. After carefully double-checking to ensure proper psi, Yeager asked Frost, now in position below and behind the B-29, to verify emergency jettison. Frost observed the procedure and notified Yeager of its satisfactory function. At that moment, Cardenas warned there was only one minute to drop.

Slowly the passage of time brought the steady voice of Cardenas back on the air. Starting the countdown at 10, the B-29 pilot slowly intoned the individual digits maintaining his record for always missing one (number 7 this time). At 21,000 feet and an IAS of 255 MPH, Yeager heard the magic word "drop" and felt the lurch as the XS-1 dropped from the bomb shackles two hooks. Separation was clean, although Yeager immediately noted the difference in aircraft weight from the glide flights, just as Goodlin had done nearly one year before. For seemingly endless seconds the XS-1 slowly drifted and then began to fall gently away. Blinded momentarily by the bright sun, Yeager allowed the XS-1 to drop nearly 500 feet. Almost 15 seconds had elapsed before he tripped the first switch for the number 1 engine chamber. The rocket plane streaked forward. Swiftly, as he had practiced on the ground runs, Yeager sequentially fired and then shut down the remaining chambers. It was by the book. The Machmeter quickly revealed Mach 0.70 during the ignition/shutdown sequence. But what accompanied the routine was not on anybody's flight card—while the number 3 chamber was ignited, Yeager put the XS-1 through a slow roll. Frost, flying chase, was taken by surprise and called out, "My God . . . That's not in the flight plan, Yeager!" Frost knew the roll could shut down the engine by unporting the lox tank outlet. The warning, however, came too late to register with Yeager, who had found a demon to retain his attention. The rolling maneuver had created a zero g condition in the XS-1. Just as Frost feared, the lox pressure dropped and shut down the rocket cylinder. As Yeager completed the roll, the plane dropped back into reality, which restored power to the engine. Reignition allowed the flight plan to continue.

Diving toward the desert floor, Yeager called back to the eversteady Frost that the snaking oscillation that Goodlin noted on his first flight had returned, similar to the snaking motion that Yeager had found in the XP-83 when he first flew that aircraft. Yeager was certain it was merely fuel sloshing (as Goodlin had concluded) because the liquidometer's fuel and lox needles oscillated in resonance with the "snaking." Yeager radioed Frost, who concurred, but Frost had a much more pertinent issue on his mind. Yeager was diving like a bat to the deck and Frost didn't know why. "Hey! Where are you going now?" he asked. Back came the reply, "To show the brass down there a real aeroplane."

At 6,000 feet Yeager began to level out. By 5,000 feet, he was only 2,700 above the high desert floor. Frost and Hoover were far above and behind. In fact, Frost was diving so hard that his P-80 began to shake with the first signs of compressibility as his speed edged up to Mach 0.84. (Waggish observers could well ask which plane—XS-1 or P-80—was the test plane on this day?) In the little saffron-colored XS-1 Yeager leveled out and closed the cabin-pressure regulator. What came next almost washed him out of the program. With only one chamber running, Yeager streaked by the tower and pointed the XS-1 upward at a 15-degree angle. Rapidly he fired chamber 2—then 3—and finally all four were running. The thrust slammed Yeager back into his seat as the full power allowed the airplane to zoom up from 9,000 feet to over 30,000 feet in an instant. At 15,000 feet, Yeager flashed by Frost in the now also steeply climbing P-80. The XS-1 looked like a skyrocket with an enormous blast of flame coming out its tail. With all chambers firing, Yeager had to increase the XS-1 angle of climb to almost 90 degrees. As Yeager rocketed nearly straight up, he struggled to no avail to hold the aircraft under Mach 0.80. Mach 0.80 came and went quickly and the Machmeter kept rising. The XS-1 climbed so steeply that Yeager couldn't see the ground. At Mach 0.83 Yeager shut down the number 4 chamber and executed the last half of a barrel roll by pulling the nose on through to a 45-degree angle as in a loop. Finally shutting down all cylinders, Yeager realized that he was all alone in the clear sky. At Mach 0.85 the XS-1's speed ceased its rise and Yeager slowly angled over to begin a descent. On his way down, he executed several 3–4g accelerated stalls at observed Mach 0.70. The plane had handled perfectly, but as Yeager later reported, he possibly was too excited to note every slight condition that the plane might have experienced.

The XS-1's descent to the desert lakebed and landing were mercifully uneventful. On the ground, Colonel Gilkey immediately notified Wright Field of the mission's success, respectfully omitting any mention of Yeager's extracurricular flight maneuvers. The flight itself certainly provided reason for some Air Force pride. Mach 0.85 at 30,000 feet was better than anything the program had achieved previously; and this was the *first* powered flight! Yeager had covered one-quarter of the distance to the unknown in his first try. An accelerated program was certainly what this effort seemed to be. Much remained to be done, but the program appeared to be moving again after the long pause caused by the transfer of ownership from Bell Aircraft. Writing to a former flying buddy, Yeager summed up the sensation: "I was so darned excited, scared, and thrilled (you know that first kill in Germany feeling) I couldn't say a word until the next day."

The joy over the completion of the first powered Air Force flight was, however, quickly tempered by the ever-present problems. First, the telemetering equipment didn't function during the flight. Consequently, no data was obtained. Second, the film in the camera behind the pilot's head was found to be overexposed and thus useless. Finally, everyone was sure an explosion of anger named Albert Boyd was bound to come from Wright Field. Few believed he would let Yeager's unauthorized flight program go by without a comment. When Frost called Larry Bell to inform him of the success, Bell only chuckled and said, "Wait until Al Boyd hears about this, . . . He'll either pin a medal on that boy or give him hell."

At Wright Field, Colonel Boyd read the flight test report with the intense eye of a commanding officer. Although momentary anger erupted while reading the final paragraphs of the report, Boyd still couldn't help suppressing a smile. Pilot to pilot, Yeager had done a good job. But—this first flight was not the approved program. A congratulations couched in stern terms was the result. The punch line of the Boyd memo to Yeager was the statement

> I would personally like to have an explanation from you . . . as to your reasons for exceeding the authorized Mach number on this flight . . . Please remember, . . . the instructions I passed on to you personally here at Wright Field with respect to the value of the pilot and the plane to the Air Force . . . The Air Force does not consider you or the plane expendable, so please approach higher speeds progressively and safely to the limit of your best judgement.

Back at Muroc, receipt of the Boyd letter caused Yeager some unease coupled with feelings of pride because Boyd sent it to him personally. The closure "Sincerely" took additional sting out of the otherwise stern sentences. Yeager and Ridley prepared a response that covered, as best they could, the wild streak that had infected Yeager on the flight. In the letter, safety and the professionalism of the Air Force were stressed. So also was the judgment that all was okay up to Mach 0.85. Finally, Yeager promised that violations of the approved flight plan "will not be repeated." Little further was heard about the stunt.

The NACA personnel were not quite as pleased as everyone in the Muroc station Air Force team. The NACA's mission was to get data, and the loss of telemetering data meant a less than perfect flight. The NACA wanted to know more than the pilot's opinion of what happened between Mach 0.80 and 0.85. The NACA always had to remind the pilots that more than just their opinion mattered . . . data told the *truth*. Even Yeager admitted in his pilot's report that he might not have noticed the small things occurring around him. In going up to the sonic barrier, the insignificant little changes were the ones that might mushroom into full-fledged catastrophes at supersonic speed. The NACA believed that another flight had to be conducted to find out what really was going on with the airplane in the mid-Mach 0.80 range. But repeating the test was not as easy as the NACA deciding to do so. Based on this first flight, the NACA had reservations about Yeager's judgment or desire to fly the flight program, and they did not hide those reservations. Yeager noted their attitude and believed that the NACA's condescension had as much to do with his lack of an advanced educational degree as with any lack of formal test-pilot training.

On Tuesday, September 2, the NACA engineers set to work to repair the malfunctioning instrumentation. September 3 was lost as a test day because the NACA technicians couldn't finish the instrumentation repair work. By September 4, the XS-1 was ready for its second powered flight test.[29] The purpose of the flight, as recorded by Ridley, was to obtain stability data in the Mach 0.80–0.85 range at 30,000 feet pressure altitude. In short, this was to be a repeat of the previous flight but, they hoped, with the NACA recorders operating to obtain data. Drop was to be at 20,000 feet at 250 MPH. After climbing to 30,000 feet, Yeager was to make a one-cylinder speed run and a 2g turn. Next, he was instructed to make a two-cylinder climbing speed run to Mach 0.85. At the end of the stabilized run, Yeager was to execute a turn and tighten it until buffeting was reached. After jettisoning fuel, Yeager was to execute an unaccelerated stall and during descent to perform turns at 3–4g positive acceleration and speeds of Mach 0.70. The plan promised a busy flight if all went well.

With final briefing and preparations ready, Yeager took his customary position in the B-29 and the team was ready to go. Once again Cardenas guided the bomber down the runway and into the air above Muroc. At 7,000 feet, Yeager entered the XS-1 and methodically moved through the flight checklist making sure that all was in readiness. No cabin pressurization would be used on this flight. Yeager set the stabilizer in the one degree down position (set one degree above neutral) and the cabin pressure regulator in the dump position. At 21,000 feet Cardenas initiated the dive to began the drop sequence. At approximately 20,000 feet and an IAS of 250 MPH, the XS-1 was released from the bomber. Separation was smooth and clean.

Immediately after clearing the bomber area, Yeager ignited the XS-1 number 1, 2, and 3 chambers. Quickly the pilot changed the stabilizer trim setting to push the XS-1 two degrees nose-down. A 40-degree climb was initiated and Yeager guided the streaking rocket plane through its paces. At 24,000 feet, he turned off the number 3 chamber. At 27,000 feet he shut down the number 2 chamber. Speed buildup continued slowly as the airplane reached Mach 0.83. With only one chamber firing, the climb finally leveled off at 30,300 feet. Nosing over and dropping to 30,000 feet, Yeager stabilized the XS-1 and began a level speed run to check the stability of the craft. For almost one minute Yeager held the XS-1 in a level run at Mach 0.84. No indications of buffeting or trim change were noted.

With the stability test finished, Yeager ignited the number 2 chamber and the XS-1's speed increased very quickly to an indicated Mach 0.865. The number 1 chamber was shut down to prevent the XS-1 exceeding this speed. An accelerated stall was simulated by pulling into the buffet boundary at Mach 0.85. A 2g acceleration was recorded on the accelerometer. Having finished one test, Yeager gently eased the XS-1 through another accelerated stall at Mach 0.80. The XS-1 behaved in a docile manner. Yeager found the buffet to be very mild. The XS-1 maintained good aileron control throughout the stall. The pilot discovered no indications of changing elevator stick force. If a supersonic demon was out there, Yeager still hadn't found him.

With the tests nearly finished, Yeager ignited the rocket engine's number 3 and 4 chambers and put the XS-1 in a 30-degree climb. The XS-1 soared up to 41,000 feet with an indicated speed of Mach 0.80. At that point, Yeager decided to call it quits. Remaining fuel was jettisoned and the XS-1 started its descent. During the drop through 30,000 feet, Yeager executed one unaccelerated stall. No unacceptable flight characteristics were noted and Yeager continued the descent. Landing and rollout were normal. Total flight time was a very lengthy 18 minutes.

The satisfaction of the second flight was again quickly dispelled upon examination of the NACA instrumentation. Once more, the NACA telemetering equipment failed to function. Thus no stability data was recorded. The good news was the airspeed, altitude, and acceleration data were recorded for later examination. After initial review of the second flight's data, the NACA/Air Force personnel discovered that Yeager had actually gone Mach 0.9015 rather than the observed 0.865. This was even better news, given the continued superb handling of the plane at the higher speeds. Yet haste had a surprise in store. After further analysis of the data, Muroc had to notify Wright Field that its initial report was in error. The XS-1 had only achieved a top speed of Mach 0.89. The September 4 flight again demonstrated the excellent aerodynamic characteristics of the XS-1. More important, the flight had revealed no surprises up to Mach 0.89. The distance to the sound barrier was rapidly closing, but nobody was yet ready to declare smooth sailing to the finish.

On September 5, Soule, Herb Hoover, and Walt Williams gathered at Muroc to meet with the Air Force project team to discuss the NACA's XS-1 data gathered from its falling body tests, wing-flow tests, and research in the eight-foot high-speed wind tunnel. The NACA presented data that could possibly give some indication of XS-1 flight characteristics in the transonic region. Since the Air Force flight program was based on exceeding the sound barrier as rapidly as possible, NACA believed the Muroc team was entitled to see all data as quickly as it could be assembled, since the upcoming tests would carry the XS-1 into the region of greatest unknowns. However, the NACA data stopped around Mach 0.93, just at the area of greatest uncertainty.[30]

Even without data, the Air Force team was not flying entirely blind. For the XS-1 tests, Ridley, Hoover, and Yeager had developed a flying technique that provided a window through which to view the unknown just beyond the current flight. Specifically, Yeager and Hoover had been testing P-80 and P-84 jets to speeds where they would buffet while flying straight and level. They found that if they slowed down, rolled the plane over, and pulled 3 g's at the lower speed, the plane would react in the same manner as at the higher speed. Thus, Yeager decided to use this information on the XS-1. Whenever the XS-1 had reached the specified Mach number for the current flight, he would roll the XS-1 and pull several g's at that speed to give him a good idea of the flight conditions to be found at a higher Mach number. It was a technique that Yeager was to rely on strongly in the coming flights.[31]

On Monday, September 8, the Muroc team was ready to try another powered flight in XS-1 number 1.[32] Since the previous flight had produced no stability data, the NACA team insisted that a repeat of that flight program was necessary before further advances could safely be authorized. The Air Force team agreed and flight preparations allowed the B-29 to try that day. The purpose of the flight was listed as stability check at an indicated Mach 0.87/0.88. Such a speed, if achieved, would equal approximately Mach 0.91 in real terms. Drop would come at 20,000 feet at an IAS of 250 MPH.

The B-29 launched the XS-1 and Yeager allowed the two planes to separate. Once safely apart, Yeager fired off three XS-1 engine chambers and climbed to 30,000 feet. Speed was recorded as Mach 0.80. Once altitude was achieved, Yeager shut down all chambers and slowed to an observed Mach 0.70. Having stabilized the craft, Yeager reignited chambers 1 and 2 and speed rapidly built up to Mach 0.88. Yeager shut down chamber 2 and conducted an accelerated stall, using full up elevator, into the buffet boundary at 2.5 g's. Yeager again reported buffeting as normal with the XS-1 exhibiting good lateral control throughout the stall. Once that test was completed, Yeager again reignited chamber 2 and climbed to 35,000 feet. A repeat of the previous test was conducted once the airplane was stabilized at the new altitude. Yeager again found stick force to be very light. Indeed, a trim change could be noted only by the different stick positions required for level flight.

As a conclusion to the aerial tests, Yeager turned on two chambers and put the XS-1 into a climb. He maintained this flight profile until fuel was exhausted. Descent and landing were routine. However, the bad news again surfaced. When the NACA engineers checked their instruments, they found no results to collect. (They privately believed that Yeager had simply forgotten to trip the switch.) In any event, whether due to error or malfunction, the NACA had no results to measure. Fortunately, the oft-troublesome telemeter functioned and revealed interesting results from Yeager's stalls. While Yeager was correct in stating that buffeting seemed mild in level flight, telemeter data indicated that heavy buffeting seemed to appear above 2 g's in the turns. Further, the data revealed that while Yeager had already noted a XS-1 tendency to experience a nose-down trim change, at Mach 0.88 a new and very worrisome phenomenon appeared. The XS-1 seemed to develop a tendency to want to nose up. Yeager had not noted serious trim changes, in spite of the significant elevator angle trim changes, because the forces involved appeared negligible. The NACA engineers were not sure what this data meant. It might be just a controllable trim change requirement, or it might be that the region around the sonic barrier caused aircraft to develop dangerous instability with rapid changes in trim throwing the nose first down and then up and then out of control. No one was sure, and no one wanted to risk premature discovery.

After the flight, Cardenas notified Wright Field that operations would pause briefly while the Air Force technicians checked out the XS-1 motor and pressure regulators.[33] For some time Frank Davis, the RMI field representative at Muroc, had been verbally reminding Frost, the Air Force, and the NACA that the RMI rocket engine had a finite service life. RMI estimated that life to be about one hour between overhauls. The XS-1 number 1 engine had now exceeded that parameter by 30 minutes. RMI wanted the engine removed and shipped to New Jersey for inspection and overhaul. Although this procedure would certainly slow the Air Force accelerated program, RMI reminded the Air Force that it "cannot assume responsibility for consequences due to probable malfunction of the engine if the above recommendations are not followed." Failing to elicit any action from the verbal warnings, RMI decided to "put it in writing." Now there was a clear record that RMI would not be responsible if the engine "malfunctioned"—a nice little euphemism for a very healthy engine explosion or maybe just a little midair fire from which there might be no escape for the pilot. In any event, Cardenas decided that the engine was due for a field inspection.

Engine work was not the only problem facing the Muroc team. In the flights to date, the curve of the elevator angle had closely matched wind tunnel predictions. Yeager had noted the decrease in elevator effectiveness. Ridley and Williams had huddled to discuss the heavy buffeting found in the NACA data from the most recent flight. They recommended that Yeager change the stabilizer settings in flight to test whether trim changes could cure the problem.[34]

On the morning of September 10, the NACA/Air Force team was ready to try for the fourth powered flight of the XS-1.[35] The procedure for this flight was to repeat the tests of the first several flights. Drop altitude was scheduled to be at 20,000 feet at an IAS of 250 MPH. The purpose was to obtain the still missing data on static longitudinal stability, determination of drag coefficient, and the investigation of the stabilizer effectiveness at the higher Mach numbers that Yeager had been flying. Takeoff and climb to altitude were shaping up as routine until the B-29 reached 15,000 feet. With Yeager already in the XS-1 and going through predrop checkout, Cardenas received a call from the NACA radar truck informing him that no one had remembered to secure the telemetering battery in the XS-1 mid-section. The heavy battery presented a potential hazard if it was loose and shifted in the airplane. While Cardenas continued to steer the B-29 upward, a hasty conference went on between all the involved participants. Most believed that the battery terminals would hold it in place. After further consultations, a decision was made to go ahead and drop the XS-1.

Once the XS-1 cleared the B-29 area, Yeager ignited the number 2, 3, and 4 chambers and put the XS-1 in a rapid climb to 35,000 feet. Once he reached that altitude, Yeager shut down the number 2 chamber and executed a stabilized speed run. But now Yeager noted a peculiarity in the speed reading. Whereas in the past two chambers firing would push the XS-1 up to Mach 0.88, now Yeager could seemingly only coax Mach 0.80 out of the research aircraft. Swiftly checking the chamber pressure readings, Yeager found they were all registering normal outputs, yet the plane seemed to lack power. In response, Yeager reignited a third chamber to try to increase the XS-1's speed. Speed certainly increased, but the lag on the Machmeter still seemed peculiar to Yeager. In an effort to play it safe, Yeager shut down power to the engine and jettisoned the remaining fuel and lox. The glide down to the lakebed and landing were routine.

The technicians searched the XS-1 for the source of the difficulties that Yeager reported. Discovery was not long in coming. Before the flight, a new 80,000-feet altimeter had been installed in the XS-1 and now had a sizable leak in the case. Since the cabin was pressurized, the static pressure it measured was not that of the free stream, but rather the cockpit. Consequently, all altimeter and Machmeter readings were wrong. A review of the NACA data revealed Yeager had actually achieved speeds up to Mach 0.92 (including position error) at 36,200 feet. That was the good news. The bad news was that the NACA data revealed that the XS-1 had exhibited a definite tendency for the nose to tuck under.

The defective altimeter was replaced on September 11. Technicians also repaired a slight leak in the XS-1 fuel jettisoning valve. A new jettisoning switch was installed at the same time. Preparations began for an additional flight on Friday, September 12, the forty-fifth flight in the XS-1 program. The purpose of the flight was to calibrate the static pressure source. Cardenas notified Wright Field that this would be the last test until a new three degree per second stabilizer actuator could be installed in the XS-1. In the interim, Yeager would again attempt to test the different stabilizer settings recommended by Ridley and Williams before the September 10 flight.

On September 12, Cardenas took the B-29 aloft for the fifth Air Force test.[36] The XS-1 launch was again perfect. Once clear of the B-29, Yeager ignited the engine and climbed to 30,000 feet. He began a stabilized flight to obtain data from Mach 0.75 through Mach 0.91. A final data-recorded speed of Mach 0.925 was achieved. Complete data was obtained in that range with the stabilizer set at plus two degrees. A separate run at plus one degree secured data at speeds from Mach 0.75 to 0.85. However, the pressure altitude tests using a trailing bomb from the B-29 failed when the cable snapped and the bomb dropped over Rosamond Dry Lake. A rough calibration was made using the service altimeter and airspeed indicator and comparing it to the calibrated P-80. Landing and rollout were again normal.

With the fifth flight completed, Yeager and Hoover headed back east. They were due to arrive in Dayton on September 16 so that Yeager could brief Colonel Boyd and be fitted for his pressure suit. After the stopover at Wright Field, Yeager went home to Hamlin to pick up his son. Hoover stayed at Dayton to be married. All together, they drove back to California.

The NACA Flight Program

With the completion of five powered flights in XS-1 number 1, program focus shifted back to the still absent NACA airplane. The XS-1 number 2 arrived at Muroc aboard the B-29 on July 27,[37] and the NACA technicians began modifications on it the next day. The original four Bell Aircraft instrument panels were replaced with the four newly designed NACA panels, strictly to satisfy Herb Hoover. Further, the NACA decided to undertake additional changes to the plane, consisting of installation of a new emergency flaps actuation system and fuel jettison system. Also, after the near disasters involving fire, the NACA team wanted to utilize a fire-protection system in the plane and proposed installing an automatic fire extinguisher and automatic engine cutoff, and providing a new manual cutoff switch. During the initial Air Force contractor glide flights, there had been little time to work on the number 2 airplane. The NACA technicians had to wait until after the delay caused by the B-29 25-hour inspection procedure to prepare XS-1 number 2 for flight.

Once the decision was made to acquire the XS-1 number 2, the NACA faced the same problems Bell Aircraft had discovered in rocket powerplant utilization. In the past, the NACA had not believed the rocket engine was the powerplant of first choice. Now their program depended on this howling, dangerous, and temperamental propulsion system. Since April, Joseph Vensel of the AERL had been at Muroc helping with the engine work. He was expected to remain available, but the magnitude of the project required additional help. To redistribute the workload, a conference was held at the AERL on July 28. During the meeting, it was agreed that the AERL would be responsible for developing procedures for operation and maintenance of the RMI engine. In order to assist that effort, RMI would be asked to furnish two engines to the laboratory, both with stainless-steel propellant lines. One engine would go to Muroc for testing and the other engine would remain at AERL to be tested. Formal approval for this program came from the NACA HQ on August 12.[38]

By mid-August, the NACA activities gained speed as the tempo of operations picked up. Henry Pearson, then serving as chief of the aerodynamic loads section at the LMAL, visited Muroc to observe procedures and make recommendations for NACA future activities. During his stay, Pearson reviewed the staff situation with Williams and De Beeler, who were informed that the NACA HQ had decided to convert the temporary status of the Muroc station to permanent status as of August 24. (Formally, the change to the Muroc Flight Test Unit occurred on September 7.) As a result, the Muroc station expanded to twenty-seven people by October. Of this total, nineteen (including two Air Force civilian personnel) were assigned to the XS-1 program, including both Jack Russell (on loan from the Air Force) and Mac Hamilton. Another old hand from the Pinecastle days had returned to the program when Gerald Truszynski moved to Muroc in the spring to serve as an instrument project engineer. On August 15, the NACA's two proposed XS-1 pilots, Herb Hoover and Howard Lilly, arrived at Muroc to begin their familiarization work on the plane. Hoover was an experienced test pilot from LMAL who had already attended some of Frost's earlier workshops on the XS-1 project. The NACA had selected him to fly the initial agency flights in the number 2 aircraft. After those flights, Hoover would turn the program over to Lilly, who was assigned to Muroc from the AERL. It was anticipated that Lilly would be prime NACA pilot on both the D-558 and the XS-1 aircraft. But as fast as some of the NACA personnel came on board, the harsh climate and primitive conditions at Muroc ended others' stay. As Hoover observed, most of the difficulties with life at Muroc could be measured against personal attitude.[39]

When the XS-1 number 2 aircraft was finally ready for flight testing, the question of the NACA acceptance of the plane again came to the forefront. After the Bell Aircraft repairs of the fire damage from the June 5 accident, the NACA demanded a demonstration flight before taking control of the plane. Since neither Hoover or Lilly had flown the aircraft (and neither could fly the aircraft until the formal NACA acceptance) the task of proving flight reliability of the number 2 plane fell to Yeager.

Initial planning called for Yeager to acceptance-test the XS-1 number 2 for NACA on September 24. However, a variety of problems hampered efforts to keep that schedule. First, the aircraft's nosewheel shimmy damper failed. Then, two turn and bank indicators were destroyed when they were inadvertently connected to the wrong pressure line. As a result, the XS-1 number 1 aircraft was cannibalized to make the number 2 airplane airworthy. Technicians stripped the number 1 airplane of the following parts: turn and bank indicator, liquidometer transmitter, and nosewheel fork. Preparations for further flights on the number 1 aircraft—the accelerated program—came to a halt, which did not sit well with the Air Force team. Walt Williams had to resort to a little "stalling" with the military group in order to get the NACA flight program in operation.[40]

On September 25, the B-29 rolled down the runway with the XS-1 number 2 aircraft firmly secured in its belly.[41] Ironically, the first NACA flight would be an Air Force show. XS-1 entry, checkout, and separation were normal. However, from the moment of separation, Yeager experienced difficulty maintaining constant tank pressures. During the flight, the number 4 chamber of the rocket motor burned out due to a lean fuel mixture. In addition, the loss of source pressure during the flight was very high, hence the required 8g pull-up demonstration was performed without power. During the flight, Yeager reported only mild buffeting consistent with data obtained with the number 1 airplane and its thinner wing. Touchdown at Muroc Field and rollout were uneventful. After the landing, the NACA technicians searched for the cause of the changes in tank pressure. The propellant valves had not failed. However, a cracked fitting in the line between the first- and second-stage regulator was discovered, which accounted for the loss of source pressure and was thought to be responsible for Yeager's inability to maintain tank pressure. In spite of the problems, the flight was dubbed a success and NACA formally accepted the number 2 aircraft.

Preparations for the next Air Force flight commenced with the conclusion of the NACA acceptance flight. The most significant modification on the number 1 aircraft was a "final" fix for an old problem.

The horizontal stabilizer actuator was reworked to provide a rate of three degrees per second of change. However, to install the new actuator motor meant the XS-1 engine had to be removed from the plane. This procedure caused the Air Force to decide to ground-test the engine to ensure proper reinstallation and reliability. But simple replacement of the actuator motor proved not so simple.

On Monday, September 29, the scheduled number 1 aircraft engine test was scrubbed because of difficulty with the new stabilizer actuator controller. The existing pressure system was inadequate to operate the stabilizer satisfactorily. After two days of tinkering with the system, Frost unilaterally made the decision to have a new set of springs made in Los Angeles which would allow the controller to center the stabilizer. After contacting numerous friends and industry representatives, Frost found a company that could manufacture the specialty springs. The newly machined springs allowed the stabilizer to respond more precisely. The new parts required the XS-1 engine to again be removed from the plane, thus increasing the program delay.[42]

Racing with the Wind

On the morning of October 3, the Air Force team was still not ready to begin XS-1 testing.[43] After all the serious difficulties of the recent past, this problem was mundane. The Air Force needed to patch the XS-1 nosewheel. No spare nosewheels or tubes were available at Muroc and temporary repairs were initially inadequate. As a result, takeoff was delayed until the late afternoon. A group of special guests from the East were to witness the flight. John Stack, the assistant chief of research at LMAL, had flown out to watch the proceedings and report on the XS-1 project's progress. He was accompanied by other NACA officials and Air Force officers.

The purpose of the flight was to make two runs at 35,000 feet up to a Mach number of 0.92, each run featuring a different stabilizer setting. Following that procedure, Yeager was instructed to make a climb to 40,000 feet at a Mach number of 0.93 or 0.94. Takeoff and climb to altitude were normal. Immediately after drop, Yeager turned on the number 2, 3, and 4 chambers of the rocket engine. He executed a swift climb to 25,000-feet pressure altitude where he turned off one chamber. During the climb, Yeager quickly noted the problem for this day.

The late hour of launch provided unusual sunlight refraction through the canopy glass. This effect and the growing darkness made reading the instruments difficult for the pilot. He simply could not see the altimeter clearly. As a result, Yeager stopped the planned climb to 35,000 feet and executed the first speed run at 25,000 feet, an unstabilized run in level flight. Speed built up to Mach 0.84 and Yeager noted the increased stick force required to maintain level flight. (At Mach 0.84, it was 10 pounds.) However, Yeager also noted an unusual phenomenon. At Mach 0.85 the stick force required actually lessened. But at Mach 0.86 a light buffet started which could be felt throughout the entire aircraft and stick force pressure began to rise once again.

At this point, Yeager turned off another chamber and put the XS-1 into an accelerated 4g stall at Mach 0.85. With the fuel load still exerting a tail-heavy condition on the XS-1, Yeager found it was only necessary to release the stick pressure to approach the stall. To finish the process, however, Yeager had to pull back on the stick and hold through the stall. Coming out of the stall, Yeager noted speed had decreased to an indicated Mach 0.82. He quickly changed the stabilizer setting to two degrees nose-down. The reworked actuator functioned so rapidly that Yeager misjudged by one-half degree the rate of change and promptly experienced a negative 4g condition, which caused such a rapid loss of chamber pressure that he immediately shut down the remaining XS-1 engine cylinder.

Yeager reset the stabilizer to two degrees nose-down position and quickly attempted to refire three cylinders. The numbers 2 and 4 chambers instantly responded, but the number 3 chamber failed to ignite. Yeager made the decision to conduct the second speed run at 25,000 feet. At Mach 0.86, Yeager noted the XS-1 exhibited a right-wing heaviness. This condition increased with speed up to an indicated Mach 0.88, at which time Yeager was required to use about 50 percent available aileron simply to hold the wings level. At Mach 0.87, the XS-1 began a moderate Dutch roll, similar to previous ones. Yeager still believed this was from fuel sloshing in the aircraft and not some inherent problem with the airplane or atmospheric conditions. From Mach 0.86 through the maximum observed speed of Mach 0.88, the XS-1 exhibited light buffeting.

By the completion of the second run, Yeager had experienced enough of the shadows and reflections. Jettisoning remaining fuel, he glided back to base and executed a normal landing on the lakebed. No problems with the nosewheel were discovered. Once back at the hangar, Yeager reported that based on his judgment a two degrees per second rate of stabilizer change was about all the plane could handle. The NACA technicians proceeded to remove and begin evaluation of the data records. No further flights were planned until the nosewheel patch could be repaired. The NACA data did little to alleviate the concerns over the continuing rapid trim changes in the XS-1 handling. Speed changes seemed to clearly coincide with continuous trim changes, which seemed to indicate that any attempt to break the sound barrier was still somewhat distant. Yeager estimated the time frame to be within five flights, or about two months. The accelerated program appeared to be slowing and, as a result, the Air Force team flew to Wright Field to discuss the situation with the boss.

On October 6, Ridley and Yeager met Colonel Boyd at his office at Wright Field.[44] It was not a pleasant experience. Boyd questioned the progress of the flights. He pointedly asked what would happen over the last eight-tenths of a Mach to supersonic flight. Ridley responded in his drawl that "the airplane will just go eight-tenths faster." Yeager concurred and added, "I don't really anticipate a great deal of difficulty, Colonel. It's been easy, so far, easier than we all expected. Buffeting has been mild, and I've always had control of the airplane." But, realizing he had perhaps oversold the program, Yeager backed off somewhat with an added "Well, . . . most of the time." Boyd, however, was not so ready to concede the conquest of the unknown. He indicated his uneasiness with the results. "There is nothing to indicate that you weren't just lucky." He also verbally pinned Yeager with a catalog of flight variations, failures to detect, and pilot excitement statements that in the brevity of his remarks lends credence to his earlier comment that maybe the XS-1 program had just been lucky to have had so few accidents. Boyd summarized his concerns with the statement that success from Mach 0.82 to 0.92 was no assurance that disaster might not be waiting with the next speed increment. All the flight reports added up to increasing problems. Maybe they meant little in the overall picture of the XS-1 going supersonic, but, Boyd indicated, maybe they were alarm bells designed to keep hot-shot pilots from making foolish mistakes.

At the end of the meeting, Ridley asserted himself once more when Boyd remarked that the airplane seemed increasingly unstable as speed rose. Boyd rhetorically asked the pilots, "What if you lose complete control and the buffeting gets worse than it's been?" Ridley responded with the assurance of the brave. "Well, Colonel . . . the buffeting may decrease . . . Let's just go back and see." (Actually, the comment was more than bravado. The NACA wind tunnel data had revealed a decrease in the compressibility loss of lift around Mach 0.90.) At that point Boyd seemed satisfied that his two pilots had passed his verbal test. He cautioned them to take the future steps to Mach 1.0 at a slow and measured pace—no errors. Boyd then invited them to join him at lunch with some other guests.

At lunch with Gen. Ben Chidlaw, the commander of the AMC, both pilots were again brought back to reality when Boyd stated to the assembled group of generals at the table, "Couple of hot pilots here. Want to go back tonight and crash the sound barrier." The small joke elicited polite laughter. Yeager and Ridley realized that Boyd was still making his point. "No rudder, no elevator, buffeting getting severe at one speed, mild at the next. Nose up at Mach 0.87, nose down at .90. That aeroplane is liable to go in any direction, or all of them at once. But Captain Ridley and Captain Yeager here," Boyd smiled for effect, "anticipate no difficulty in, ah"—Boyd paused to cough to cement the effect—"attaining Mach 1 over the weekend." As if this litany of risk was not bad enough, now Chidlaw closed the story. "Lucky to make it at all." None of this information was intended to please its listeners. But the purpose of this exercise was not to stop the XS-1 tests, so Boyd now stepped in to save his pilots. Following up Chidlaw's "lucky" comment, Boyd replied, "That's it . . . They're just that lucky, General. They might just make it. So I've told them to slow down."

The two Muroc veterans sat through the rest of lunch with less enthusiasm than they arrived with. Brutal as the verbal body blows were, it was a necessary exercise. Boyd was reasserting reality into the joys of flying the hottest ride in the air. Yeager and Ridley knew the risks, but Boyd wanted to ensure that familiarity did not breed mistakes. Boyd believed in the program and he trusted his two pilots' judgment. He reminded the duo of what they thought they knew and what they really were just assuming. Yeager and Ridley departed Wright Field with a little less cockiness.

The mental notes of the Boyd conversation stayed with Yeager and Ridley on the long flight back to Muroc. Yeager did not need any prompting about safety. He already had experienced vivid reminders of the perils of experimental flying. Two somber messengers had been visiting him regularly in his dreams. One concerned the danger of fire. It was the horrible dream, that came night after night. Each time the manifestation was the same. The XS-1 was on fire and there was no way out. The other dream was equally as deadly, if not quite as mentally tormenting. This time the adjustable stabilizer was the villain. The Air Force had recently changed the three degree per second controller to a two degree per second rate. The previous abundance of stabilizer control might have been resolved by the new springs. But— there was always a but in experimental flying—Yeager dreamed about climbing out onto the fuselage and sliding back to the tail to fly (guide) the XS-1 to a landing. In any event, Boyd's speech could not tell Yeager anything new to fear.[45]

In the days after September 25, work continued on the NACA's XS-1 number 2 aircraft. The crack in the pressure fittings discovered after the Yeager acceptance flight caused the NACA to ground the plane until new fittings could be manufactured in the Muroc machine shop. But the new fittings proved difficult to make and thus it was not until October 13 that the plane was ready for an engine ground check. However, as was seemingly always the case with the X-craft, progress brought additional setbacks. As soon as the lines were pressurized on the number 2 airplane, the NACA technicians discovered leaks around the new fittings. When these were repaired, new leaks were discovered. As fast as the NACA could find and repair the leaks, they discovered new ones. In any event, the NACA airplane was out of commission and other, more pressing Air Force events were grounds for letting the work slip.[46]

On October 8, Cardenas guided the B-29 down the runway for the forty-eighth flight of the XS-1 program.[47] It was the seventh powered flight of the Air Force accelerated program. During the B-29's climb to drop altitude, Yeager entered the XS-1 at 7,000 feet. On this flight, Yeager would again pressurize the XS-1 cabin, by discharging nitrogen from the attitude gyro into the cockpit cabin. After cockpit checkout, he signaled to Cardenas that he was ready. At 20,000 feet and an IAS of 240 MPH, Cardenas finished the countdown and released the experimental aircraft from the B-29's grip.

Almost immediately, trouble appeared. The XS-1 emerged from the drop sequence in a near stall condition due to the lower B-29 drop speed. With the stabilizer set at one degree nose-down, Yeager immediately flipped the switches for all four engine chambers in rapid sequence. The roaring rocket engine caused the XS-1 to leap forward in a steep climb, so steep in fact that the center of gravity shifted tailward and caused Yeager to have to adjust the stabilizer trim to two degrees nose-down to compensate. Once again, Yeager experienced such rapid change in the trim setting that the XS-1 momentarily suffered negative acceleration and the line pressures dropped 100 psi! This time, however, the engine continued to function and Yeager took no remedial action. Within seconds, all pressures began to rise to normal levels.

Yeager continued to climb to 35,000 feet. Mach speed varied from an indicated 0.85 to 0.88. As he passed through 35,000 feet, Yeager began to shut down the engine chambers. He coasted through 40,000 feet with speed falling off to an indicated Mach 0.82. At this point in the flight, the XS-1 still retained 60 percent of its fuel. With this fuel weight and slower speed, the aircraft once again entered stall conditions. Yeager quickly reignited first one cylinder and then another to increase speed back to an indicated Mach 0.88. At that speed, Yeager put the XS-1 through a 2g turn to achieve an accelerated stall. However, in spite of pulling the stick full back (and using full up elevator), Yeager could not get the XS-1 to stall. Finally giving up, Yeager shut down one cylinder and leveled out the XS-1 for a speed run. In level flight at 40,000 feet, the XS-1 speed built up to an indicated Mach 0.925. The airplane again exhibited the right-wing heaviness that Yeager experienced on the last flight. After finishing the stabilized speed run, Yeager put the XS-1 in a climb and quickly reached Mach 0.938. As Yeager increased speed up to the recorded higher Mach number, the buffeting and wing-heaviness conditions seemed to disappear. No Dutch roll was noticed and all conditions seemed to be of lesser magnitude than Yeager previously remembered. Also, Yeager reported two other characteristics. The various phenomena were the same whether the aircraft was increasing or decreasing speed through the Mach 0.86–0.88 range. Second, at about Mach 0.90, the XS-1 seemed to stabilize its speed. Then suddenly, with no visible change in engine performance, the aircraft increased acceleration, a condition Yeager was to experience more dramatically very soon.

After jettisoning fuel, Yeager began the glide back to Muroc. Amid all the activity, Yeager had forgotten (at 10,000 feet) to switch from normal to the emergency breathing supply. As a result, he suffered a noticeable oxygen deficiency in the descent to landing, an extremely dangerous condition with sometimes fatal aviation consequences, but not on this day. The XS-1 landed cleanly, but the front tire was badly out of round with the temporary patch. Yeager used full up stabilizer to hold the wheel off the ground as long as possible. Final contact occurred at 80 MPH and the remaining rollout was normal. Preparations began immediately for the next flight. The NACA XS-1 records revealed a higher recorded speed of Mach 0.945. The pilot report confirmed the good news about the increasing stability in the handling of the plane. Good progress and better handling made for excellent reports for visiting dignitaries.

One visitor was not there on a social call, however. When word of the elevator ineffectiveness reached Wright Field, Colonel Boyd decided to fly to Muroc on October 9 to meet with his team. Although stability and buffeting appeared better at Mach 0.94 than had previously been seen around Mach 0.92, the lack of elevator effectiveness revealed in the 2g turn seemed to promise serious problems when Yeager tried to fly level at a higher Mach number. Also on October 9, Secretary of the Navy J. L. Sullivan and a group of Navy admirals came to Muroc to observe the proceedings. They met with Colonel Gilkey and spent time observing the Navy research projects. On the tenth, President Truman's entire Air Policy Board visited the base aboard the president's airplane "Independence" to watch the latest XS-1 flight. The board also witnessed flight tests of every plane at Muroc except the XB-46 and the XP-86. Progress at Muroc on other aircraft provided increasing distractions for some personnel. The former focus on speed records and new aircraft had lost some novelty. But nobody could deny that the assault on the sonic barrier continued to capture the attention of most everyone. The NACA certainly had their hands full trying to keep up with the XS-1, but even they were distracted. Unfortunately, it was not the flights—rather, it was the lack of flights that caused concern.[48]

On October 10, the Air Force team was ready to take the XS-1 aloft for another flight.[49] In many ways this was a far from settled program. Everyone believed the plane was getting close to something—the question was what. But the program's progress was not obvious from the press coverage—there wasn't any. It remained a sore spot with Yeager when he read of the Navy and the British bragging about their efforts, as if the Air Force wasn't doing anything, when in reality they were leading the way. Nor could Yeager and Ridley quite shake the recent Boyd lecture out of their minds. Everyone was positive that the sound barrier could be broken; the question involved what would be found. As Williams remembered, the group "developed a very lonely feeling as we began to run out of data." The NACA team continued to caution safety first and foremost. The Air Force was ready to go. But, the stated purpose of the flight—"Determination of airplane drag coefficient vs Mach number; investigation of effects of increased Mach number"—provided a glimpse of the uncertainty everyone felt regarding the flight. The current program was to fly—fly fast—and see what happened. Cardenas would again pilot the bomber. Frost and Hoover would fly chase.

After takeoff, Yeager climbed into the XS-1 at 7,000 feet. After he checked out the experimental craft he signaled Ridley all was ready. At 20,000 feet, Cardenas began the countdown. After the near stall of the previous drop, Cardenas decided to keep the B-29 at a higher launching speed. The B-29 initiated separation from the XS-1 at 260 MPH rather than the previous 240 MPH. This seemed to work fine and Yeager reported no problems with the launch. Immediately after clearing the B-29, Yeager fired all four cylinders in rapid sequence.

Yeager put the XS-1 into a steep climb to 35,000 feet. Once again, the full power of the rocket engine pinned him into the seat. During the climb, he vented the attitude gyro into the cockpit to achieve cabin pressurization. Simultaneously, Yeager reset the stabilizer to 1.8–1.9 degrees down to counteract the tail-heavy condition experienced in the last flight. At 30,000 feet and with speed approaching Mach 0.88, Yeager sequentially turned off three cylinders and continued to climb. By 38,000 feet speed had dropped off to an indicated Mach 0.80. Yeager smoothly leveled out and executed a stalled turn. Performing the maneuver, given the XS-1's weight, speed, and altitude, required almost full up-elevator and produced 1.9 g at stall. Satisfied with the initial test, Yeager descended and leveled out at 36,500 feet. The pilot reignited a second cylinder and a stabilized speed run was made to Mach 0.915. Having provided time to record the data, Yeager ignited a third cylinder and began to climb once again. At 40,000 feet, Yeager observed an indicated Mach 0.94. The airplane felt right. The thought crossed his mind to go ahead, let it rip, and get it over with. But reality intruded as the engine noise changed pitch, which indicated rapid onset of fuel exhaustion. Yeager quickly shut down the engine to prevent damage, but the XS-1 knew no bounds at this instant. Still climbing, the plane soared past 45,000 feet before it nosed over and began the glide back to earth.

Something else had disrupted Yeager's pleasant thoughts of just the moment before. During the climb and the tests, buffeting and lateral/longitudinal instability had been as observed previously: light, sometimes sharp, but controllable. Above Mach 0.90, it had disappeared. Now, something everyone had recently speculated about made a dreaded reappearance. Between Mach 0.70 and 0.87, NACA data revealed that elevator effectiveness declined 50 percent. Once above Mach 0.90, elevator effectiveness decreased even more appreciably. By Mach 0.94, Yeager could move the control column each way with little force, but also with little appreciable effect. Luckily, the ailerons were still effective. But no elevator to correct the nose-up/nose-down pitch seen earlier could be deadly—if the stabilizer could not control the aircraft. Fortunately, the plane remained stable and no evidence of Dutch roll was observed (since the propellants were gone).

During the descent, Yeager executed a 1g stall at 41,000 feet in a shallow descent and then allowed the XS-1 to continue to glide toward the desert floor. His thoughts remained focused on the sequence of events at Mach 0.94. The difficulty with elevator control boded little good for the program. But future problems were quickly dismissed for more immediate concerns as Yeager noticed the cockpit was extremely cold. The instrument panel was so cold that vapor emanated from it. Worse, it was condensing on the canopy. The frost collected everywhere as Yeager frantically tried to wipe it off the windshield glass. He reported, "Can't see a thing," to the Muroc tower. As the descent continued and the frost collected on the canopy, the issue became landing the XS-1 safely. There were no blind-flying instruments (other than the attitude gyro) on this research craft. Quickly, Frost moved alongside the XS-1 as Hoover continued to approach. All three were in voice communication. Frost indicated to Yeager that he would talk the XS-1 down. With sure precision, Frost became Yeager's eyes as he slowly and smoothly guided the XS-1 back to the high desert floor for a gentle touchdown on the dry lakebed. Another crisis was averted. Yeager was exhausted after the dramatic finish to the flight. He had experienced enough excitement for that day and headed for home.

The confirmation of the elevator effectiveness problem presented a major disruption in the program if the movable horizontal stabilizer could not be used to trim the plane. Discussions on the problem continued for the rest of the day. However, the disquiet over the elevator problem temporarily masked the discovery of a bigger surprise. After the NACA technicians performed a preliminary data analysis from the flight, they provided stunning information. The high static pressure effect on the nose-boom's pitot head—the one used by the cockpit airspeed and Mach number indicators—at the increased Mach numbers continued to mask the XS-1's true speed. The preliminary data reduction late on the evening of October 10 confirmed that the XS-1 has flown faster than any airplane in history. The little saffron-colored aircraft's cockpit Machmeter may have indicated Mach 0.94, but in reality the XS-1 had streaked to a new unofficial world speed record of Mach 0.997 at 37,000 feet. It might even have exceeded the speed of sound, but no one was quite sure with the data showing such a close pass at the mark. Everyone agreed the thing to do was repeat the flight and go for clear penetration of the Mach 1.0 barrier.

A Leap into the Unknown

On Sunday, Yeager decided to take his wife, Glennis, out to dinner at Pancho's bar on the edge of the base. Florence L. "Pancho" Barnes's Rancho Oro Verde Fly-Inn Dude Ranch was the local place to gather. After several rounds of drinks, both had the idea that a moonlight horseback ride would be a fitting conclusion to the day. While returning the horses to the stables at breakneck speed, Yeager failed to notice that the corral gate was closed. The horse, however, noticed. Bolting at the closed gate, it threw Yeager solidly to the ground. A sharp pain in his side and a little loss of dignity were the immediate results. On Monday, Yeager secretly traveled to Rosamond to visit a doctor. The diagnosis—two broken ribs. Two broken ribs before the most important mission of his career. Until the next day, Yeager sweated out the pain and worried that the mission might be scrubbed or, worse, that Hoover would get the call to fly the flight.[50]

On Monday, October 13, the NACA and Air Force teams gathered with outside specialists to review the preliminary data and discuss Yeager's report of flight number 8. The NACA data revealed that the elevator ineffectiveness was due to shock waves forming right at the hinge-line. The shock wave first revealed itself along the wing at an indicated Mach 0.88, and by Mach 0.94 it had moved back to the tail. The result was a loss of elevator effectiveness and serious questions about what to do next. The NACA team believed that the lack of effective elevator control meant Yeager would face severe difficulties if another attempt was made to break the sound barrier. Review of the NACA data revealed the nose-up trim change around the recorded Mach 1.0 had been much larger than earlier ones. Ridley argued that the movable horizontal stabilizer could be adjusted as a substitute for the lack of elevator effectiveness. Prior to this the stabilizer settings were set at lower speeds before the tests were executed. Although Goodlin and Yeager had already demonstrated the effectiveness of the moving stabilizer at slower speeds, what Ridley was proposing was something entirely different. Ridley suggested that Yeager trim the stabilizer *while* traversing the transonic zone near Mach 1.0. The NACA/Air Force debate revolved around the what-ifs. No one in the room was sure what the effect might be at speeds nearer supersonic. What if the turbulence prevented the tail from pivoting at high speed? Answer—Yeager would be no worse off than the last flight. What if the turbulence tore the moving tail off?—Yeager had better check his life insurance.

NACA and Ridley continued the debate to a verbal draw. Ridley's solution was to ground-test the stabilizer to find out if precise changes were possible under loads. The decision was made to test Ridley's theory since no one seemed to have an alternative. Colonel Boyd was notified of the proposed solution and gave the go-ahead. Yeager and Ridley ground-tested the stabilizer remedy in increments of one-quarter and one-third degrees and satisfied themselves it could work. The new two degree per second stabilizer trim springs gave them just enough control without the previous overcontrol from the three degree per second actuator.[51]

On October 14, Yeager and Glennis traveled to Muroc just as the sun started to come over the horizon.[52] Yeager implied that Glennis should return in time for the flight because something—subject undetermined—might happen that day. She dropped him off at the operations shack and vowed to return just before 10 o'clock. Yeager and Ridley huddled over last-minute details as the XS-1 mechanics continued their work. On Monday afternoon, Yeager had secretly advised Ridley of the riding accident and the painful physical result. He informed Ridley that he could fly the airplane but might not be able to snap the XS-1 hatch door shut. Ridley had gone to the maintenance hangar and provided Yeager with a simple solution: a sawed-off broom handle. The broom handle currently lay hidden in the XS-1.

Out near the XS-1 pit, the routine loading and fueling procedure seemed to have extra significance on this day. Slowly, the normal activity level around the base increased. The NACA and Air Force teams arrived and conferred with Ridley, Yeager, and Frost. Russell advised Yeager that the Drene shampoo he requested would be used to coat the windshield, an old trick Yeager had used in the past to prevent windshield frosting, proving once again that not everything new under the sun is high technology. Although project manager Cardenas knew of Yeager's broken ribs, he decided to say nothing and not scrub the mission, figuring that Yeager was professional enough to indicate if he had a serious problem with flying the aircraft. Nor did Walt Williams say anything, although Yeager had also confided in him. After one last checkover everyone climbed into the B-29 for launch.

Cardenas called out "Muroc Tower, Air Force Eight Zero Zero taxi instructions." The tower responded, "B-29 Eight Zero Zero cleared Runway Six. Wind out of the East, seven miles an hour." Finally Cardenas asked the pertinent question. "We cleared to roll?" The tower provided the affirmative and at 10:00 A.M. the B-29 rolled down the runway and off into the high desert sky. After takeoff, Cardenas checked with the B-29 flight engineer to ensure the bomber was okay and received a positive response. Next he checked with Muroc Tower to ensure that they could hear the B-29. Once again, Cardenas received an affirmative. The crew settled into its drop routine. Tense and businesslike was the attitude, no idle chatter on the intercom.

After passing 5,000 feet, Yeager and Ridley moved back toward the belly of the B-29. Opening the access door to the bomb bay, the pair caught sight of the bullet-shaped aircraft hanging in the bowels of the bomber. The slipstream howled loudly as Yeager and Ridley made their way along the catwalk to the ladder. Standing on the ladder's platform, Yeager bounced on the structure to start the descent. Now came the tricky part. Sliding through the cockpit hatch with broken ribs was no walk in the park. Twisting and sliding, Yeager managed to enter the cockpit and begin buckling up. Ridley lowered the door into position and Yeager used the broom handle as leverage to snap the door handle lever locked. Cockpit checkout replaced idle thoughts. The wonder of the adventure was immediately replaced by the necessity of routine.

Yeager carefully set the horizontal stabilizer to one degree nose-down. All the months of planning, of testing, of rehearsing, of thinking, were now down to the next few minutes. Yeager was all alone with his thoughts. The moment of performance was upon the XS-1 program. If there was a beyond on the other side of the sound barrier, he was about to experience it. Engineers had their slide-rule calculations but they were all safe on the ground. A pilot was about to test the theory.

The headphones suddenly crackled and an oddly distant voice pierced the silence. "Yeager, this is Ridley. You all set?"

Back came the reply, "Hell yes, let's get it over with." Ridley gave the parting warning, "Remember those stabilizer settings."

"Roger," Yeager replied and the silence returned. High above the desert floor, the lumbering silver and black B-29 executed a shallow dive and then began a slight leveling. A disembodied voice came over the earphones as Cardenas slowly intoned the drop count. 10-9-8-7-6-5-3-2-1—Drop (once again he missed a number.) At 20,000 feet and at an IAS of 250 MPH, the XS-1 began its historic flight.

For a second the snap of the two shackles as the XS-1 left the bomb release hooks seemingly produced no effect on the research craft. Then the fall gathered momentum. As the experimental plane dropped away from the B-29, a parting vestige of the mother ship remained on its left wing: the B-29 engines had again spewed oil on the upper surface. As the XS-1 cockpit cleared the darkened shroud of the bomber, the dazzling sunlight momentarily blinded Yeager. Shortly after drop, Yeager ignited all four cylinders in rapid sequence. Engine performance was the same as the last flight. For the pilot, that was a good omen. The bad news was that for the first 185 seconds of this journey into history, there was no NACA data recorded by the XS-1. Yeager forgot to turn on the instruments.

At 20,000 feet, and with cylinders 1 and 3 running, Yeager put the XS-1 into a climb. Speed varied between an indicated Mach 0.85 and 0.88. On the way up, Yeager adjusted the stabilizer setting from the predrop one degree nose-down to two degrees nose-up. In anticipation of the decrease in elevator effectiveness above Mach 0.93, longitudinal control was investigated in increments of one-third and one-quarter degrees. Checks of effectiveness were made at Mach 0.83, 0.88, and 0.92. The stabilizer retained effectiveness at all speeds and the plane smoothed right out. During the climb, Yeager reignited the two other cylinders and continued to race toward 35,000 feet. After passing 35,000 feet altitude, Yeager shut down two cylinders to hold speed around an indicated Mach 0.92. At 42,000 feet, Yeager nosed the XS-1 over and detected a slight decrease in lox line pressure. The result was a drop in engine chamber pressure from the rich mixture. G force change was negligible and the condition quickly passed.

At 42,000 feet in level flight, Yeager reignited chamber number 3. Acceleration was rapid, but so was the onset of mild buffeting. So far the XS-1 had experienced just the usual instability. But, as speed increased, Yeager was alive to any further stability changes. As in the past, at Mach 0.94 elevator effectiveness sharply decreased. However, as the plane passed an indicated Mach 0.96, Yeager quietly received an encouraging signal. "Say Ridley, make a note here. Elevator effectiveness regained." "Roger. Noted." A small grin must have been lurking on that Oklahoman's face. His calculations were correct so far.

As the XS-1 smoothed out to normal flying characteristics, Yeager registered an added level of confidence that the XS-1 was not going to bite him this day. However, as the XS-1 reached an indicated Mach 0.98, Yeager felt a sudden increase in acceleration. The needle of the Machmeter fluctuated momentarily. No other human had been this fast before. Now this day would tell whether Yeager would live to explain the event. Straining at the leash, the XS-1 took its leap into the unknown. Suddenly the Machmeter paused as the effect of the shock waves passed over the static source. Then—the needle jumped off the scale.

Far below the streaking saffron-colored fireball, NACA and Air Force ground personnel gathered around their various radar, radios, and instruments. Waiting for the unexpected was always tense. For those in the know who could hear it, the next sound was one of pure joy. Suddenly, the air was rent by the sound of what seemed to be the distant double-crack of rolling thunder. As von Kármán with his theory of sonic boom had predicted, the XS-1 had audibly confirmed its trip of destiny. As the noise reverberated across the dusty flats, those who knew could only smile. Far up in the sky, no sound could reach the streaking pilot's ears. Only the Machmeter, now pegged past Mach 1.0, revealed what had just been left behind—human ignorance of the spatial unknown. Chuck Yeager had broken the sound barrier. He was the fastest man alive.

"Ridley!—Make another note. There's something wrong with this Machmeter. It's gone screwy!" In the B-29 Ridley heard the sentence and replied in kind, "If it is, we'll fix it. Personally, I think you're seeing things." The secret aviation code of two good ol' country boys now quaintly passed into history. "I guess I am, Jack." Approximately 30 percent of fuel/lox still remained and Yeager held the speed steady to allow for better NACA data recording. Finally, he notified Ridley that he was prepared to shut down the engine. Speed was constant and the XS-1 continued to climb like a bat. During deceleration, Yeager noticed the same reverse sequence of physical events. At Mach 0.98 a single sharp impulse, like a turbulence bump, was all that occurred to remind Yeager of where he had just ventured. After jettisoning fuel/lox he continued the ride to the top of nose over. (Although later flights would reveal the scenery of the stars and moon, on this first flight Yeager was simply too busy to see the fruits of his eternal triumph.) Gliding over, he began the descent to Muroc. For the last several minutes of his flight he performed victory rolls and wing-over-wing aerobatics in personal celebration of the end of the supersonic journey. Landing and rollout were routine. When the plane stopped, an era of aviation history also came to an end. On the ground, Yeager hitched a ride on the base firetruck. Dick Frost immediately called Larry Bell. Ridley and Yeager returned to the office to call Colonel Boyd. At the same time, Colonel Gilkey officially notified AMC that the XS-1 had broken the sound barrier. It was a spectacular finish to a much-anticipated event. But Yeager was so tired and sore because of his ribs that he simply asked Glennis to take him home. This was not to be as Frost, Ridley, and Hoover planned an immediate celebration. However, plans for a party at Pancho's later that day quickly went astray.

The NACA data reduction later confirmed the supersonic achievement of the October 14 flight. The true recorded speed was officially pegged at Mach 1.06. When the Air Force formally took over the program on July 10, Wright Field had estimated an accelerated program of sixty to ninety days to complete. Ninety-six days later, Yeager fulfilled that expectation. It was news for the whole world to share—the fearsome sound barrier had been broken. But, within hours of the notification call to Colonel Boyd the word came down from Wright Field: no comment on this success was to go beyond the flight line. Unfortunately, the visible picture of celebration at the base that afternoon had already tipped the story to more than the few who originally knew of the event. Private parties at Yeager's and then at Dick Frost's house were the only initial celebrations to honor the fall of the fearsome sonic barrier. On October 17, Wright Field informed Bell Aircraft and the NACA that all data above Mach 0.92 was to be regarded as secret. The military brass had decided that the flight would remain in deep cover.

Beyond the Wall

The quest to open the supersonic horizon had concluded and the Air Force discovered a whole new universe on the other side. But the ability to recross that barrier did not come easily. On the very next attempt on October 27, the old familiar pattern of flight difficulties reemerged. Yeager was forced to execute a glide flight due to an electrical failure. On later attempts, the program experienced engine-cylinder failures, delays over lost parts, and the anticipated B-29 over-hauls, all misfortunes that Slick Goodlin knew only too well. In all, between October 14, 1947, and January 1948, Yeager piloted the XS-1 only seven additional times. The plane managed to go through the sound barrier on most of those flights. But rather than a great leap forward, the new flights continued the careful, measured probing that had been the hallmark of the XS-1 program. It was a program that worked well.

The lid of security may not have kept everyone at Muroc from knowing of the supersonic flight, but it certainly limited the numbers. Nor did the press immediately pick up on the story. On November 3, *Aviation Week* announced that the Douglas D-558-1 had become the newest world's speed record holder when pilot Gene May flew the aircraft to a speed of approximately 680 MPH. This was 30 MPH faster than Maj. Marion Carl's flight on August 25. May reported no indica-tion of adverse compressibility in the plane's behavior. It was expected that tests at 35,000 feet would begin soon and would signal the real attempt at supersonic flight. However, if the D-558-I failed to achieve the goal, the Navy had a new entrant in the race ready to participate. In mid-November, the security wraps came off when Douglas Aircraft unveiled their new D-558-II Skyrocket. The latest challenger for the supersonic goal was a highly modified version of the D-558-I Sky-streak, incorporating many of the features once considered for the XS-1 and later demonstrated by research and wind-tunnel testing. The D-558-II combined rocket and turbojet power, incorporated thin swept wings of low-aspect ratio, and a swept tail. It also featured a long thin nose for low fuselage drag and flush air inlets unlike the blunt nose of the D-558-1. *Aviation Week* quoted Navy, Douglas, and the NACA engineers as being confident that supersonic speeds were now only a matter on getting the D-558-2 into the air and making the flight. The impressive story could only lead to smiles at Muroc and in Buffalo. On November 6, Chuck Yeager piloted the XS-1 on its fifty-eighth flight and reached a speed of Mach 1.35/905 MPH. No word of these achieve-ments would come from Bell Aircraft or the Air Force. While the news media looked for the first supersonic flight, the Air Force and Bell Aircraft began the second phase of their program. On November 14, Wright Field authorized the initiation of studies that would lead to a contract on December 11 for three new planes.[53] In the meantime, the NACA continued its efforts to start its own program.

On October 20, the NACA XS-1 number 2 finally made a successful ground test of its engine, after being plagued by mechanical delays since October 3. A glide flight was scheduled for October 21, when the NACA's Herb Hoover flew the number 2 airplane on the first pure NACA flight test. The B-29 departed Muroc late in the afternoon. The glide flight lasted approximately eight minutes and Hoover attained a speed of Mach 0.84. However, on landing, he experienced the tricky nature of the XS-1's handling characteristics. After porpoising the aircraft over four touchdowns, #6063 experienced a nosewheel collapse. The number 2 airplane skidded to a stop within 2,000 feet. Hoover took responsibility for the accident but, given the nature of the XS-1's landing characteristics, it was not a completely unexpected problem. The NACA plane was out of commission awaiting parts until November 24. Unfortunately, on that same day, the NACA engineers discovered that Bell Aircraft had left spare parts for the nosewheel in the Muroc warehouse. It was a fitting new chapter to the same old supply problems.[54]

Finally, with the plane now ready to resume flight testing, the rains came to Muroc. The two-day (December 4–5) downpour postponed any opportunity for the NACA to fly for 11 days. On December 16 and 17, Hoover managed to get the number 2 airplane into the air for two additional checkout flights. As the program continued to gather momentum, the NACA expanded its plans to increase its flight station to handle the evolving workload necessitated by the present and future aircraft at Muroc (XS-1's number 1 and 2, D-558-1-2 and 1-3, the XS-4 due January 1, 1948, the D-558-II-2 due February 1, 1948, and a special P-80 and C-45). But, in spite of the NACA progress, other events at Muroc quickly distracted everyone's attention.

While the NACA research efforts at Muroc continued during November in a "routine" manner, the walls of silence regarding the first supersonic flight were crumbling rapidly. Slowly at first, and then with increasing momentum, the story began to surface. Press security initially kept the headline in check. However, by December the Air Force was fighting a losing battle to control the release of the information. Although newspaper editors initially agreed not to release the details until such time as it became public knowledge, the very flexibility of that term meant that the release date was not long for the future. The future came suddenly in mid-December.

In an issue prepared earlier but dated December 22, *Aviation Week* magazine made public the astonishing story of Yeager's supersonic flight in the XS-1. It was to be the magazine's biggest news break and came at a time when the Air Force's security policy on this story was already in tatters. National newspapers such as the *New York Times* and the *Los Angeles Times* picked up the story dated the same day. Not surprisingly, with a "secret" program, the stories missed several points. Most significantly, they reported that the NACA's aircraft had also flown supersonic. On December 22, the NACA was still awaiting that event. Also, *Aviation Week* reported the turbine pump had been installed in the XS-1, greatly increasing efficiency. (This old story about the elusive turbine pump lived for several more years.) At USAF HQ in Washington, official spokesmen maintained complete silence. At Muroc, when the press found Colonel Gilkey, he declined to verify the story. When pressed, Gilkey admitted that he had only "indirect supervision" over the test flights, as if to confirm "something" might have happened. However, he attempted to maintain the security cover by adding, "I think I would have been notified if such flights had been made." Colorfully, Gilkey stated, "It sounds to me like another flying saucer story."[55] The quote proved entertaining, but no amount of levity could put the secret genie of supersonic flight back into its security bottle.

On January 9, 1948, Wright Field convened a conference of all the primary participants of the XS-1 program.[56] A total of 122 people were officially cleared to attend the meeting. The list of attendees reads as a Who's Who of the Air Force, Bell, NACA, and the aviation industry. The stated purpose was to provide reliable data on the results of the XS-1 tests in a form suitable for industry use. The report featured four presentations by Bob Stanley/Roy Sandstrom, Walt Williams, Chuck Yeager, and Charles Hall. Each covered their area of expertise in the development of the program. Appended to the report were actual data points for reference. Unfortunately, the report contained important errors that became the foundation for later historical misinformation about the program. Although supposedly top secret, the document leaked and became the basis for the *Aviation Week* story on July 26, 1948, that publicly revealed many of the details of the program.[57] But, in the interim, the U.S. government compounded the damage caused by the leak by seeking a villain.

With the initial wave of publicity came additional press inquires. The U.S. Justice Department only compounded the attention by announcing an FBI investigation of *Aviation Week* magazine for an alleged breach of national security. That case dragged on fitfully until May 27, 1948, when the Justice Department finally announced that it could not prosecute the magazine because no federal laws had been broken. It was not illegal to disclose top secret information in America. Nor had any U.S. espionage laws been broken. In short, there was no case. While public attention was focused on that issue, President Truman declined to comment, but Air Force sources leaked the story that the secretary of defense, James Forrestal, was continuing to hold the XS-1 story confidential in spite of the press coverage to relieve a public relations problem. The allegation was that Forrestal, a former secretary of the Navy, was taking this action at the request of the Navy to avoid embarrassing the service. This very contentious issue only fueled the public debate over the necessity to disclose the exact nature of the situation regarding the supersonic flight.[58]

Finally, on June 15, 1948, Air Force Chief of Staff Gen. Hoyt Vandenburg officially confirmed that the sound barrier had been broken. The long delay in confirmation had caused some harsh words and hard feelings to be distributed among the government, the press, and industry. It had removed the immediate glamour of the event in a vain attempt to fulfill the demands for long-term security. But the secrecy had also cloaked one major benefit of the classified program. The success of the flying tail, the movable horizontal stabilizer, remained unappreciated for several additional years before it finally became accepted aviation practice in Russian, British, and French aircraft. In the interim, the moving stabilizer provided an important technological advantage for U.S. combat aircraft.[59]

Epilogue

The public acknowledgment of the breaking of the sound barrier provided the basis for recognition of the grand aviation achievement of supersonic flight. The prestigious National Aviation Association voted to award its 1948 annual prize for greatest achievement in aviation during the previous year to the veterans of the XS-1 program. In a presentation ceremony at the White House, John Stack, Larry Bell, and Chuck Yeager jointly received the Robert J. Collier Trophy from Pres. Harry S. Truman on December 17, 1948. For Stack and for Larry Bell, it was certainly an ironic and perhaps even bittersweet victory. Larry Bell had wanted the supersonic honors for his company. It was not to be except in the sense that all aircraft manufacturers share in the glory of their products.

For John Stack the situation was even more complex. Stack's perseverance and technical knowledge had allowed the NACA to play a primary initial role in the creation of the high-speed project. Many of the original design features of the XS-1 were based on NACA concepts. However, it appears very ironic in hindsight that the plane that broke the sound barrier was not the one that Stack desired to test. Stack had opposed the Army plan to build a very thin-winged, air-launched, rocket-powered aircraft. He had favored a more conservative approach for safety and data. In retrospect, the XS-1 provided the data that Stack and his colleagues wanted from the new research aircraft program while the ground-launched, jet-powered, conventional D-558 experienced an engine failure on takeoff that killed Howard Lilly in May 1948. The favored D-558-I that NACA so strongly supported turned out in the end to be an "unnecessary" airplane since production Air Force aircraft in 1948 rapidly provided the same data below the Douglas aircraft's Mach 1.0 limit.[1] Nevertheless, the myth continued to grow that the XS-1 and the D-558 were complementary aircraft. As seen from the historical debate, they were never intended to function in that role.

Nor is it completely objective to believe that the AMC, NACA, and Bell Aircraft "cooperated" on the XS-1 program. It is true that they participated in the same project. However, it is more precise to say, as revealed in this work, that they exhibited a desire to function under the same program umbrella while pursuing different program objectives.

The XS-1 number 1 continued to fly for three more years. When it retired on May 12, 1950, it had made eighty-three flights with ten different pilots (including Ridley and Boyd). On August 26, 1950, the number 1 airplane became a permanent resident of the National Air Museum (now the Smithsonian National Air and Space Museum).

The XS-1 number 2 had an even longer career. The aircraft flew until October 23, 1951, when it was retired to be rebuilt. It returned on December 12, 1955, as the new X-1E and flew as an NACA research aircraft until November 6, 1958. It was the last of the X-1 series of aircraft to remain flying. It completed a total of 100 flights (XS-1 = 74; X-1E = 26) with eleven pilots. It now is on display at the Dryden Flight Research Center at Edwards Air Force Base.

The oft-delayed XS-1 number 3 with the much-anticipated turbine pump was finally completed in 1951 and made its first glide flight on July 20, 1951, with Joe Cannon as pilot. Its next test on November 9, 1951, became its last one as a one-hour captive flight ended with an explosion and fire on the ramp while preparing to jettison propellants. Both the number 3 airplane and the B-50 carrier plane were destroyed and Cannon was severely burned. It was a tragic end to the final airplane of the original contract.

The success of the XS-1 demonstrated the feasibility of supersonic flight. However, the XS-1 program was more than simply flight tests of another prototype airplane. Rather, its distinction rests as the first major American experimental research program. During the presentation ceremonies for the number 1 airplane at the National Air Museum, Air Force general Hoyt Vandenburg provided a fitting conclusion to the aircraft's role in history. The XS-1's flight through the sound barrier, Vandenburg stated, marked "the end of the first great period of the air age, and the beginning of the second."[2]

The XS-1 not only proved that humans could go beyond the speed of sound, it reinforced the understanding that aviation's "absolute" technological barriers could be overcome. The XS-1 possesses a rare distinction as one of the few aircraft to complete its test program without a significant aerodynamic or structural alteration. The XS-1 airplanes laid the foundation for a legacy of knowledge, not only on government and industry teamwork, but also by providing continuing information on rocket powerplants, aircraft structural design, and aerodynamics. In any immediate evaluation of the project, NACA obtained its basic research data for application to all future supersonic plane design; the Air Force acquired its technological advance over all opponents, and the prestige and experience of its first piloted supersonic plane; and Bell Aircraft continued its reputation on the cutting edge of aviation science.

In the final analysis, the XS-1 project provided the impetus for the increased congressional funding necessary for the Air Force and the NACA to develop the technological basis for American aviation supremacy in the latter half of the twentieth century. The union of industry capabilities, scientific skills, and military needs had triumphed. By the end of the XS-1 program, the pattern of high-speed/high-altitude aviation research was set and the foundation of cooperation necessary to make it function had been established: a foundation that would later carry America safely into outer space and back. It was a remarkable achievement for a three-aircraft production contract that no one initially wanted to build.

Appendix

XS-1 Flights

Date	Flight No.	Aircraft	Pilot	Remarks
1/10/46	—	#1	———	Captive flight
1/21/46	—	#1	———	Captive flight
1/25/46	1	#1	Woolams	1st glide flight
2/5/46	2	#1	Woolams	
2/5/46	3	#1	Woolams	
2/8/46	4	#1	Woolams	
2/19/46	5	#1	Woolams	
2/25/46	6	#1	Woolams	
2/25/46	7	#1	Woolams	
2/26/46	8	#1	Woolams	
2/26/46	9	#1	Woolams	
3/6/46	10	#1	Woolams	Last Pinecastle flight
9/26/46	—	#2	Goodlin	Captive flight
10/9/46	—	#2	Goodlin	Aborted
10/11/46	11	#2	Goodlin	1st Muroc glide flight
10/14/46	12	#2	Goodlin	
10/17/46	13	#2	Goodlin	
12/2/46	14	#2	Goodlin	
12/6/46	—	#2	Goodlin	Aborted
12/9/46	15	#2	Goodlin	1st powered flight
12/20/46	16	#2	Goodlin	
1/8/47	17	#2	Goodlin	
1/17/47	18	#2	Goodlin	
1/20/47	—	#2	Goodlin	Aborted (pressure regulator)
1/21/46	—	#2	Goodlin	Aborted (lox regulator)
1/22/47	19	#2	Goodlin	
1/23/47	20	#2	Goodlin	

Date	Flight No.	Aircraft	Pilot	Remarks
1/30/47	21	#2	Goodlin	
1/31/47	22	#2	Goodlin	
2/5/47	23	#2	Goodlin	
2/7/47	24	#2	Goodlin	
2/19/47	25	#2	Goodlin	
2/21/47	26	#2	Goodlin	
4/10/47	27	#1	Goodlin	1st flight of 8% wing/6% tail.
4/11/47	28	#1	Goodlin	
4/29/47	29	#1	Goodlin	
4/30/47	30	#1	Goodlin	
5/1/47	—	#1	Goodlin	Aborted (pressure relief valve)
5/2/47	—	#1	Goodlin	Aborted (pressure relief valve)
5/5/47	31	#1	Goodlin	
5/15/47	32	#1	Goodlin	
5/16/47	—	#1	————	Captive flight
5/19/47	33	#1	Goodlin	
5/21/47	34	#1	Goodlin	
5/22/47	35	#2	Johnston	
5/29/47	36	#2	Goodlin	
6/5/47	37	#1	Goodlin	Last contractor flight
8/6/47	38	#1	Yeager	1st AF glide flight
8/7/47	39	#1	Yeager	
8/8/47	40	#1	Yeager	
8/29/47	41	#1	Yeager	1st AF powered flight
9/4/47	42	#1	Yeager	
9/8/47	43	#1	Yeager	
9/10/47	44	#1	Yeager	
9/12/47	45	#1	Yeager	
9/25/47	46	#2	Yeager	1st "NACA" flight
10/3/47	47	#1	Yeager	
10/8/47	48	#1	Yeager	
10/10/47	49	#1	Yeager	
10/14/47	50	#1	Yeager	1st supersonic flight

Notes

Chapter 1. In the Beginning

1. James Hansen, *Engineer-in-Charge* (Washington, D.C.: NASA, 1987), 253. Richard Hallion, *Supersonic Flight: The Story of the Bell X-1 and the Douglas D-558* (New York: Macmillan, 1972), 11.

2. Details on the Congress are found in Theodore von Kármán and Lee Edson, *The Wind and Beyond* (Boston: Little Brown, 1967), 216–23.

3. Hansen, *Engineer-in-Charge*, 253–60. Kotcher was a senior instructor at the engineering school at Wright Field. He was recalled to duty after Pearl Harbor as a captain. Hallion, *Supersonic Flight*, 12–13 and 20. Also see von Kármán and Edson, *Wind and Beyond*, 233–34. The prestigious board, whose members included Charles Lindbergh and Lt. Col. Carl Spaatz under the direction of Brig. Gen. Walter G. Kilner, was charged with defining the needs of future U.S. aircraft. The board proposed a major research and development plan to support the production of such aircraft.

4. W. F. Durand to H.J.E. Reid at Langley, NACA letter R.A. (Research Authorization) 1347, February 14, 1944. Also see the Wolf letter and the Dr. G. W. Lewis response in News Release, "Beginnings of the X-1," Bell Aerospace Textron, updated February 23, 1977, pp. 3–4. Wolf was the Bell technical advisor for the XP-59A, America's first turbojet aircraft. Hallion, *Supersonic Flight*, 19; Hansen, *Engineer-in-Charge*, 260.

5. Staff minutes of the High-Speed Panel meeting, NACA memorandum R.A. 1347, March 2–3, 1944.

6. The discussion of the Mach 0.999 study is from the Kotcher letters to Hallion. See *Supersonic Flight*, 20–21.

7. 2nd Lt. G. H. Crocker, AAF Materiel Center, Engineering Division ENG-51-4589-1-5, "Memorandum Report on Compressibility Conference at NACA Langley Field," April 10, 1944, in NACA memorandum R.A. 351. John Becker, *The High-Speed Frontier: Case Histories of Four NACA Programs, 1920–1950* (Washington, D.C.: NASA, 1980), 91. The *Air Force Supersonic Research Airplane XS-1 Report No. 1*, January 9, 1948 (hereafter *AFSRA XS-1*), p. 5, erroneously lists the date of the meeting as March 15, 1944. The document may be found in the NASA HQ Historical Archives, Washington, D.C.

8. Capt. F. Oranzio and Capt. G. Bailey, AAF Materiel Center, Engineering Division ENG-51-43011-1-2, "Preliminary Design of Transonic Flight Test Airplane," July 17, 1944, in NACA memorandum R.A. 1291.

9. Personnel and budget numbers from Wesley Craven and James Cate, eds., *The Army Air Forces in World War II:* Vol. 6, *Men and Planes* (rpt.: Washington, D.C.: GPO, 1982), 243. A thorough discussion of the NACA-AAF struggle over high-speed research may be found in Alex Roland, *Model Research*, 2 vols. (Washington, D.C.: NASA, 1984), 1:194–221. Another discussion of this point is found in a variety of sources excellently summarized by Dr. James Young in Young, ed., *Supersonic Symposium: The Men of Mach 1* (Edwards AFB, Calif.: AFFTC-HO, 1990), 82–83. Also see von Kármán and Edson, *Wind and Beyond*, 225–26 on the Air Corps–NACA disagreement. However, not all AAF officers were so sure the NACA had disappointed the service. Certainly General Arnold believed this, but in an interview with the author, Lt. Gen. Laurence Craigie indicated others "who were in the know did not think NACA let us down." Craigie stated the NACA problems certainly were compounded by the lack of an imaginative jet aviation pioneer like Sir Frank Whittle (Great Britain) or Dr. Hans von Ohain (Germany), but the real issues were personnel and funding. Significant as the extra resources received by NACA during the war were, they did not compare to those received by the German aviation industry. Craigie, interview with the author, June 9, 1992. See also Brig. Gen. Laurence Craigie, "Research and the Army Air Forces" in *Aeronautical Engineering Review* 4, no. 10 (October 1945), for similar remarks. Also see Michael Gorn, *Harnessing the Genie: Science and Technology Forecasting for the Air Force, 1944–1986* (Washington, D.C.: USAF, 1988), 38.

10. See Herman Wolk, *Planning and Organizing the Postwar Air Force 1943–1947* (Washington, D.C.: USAF, 1984), 39; and Gorn, *Harnessing the Genie*, 11–20, 27–28, 46.

11. "Beginnings of the X-1," 4.

12. R. G. Robinson to Dr. G. W. Lewis, NACA memorandum R.A. 1347, April 25, 1944. (In 1948, the AERL became the Lewis Flight Propulsion Laboratory in honor of George Lewis.) One week after the meeting, the Navy's Bureau of Aeronautics (BuAer) sent a letter to NACA outlining their role in high-speed research. The Navy stated their responsibility would be mainly to provide equipment. This included radar, free-fall models, a specially modified XF8F airplane, and procurement of a very high speed airplane. BuAer letter Aer-E-241-EWC C10918, April 27, 1944. Also see Capt. W. H. Miller to NACA Washington, BuAer memorandum Aer-3-2411-CBL C28013, October 13, 1944.

13. W. J. McCann, notes on AAF Conference on Experimental Research and Development Projects for Fiscal Year 1945: May 15, 16, 17, 18, 19, 20, 1944, AAF memorandum R.A. 1347, May 26, 1944, pp. 1–2; also see Col. Carl Greene (MC Liaison Langley) to AAF Materiel Center Wright Field, Materiel Command letter R.A. 1347, June 10, 1944; Minutes of the Committee on Aerodynamics, NACA memorandum R.A. 1347, May 1944.

14. Russell Robinson to the Chief of Research, NACA memorandum R.A. 1347, August 12, 1944.

15. Hallion, *Supersonic Flight*, 34.

16. Eugene Draley to Langley Chief of Research, NACA memorandum R.A. 1347, January 17, 1945. (The memo lists a Captain Carl which I believe is Col. Carl Greene.)

17. The NACA argument centered on the differences in power versus weight. At 20,000 feet the rocket engines would provide 4,500 pounds of thrust for two minutes from 3,900 pounds of unit and fuels. A four-unit jet engine would provide 3,850 pounds of thrust for 13 minutes and weigh only 3,975 pounds. Since the rocket engine did not have sufficient fuel to climb to 20,000 and power a speed run, any airplane equipped with rockets must also have an auxiliary powerplant to allow it to reach test altitude. Also, NACA continued to have severe reservations about the reliability of rocket propulsion. See Becker, *High-Speed Frontier*, 91–92. Becker quotes Melvin Gough, chief of LMAL's

Flight Research Division as stating that "no NACA pilot will ever be permitted to fly an airplane powered by a damned firecracker."

18. Minutes of the High-Speed Panel meeting, NACA memorandum R.A. 1347, December 18, 1944; William Lundgren, *Across the High Frontier* (London: Victor Gollancz Ltd., 1956), 26–27. The Lundgren book is written in the first person (from Yeager's point of view) and is quite detailed in its discussion of the flights. Several of the key XS-1 participants also contributed to the preparation of the manuscript. However, there are some chronology errors and questionable events in the book. Where appropriate, I have attempted to clarify these chronology errors and details. It is interesting to note how Kotcher and Bell Aircraft were already changing the agreement (transonic is now supersonic) from the December 13–14 meeting.

19. Hallion, *Supersonic Flight*, 23–26, 56–59; Hansen, *Engineer-in-Charge*, 273–74; Becker, *High-Speed Frontier*, 91–92.

20. "Beginnings of the X-1." Also J. van Lonkhuyzen, "Problems Faced in Designing Famed X-1," *Aviation Week* 54, no. 1 (Jan. 1, 1951): 22–24. Also see Benson Hamlin, "The Design Conception of Supersonic Flight," unpublished manuscript provided by R. H. Frost.

21. John Becker to Langley Chief of Research, NACA memorandum R.A. 1347, March 21, 1945. Written biographical details of Robert Stanley supplied by Doris Stanley Proebstel, November 16, 1989.

22. Biographical details of Jack Woolams are from a six-page autobiography prepared on March 20, 1946, and supplied to the author by Mary (Mrs. Jack) Woolams. This information is supplemented by taped interviews with Woolams's associates, Don Thomson and Jack Russell. A separate contract for supersonic flight was to be issued after the completion of the acceptance tests. No formal guarantee had been issued to Bell Aircraft that it would be selected. However, Air Force statements (May 31, 1945) and later 1946 newspaper articles indicate that Woolams was the pilot expected by many to fly the craft after the acceptance flights. It was normal for the corporation to assign their lead test pilot to a program of this type. Bell retained a stable of experienced experimental test pilots on the payroll simply to fly the newest corporation aircraft. This program was significantly different than anything Bell or any other aircraft company had previously attempted. However, Woolams insisted on flying the XS-1 from the beginning of the serious discussions to bring the program to Bell. Bell agreed to pay Woolams a $10,000 bonus for the initial flight tests. However, Woolams often privately indicated that he so desired to participate in the program that he would fly the project for free. Mary Woolams, interview with the author, August 22, 1989 (hereafter Woolams interview); Jay Demming, conversation with the author, July 31, 1991; Dick Frost, conversation with the author, August 15, 1991.

23. "Beginnings of the X-1," 4. Also see Col. P. B. Klein, "Research Aircraft," *Mechanical Engineering* 69, no. 10 (October 1947): 813, says the contract letter was signed March 12, 1945. The *Project MX-653 History* (Jan. 11, 1947) prepared by James Voyles says March 10, 1945, is when ATSC notified NACA of the contract award; document supplied to the author by Ezra Kotcher. Other contract date is from *AFSRA XS-1*, 5. Details of the funding are from the AMC 1947 "Green Book" found at Edwards AFFTC-HO. The book was maintained by senior Engineering Division personnel at Wright Field and lists budget data for all AMC aircraft projects.

The XS-1 was an expensive program. The original XP-80 contract, signed in July 1943, was for only $515,000, including a $19,800 fixed fee. Likewise, the three-plane XP-84 plus spares and GFE contract, also signed in March 1945, was for $2.5 million, excluding a $99,000 fixed fee. Thus, the XS-1 was twice as expensive as these production aircraft. Surprisingly, it was *theoretically* not the most expensive contract in the AMC collection. As listed in the "Green Book" at least six aircraft (in millions: XP-87 = $9.1;

XP-88 = \$5.3; XP-89 = \$5.6; XP-90 = \$4.8; XP-91 = \$5.1; XP-92 = \$7.5) contracts were authorized at higher levels. Most were major disappointments as aircraft. The XP-80 and XP-84 numbers from Marcelle Knaack, *Post–World War II Fighters, 1945–1973* (Washington, D.C.: GPO, 1986), 1, 24. Bell model number from ENG 04005-3 PCE (Paul Emmons) to R. M. Stanley, Bell Aircraft Corporation internal memorandum, April 5, 1945. Classification information from Col. Carl Greene to Engineer-in-Charge Langley Lab, ATSC letter 1259 TSENG-3, April 12, 1945.

24. Hallion, *Supersonic Flight*, 61–62.

25. Lundgren, *Across the High Frontier*, 79, 82.

26. Roy Sandstrom to ATSC Wright Field, MX-653 Bell Progress Report, April 16, 1945.

27. Milton Davidson to Langley Chief of Research, NACA memorandum R.A. 1347, April 28, 1945.

28. Milton Davidson to Langley Chief of Research, NACA memorandum R.A. 1347, May 1, 1945; William Gracey to Langley Chief of Research, NACA memorandum R.A. 1347, May 1, 1945; Roy Sandstrom to ATSC Wright Field, MX-653 Bell Progress Report, May 1, 1945. This is the first documented mention of Muroc in regard to the MX-653 program that the author was able to find.

29. John Stack to Langley Chief of Research, NACA memorandum R.A. 1347, May 1, 1945. The ATSC later commissioned General Electric to assist with development of the turbine pump.

30. Col. George Smith to Dr. G. W. Lewis, ATSC letter TSESE-2G, May 31, 1945. At this point, the ATSC seems merely to be repeating the Bell recommendation for testing at Muroc.

31. Stanley Smith to ATSC Wright Field, MX-653 Bell Progress Report, June 19, 1945. Smith had taken over from Sandstrom as XS-1 project engineer.

32. Milton Davidson to Langley Chief of Research, NACA memorandum R.A. 1347, May 1, 1945, pp. 2–3.

33. John Stack to John Crowley, NACA memorandum R.A. 1347, June 14, 1945.

34. John Crowley to John Stack, NACA memorandum R.A. 1347, June 19, 1945.

35. Hansen, *Engineer-in-Charge*, 275–79.

36. See Milton Davidson to Langley Chief of Research, NACA memorandum R.A. 1347, June 16, 1945; and H.J.E. Reid to ATSC Langley Liaison, NACA letter R.A. 1347, June 20, 1945. Also see Robert Byrne to Langley Chief of Research, NACA memorandum R.A. 1347, October 22, 1945; Walter Williams to Langley Chief of Research, NACA memorandum R.A. 1347, November 1, 1945.

37. Draft remarks of Gen. Laurence Craigie entitled, "An-NACA-Industry Conference re High-Speed Research Program," September 6, 1945, AFFTC-HO. Also see ATSC Liaison Office LMAL to the Commanding General ATSC Wright Field, ATSC memorandum 1433 TSEXL/JAR in R.A. 1347, March 14, 1946. Hansen, *Engineer-in-Charge*, 289.

38. William Gracey to Langley Chief of Research, NACA memorandum R.A. 1347, September 17, 1945; Henry Pearson to Langley Chief of Research, NACA memorandum R.A. 1347, September 27, 1945.

39. The comment was made by Stanley to Capt. David Pearsall; Pearsall, interview with the author, September 10, 1989.

40. Milton Davidson to Langley Chief of Research, NACA memorandum R.A. 1347, October 16, 1945. Stanley Smith to ATSC Wright Field, MX-653 Bell Progress Report, October 19, 1945. The quote appears in H.J.E. Reid to NACA HQ, NACA letter R.A. 1347, May 5, 1945.

41. Walter Williams to Langley Chief of Research, NACA memorandum R.A. 1347,

November 1, 1945; Maj. Donald Eastman, ATSC Acting Liaison, to Director ATSC Wright Field, ATSC Langley letter, October 23, 1945.

42. Boeing B-29s were made under license by several U.S. aircraft companies. Each company was given leeway in construction techniques and, as a result, the finished planes sometimes exhibited slight differences in airframes. The Renton B-29 built at Boeing had a different wing attachment than the Bell B-29. As a result, the Boeing aircraft had a larger bomb bay with a flatter center surface.

43. In 1945, Cherry Point's radial-configured airfield consisted of eight runways of varying lengths between 4,800 and 7,600 feet. A special center-section jog of 5 degrees allowed the claim of 16,000 feet; information provided in a letter to the author by Command Historian Rudy Schwanda, May 28, 1992.

44. Roland, *Model Research*, 1:203–13; Gorn, *Harnessing the Genie*, 28; Wolk, *Planning and Organizing*, 44.

45. NACA budget data from Hansen, *Engineer-in-Charge*, 428. NACA Information Release, R.A. 1347, "Super-Speed Military Aircraft Program Revealed," October 29, 1945.

46. Stanley Smith to ATSC Wright Field, MX-653 Bell Progress Report, November 8, 1945.

47. Melvin Gough to Langley Chief of Research, NACA memorandum R.A. 1347, November 15, 1945. Actually, in 1945, due to previous wartime restrictions, few people outside the military knew the number and locations of many of the auxiliary airfields around the country.

48. Col. George Price to ATSC liaison Langley Laboratory, ATSC letter TSESA-7, November 15, 1945.

49. Col. George Smith, Wright Field to Dr. George Lewis, NACA Washington, ATSC letter TSESA-7, November 23, 1945. The copy in the Langley files exhibits a handwritten note that states, "I think we are getting the bag—and not what it was agreed we would have—and maybe we should drop the project!" The signature is illegible, but is dated December 13, 1945. In *Supersonic Flight*, 86, and the footnote explanation on p. 98, Hallion cites a Reid letter of December 29, 1945, that says that the ATSC made the decision. In an interview with R. Hallion, Walt Williams stated that the ATSC selected Pinecastle; Hallion, *On the Frontier, Flight Research at Dryden 1946–1981* (Washington, D.C.: NASA, 1984), 7 and 363 n. 3. As seen from the Smith letter, the ATSC accepted the Woolams recommendation. It is hard to determine to what degree the ATSC forced the Bell recommendation by the requirement for a 10,000-foot runway or similar open area.

50. Stanley Smith to ATSC Wright Field, MX-653 Bell Progress Report, December 6, 1945.

51. Information from Joseph Cannon's interview of August 23, 1989, with the author and an article by Jack Woolams, "How We Are Preparing to Reach Supersonic Speeds" in *Aviation* (September 1946): 38–39. Harold Dow is now known as "Pappy," although the nickname did not become common until a later date. Dow had been the chief test pilot at the Marietta facility during the war.

52. Details on the XS-1 are provided in Hallion, *Supersonic Flight;* van Lonkhuyzen, "Problems Faced in Designing Famed X-1"; Jay Miller, *The X-Planes: X-1 to X-31* (Arlington, Va.: Aerofax, 1988); and the article by Robert Stanley and Roy Sandstrom, "Development of the XS-1 Airplane" in *AFSRA XS-1*. Also see Klein, "Research Aircraft," and Joel Baker to Langley Chief of Research, NACA memorandum R.A. 1347, February 25, 1947. The 18g requirement is from Paul Emmons to Floyd Thompson (Assistant Chief of Research at Langley), Bell Aircraft memorandum ENG 0405-2, April 15, 1945. Hallion, *Supersonic Flight* (52 n. 20), reports that Bell, ATSC, and NACA arrived independently at the 18g requirement. However, Hallion (p. 44) states

that Bell only wanted a 7g requirement. This is confirmed by Hamlin, "The Design Conception of Supersonic Flight," 7.

53. Stanley Smith to ATSC Wright Field, MX-653 Bell Progress Report, December 19, 1945. Other details from the personal flight log of Harold Dow.

54. Stanley Smith to ATSC Wright Field, MX-653 Bell Progress Report, January 10, 1946.

55. William Gracey to Langley Chief of Research, NACA memorandum R.A. 1347, December 17, 1945.

56. Maj. Donald Eastman to Maj. Ezra Kotcher, at ATSC Wright Field, ATSC Langley telegram, December 6, 1945.

57. Walter Williams to Langley Chief of Research, NACA memorandum R.A. 1347, December 18, 1945; Maj. Donald Eastman to Capt. David Pearsall, ATSC Langley TSEXL 100, December 19, 1945.

58. John Stack to Langley Chief of Research, NACA memorandum R.A. 1347, December 28, 1945; "Participation of Langley Laboratory in Flight Tests of the MX-653," NASA Federal Records Collection, Box 310, National Archives Pacific Southwest Region, Record Group 255, from the Dryden Flight Test Center; G. H. Helms (Assistant Director of Aeronautical Research) to ATSC Wright Field, NACA memorandum R.A. 1347, January 8, 1946.

59. Hartley Soule to Langley Chief of Research, NACA memorandum R.A. 1347, January 25, 1946; also Stanley Smith to ATSC Wright Field, MX-653 Bell Progress Report, January 10, 1946.

60. Dow flight log; Maj. Don Eastman to ATSC Orlando, ATSC TSEXL-116, January 11, 1946.

61. Dow flight log; Stanley Smith to ATSC Wright Field, MX-653 Bell Progress Report, January 10, 1946.

Chapter 2. The Pinecastle Tests

1. C-45 tail number from Joel Baker's personal flight log. Biographies from correspondence and interviews by the author with Gerald Truszynski (September 15, 1989) and Charles Taylor (August 31, 1989).

2. Details on Pinecastle come from Henry Swanson, *Countdown for Agriculture* (Orlando, Fla.: Designers Press, 1975), 2–3, and Eve Bacon, *Orlando, A Centennial History* (Chuluota, Fla.: Mickler House, 1977), 147. Dick Marlowe, "Deed Marks Up Airport $1 Price Tag," *Orlando Sentinel*, July 1, 1991 (Central Florida Business), 3.

3. The Pinecastle trip date is from the Dow flight log. Cannon interview, August 23, 1989.

4. Walter Williams to Langley Chief of Research, NACA memorandum R.A. 1347, February 1, 1946. These detailed notes cover the day-to-day events at Pinecastle.

5. Details for the biographies of Williams, Hayes, Baker, and Eastman are from correspondence and personal and telephone interviews conducted by the author (August 5, September 1 and 18, and October 11, 1989), and supplemented by Hallion, *Supersonic Flight*, 86.

6. Baker flight log. Williams to Langley Chief of Research, February 1, 1946. These details are supplemented by the personal interviews conducted by the author with the aforementioned individuals.

7. The written record indicates a P-51D (#44-84953). See Maj. Donald Eastman to ATSC Wright Field, ATSC TSEXL-122, January 18, 1946. However, Williams's February 1, 1946, memo (see n. 4, above) identifies a P-51H as the chase plane. A review of the Bell Aircraft (now Smithsonian FA00204) film taken at Pinecastle reveals a P-51H (tail number illegible) landing after Woolams's flight.

8. Dow flight log.

9. Donald Eastman, telephone interview with the author, October 11, 1989 (hereafter Eastman interview).

10. Joel Baker, interview with the author, September 1, 1989 (hereafter Baker interview).

11. Few participants remember the B-17 photographic plane. Photos taken at Pinecastle attest to its existence, although the author has been unable to identify the crew. B-17 tail number supplied from the Baker flight log.

12. Assistant Director of Aeronautical Research to Director ATSC Wright Field, NACA letter, R.A. 1347, January 8, 1946, Washington HQ. The initial Langley reply is found in H.J.E. Reid to NACA HQ Washington, NACA letter R.A. 1347, December 29, 1945.

13. Details for the first flight come primarily from Jack Woolams's official Pilot Report #1 to Bell Aircraft Corporation, January 25, 1946. This information is supplemented from the Williams memo of February 1, 1946, and the Dow flight log.

14. Baker interview; Williams, taped interview with the author, August 5, 1989 (hereafter Williams interview).

15. The in-flight details were supplied by the Cannon interview and Woolams, "How We Are Preparing to Reach Supersonic Speeds," 39. The cockpit setup details are from an article written by Chalmers "Slick" Goodlin, "Twenty Minutes of Eternity," *Air Trails Pictorial* (January 1948): 22. A review of the Smithsonian film #FA00204 reveals that Woolams wore the Mae West and the white-faced helmet with his initials that is visible in many still photos of him from that period.

16. Performance details for the XS-1 come from the recording instruments in the plane. This data was compiled by NACA engineers into a bound report, "Preliminary Flight Tests of the XS-1 Airplane (Project MX-653). Part 1—Measured Results Obtained from First Glide Flight," February 13, 1946 (hereafter "Flight Test Data, XS-1").

17. The Smithsonian film #FC00214 is the Wright Field compilation film of all ten drops. It reveals that the first drop was filmed in slow motion, presumably to assist in viewing separation.

18. Woolams's official Pilot Report #1.

19. Details from the landing come from the interview participants and the original Bell Aircraft film now in the Smithsonian Institution #FA00204. The film originally was mislabeled at the Smithsonian as tests at Eglin AFB.

20. Donald Norton, *Larry: A Biography of Lawrence D. Bell* (Chicago: Nelson-Hall, 1981), 173.

21. Eastman interview.

22. The rodeo incident is from the Cannon interview; Doug Rumsey, taped interview with the author, August 22, 1989 (hereafter Rumsey interview); and Woolams interview. This information is supplemented by an article, "Kowboys, Pilot Shine At Rodeo," *Orlando Sentinel*, January 26, 1946, p. 5.

23. Baker flight log. Williams to Langley Chief of Research, February 1, 1946.

24. Walter Williams to the Langley Chief of Research, NACA memorandum R.A. 1347, February 14, 1946. Baker flight log. The Williams memorandum serves as the general source of day-to-day information for the events at Pinecastle between February 2 and 9.

25. Details of the second flight of the XS-1 come from Jack Woolams's official Pilot

Report #2 to Bell Aircraft Corporation, February 5, 1946. This information is supplemented by Williams to Langley Chief of Research, February 14, 1946, and the Dow flight log. The P-47N tail number is from the Baker flight log.

26. Details of the third flight of the XS-1 come from Jack Woolams's official Pilot Report #3 to Bell Aircraft Corporation, February 5, 1946. This information is supplemented by Williams to Langley Chief of Research, February 14, 1946, and the Dow flight log.

27. The events for February 6–8, 1946, come from Williams to Langley Chief of Research, February 14, 1946.

28. Details of the fourth flight of the XS-1 come from Jack Woolams's official Pilot Report #4 to Bell Aircraft Corporation, inaccurately dated February 11, 1946. It appears probable that this is the date that Woolams prepared the report, since the detailed Williams memo of February 14, 1946, gives a day-by-day account of events and clearly indicates February 8, 1946, for the fourth flight, as does the Dow flight log. This date is also confirmed by the Baker flight log as a day he flew the P-47. Woolams listed "ventral fin" damage in his pilot report, but from personal examination of the XS-1 damage photos, it is possible this is a typographical error and Woolams meant ventral skin of the wing.

29. Norm Hayes, telephone interview with the author, September 18, 1989.

30. Williams (in Williams interview) and Pearsall (in a telephone interview of September 26, 1989) cleared up the mystery of the oft-seen Woolams photo in the pressure suit. Photos taken minutes prior to the pressure suit photo show Woolams wearing his coveralls and leather flight jacket with the XS-1 laying on its wing in the grass.

31. Dow flight log.

32. Pearsall, interview with the author, September 10 and 26, 1989.

33. Dow flight log. Office of the ATSC Liaison Langley to ATSC Wright Field, ATSC letter 1433 TSEXL/JAR R.A. 1347, March 14, 1946.

34. Williams to Langley Chief of Research, NACA memorandum R.A. 1347, March 8, 1946. Williams apparently wrote a longer version of this memorandum while in Orlando. It is dated February 28, 1946, and is used to supplement the March 8 memorandum (National Archives Pacific Southwest Region, RG 255, January–March 1946).

35. Details of the fifth flight of the XS-1 comes from Jack Woolams's official Pilot Report #5 to the Bell Aircraft Corporation and dated February 19, 1946. This is supplemented by Williams to Langley Chief of Research, February 28 and March 8, 1946. The Dow flight log contains an error in regard to the fifth flight; Dow apparently wrote the wrong date (February 18) for the fifth flight. Baker interview.

36. Rumsey interview.

37. Details of the perceived problem and the recommended fix come from Jack Woolams's official Pilot Report #6 to Bell Aircraft Corporation, February 25, 1946.

38. Baker flight log and telephone conversation with author, October 16, 1989.

39. Details of the sixth and seventh flights of the XS-1 come from Jack Woolams's official pilot reports #6 and #7 to Bell Aircraft Corporation, February 25, 1946. This information is supplemented by the Baker flight log, the Dow flight log, and Williams to Langley Chief of Research, February 28 and March 8, 1946.

40. Details for the eighth flight of the XS-1 come from Jack Woolams's official Pilot Report #8 to Bell Aircraft Corporation, February 26, 1946. This information is supplemented by the Baker flight log, the Dow flight log, and Williams to Langley Chief of Research, February 28 and March 8, 1946.

41. Baker flight log.

42. Details of the ninth flight of the XS-1 comes from Jack Woolams's official Pilot Report #9 to Bell Aircraft Corporation, February 26, 1946. This information is supple-

mented by Williams to Langley Chief of Research, February 28 and March 8, 1946, and the Dow flight log.

43. Don Thomson, taped recollections on August 20, 1989, provided to the author.

44. Williams to Langley Chief of Research, February 28, 1946, and Williams interview.

45. Williams to Langley Chief of Research, NACA memorandum R.A. 1347, March 21, 1946, covers the day-to-day events.

46. Frost biographical information supplied by R. H. Frost; R. H. Frost to ATSC Wright Field, MX-653 Bell Progress Report, March 4, 1946.

47. Williams to Langley Chief of Research, March 21, 1946. Tex Johnston, interview with the author, October 8, 1989 (hereafter Johnston interview).

48. Johnston interview. Details for the tenth flight of the XS-1 come from Jack Woolams's official Pilot Report #10 to Bell Aircraft Corporation, March 6, 1946. This information is supplemented by Williams to Langley Chief of Research, March 21, 1946, the Baker flight log, and the Dow flight log. R. H. Frost to AMC Wright Field, MX-653 Bell Progress Report, March 4, 1946, had indicated the tests would end on March 9.

49. Dow flight log.

50. Williams to Langley Chief of Research, March 21, 1946.

Chapter 3. The Program Takes a Pause

1. For a discussion of bureaucratic imperialism see Matthew Holden, " 'Imperialism' in Bureaucracy," in *Bureaucratic Power in National Politics*, 2d ed., ed. Francis Rourke (Boston: Little, Brown, 1972), 197–214.

2. Craven and Cate, eds., *Men and Planes*, 239–40.

3. See Roland, *Model Research*, 1:211–18, for an excellent discussion of this issue. The NACA lost a key ally in 1945 when Congressman Clifton A. Woodrum of Virginia, chairman of the appropriations subcommittee that oversaw the NACA budget, retired from Congress; and Roland, *Model Research*, 2:693–95, for NACA's "National Aeronautical Research Policy," a very political and self-aggrandizing (from the AAF perspective) document. See also Wolk, *Planning and Organizing*, 44.

4. See the ATSC memorandum TSENG3/JAR/, "Financing of NACA Research," October 22, 1945, attached to ATSC letter TSEST7/JPA/ through ACAS-4, "Financing of NACA Research," no date listed, to the Commanding General Army Air Forces. The replies include Col. Barron Powers to ATSC Liaison at Langley, ATSC memorandum TSEST7B/DS/ of January 16, 1947. Also see Brig. Gen. Laurence Craigie to Subdivision Chiefs, Laboratory Chiefs, NACA Committee members, Project Officers, ATSC memorandum TSENG "ATSC Participation in the NACA Research Program," no date listed, in "Sara Clark" RD 3763, File 121.6, "Costs—Research & Development—Policy Engineering & Experimental," National Archives Washington National Records Center, Suitland, Maryland.

5. See Brig. Gen. Laurence Craigie to Aircraft Manufacturers, ATSC letter and cover memorandum, "Policy Regarding Research and Development Aircraft," January 11, 1946 (letter) and January 24, 1946 (memo), in "Sara Clark" RD 4026, File "Research Aircraft 1946-49-50." Also see R. H. Frost to Air Materiel Command, Wright Field, Bell Aircraft Corporation memorandum R.A. 1347, August 28, 1946 (Bell changed the designation ATSC to AMC on the Wright Field reports in April). Stanley was actually the author of the document as noted by the initials RMS on the memorandum.

6. See R. H. Frost to AMC Wright Field, MX-653 Bell Progress Report, May 2, 1946; Dow flight log.

7. Hansen, *Engineer-in-Charge*, 297. Hallion, *On the Frontier*, 9. The issue of who decided the question to test at Muroc is a logical outgrowth of the earlier decision for Pinecastle. Bell had considered Muroc from the first days of the program as evidenced by their June 19, 1945, Progress Report to ATSC. Stanley and Woolams were intimately familiar with the advantages of Muroc from their days on the XP-59 tests. Muroc Dry Lake was flooded during the Pinecastle tests. Once Woolams experienced the difficulties of XS-1 flight, and considering the Bell/ATSC reluctance to test near Langley, the spacious dry lakebed became the perfect location for the powered tests.

8. See the *Historical Summarization of Muroc Army Air Field, January 23, 1942– September 2, 1945*, pp. 1–2, AFFTC-HO.

9. See Richard Frost to ATSC Wright Field, MX-653 Bell Progress Report, April 1, 1946; R. H. Frost to AMC Wright Field, May 2, 1946; R. H. Frost to AMC Wright Field, MX-653 Bell Progress Report, May 14, 1946; and Williams to Langley Chief of Research, NACA memorandum R.A. 1347, May 21, 1946. Also see Williams to Langley Chief of Research, NACA memorandum R.A. 1347, July 8, 1946. Gilkey was base commander from May 1946 to August 1949. He was also, coincidentally, a survivor of an ill-fated P-38 flight that crashed due to compressibility effects.

10. R. H. Frost to AMC Wright Field, MX-653 Bell Progress Report, March 16, 1946; Frost to ATSC Wright Field, April 1, 1946; R. H. Frost to AMC Wright Field, MX-653 Bell Progress Report, April 16, 1946; R. H. Frost to AMC Wright Field, MX-653 Bell Progress Report, May 31, 1946.

11. Air Materiel Command TSESA-4-287 to Bell Aircraft Corporation, April 25, 1946, is the AMC request; see Frost to AMC Wright Field, May 2, 1946. The date for the flight is the author's calculation from the Dow flight log, which lists a C-45 flight that date to Wright Field. I believe this flight was to pick up the B-29 crew. Frost to AMC Wright Field, May 31, 1946, and May 14, 1946.

12. "Supersonic Plane and Jet Bombers Revealed by Army Air Forces," *Aviation* 45, no. 7 (July 1946): 68–69.

13. The return flight date is listed in the Dow flight log. Details for the progress in June are found in R. H. Frost to AMC Wright Field, Bell Aircraft memorandum, June 14, 1946; R. H. Frost to AMC Wright Field, MX-653 Bell Progress Report, June 17, 1946; and R. H. Frost to AMC Wright Field, MX-653 Bell Progress Report, July 1, 1946. All documents supplied by Frost. Also see Williams to Langley Chief of Research, July 8, 1946.

14. For July/August events see R. H. Frost to AMC Wright Field, MX-653 Bell Progress Reports, August 15, 28, and September 3, 1946 (all documents supplied by Frost). Also see Williams to Langley Chief of Research, July 8, 1946; Dow flight log.

15. See R. H. Frost to AMC Wright Field, MX-653 Bell Progress Report, September 19, 1946. Also see Donald Eastman to Files, AMC 2256 TSEXL/DRE in R.A. 1347, September 11, 1946. Also see the R. H. Frost flight log; Dow flight log.

16. See Frost to Air Materiel Command, August 28, 1946. The Frost/Stanley memo has a typographical error in regard to the AMC policy letter. It gives the date as February 1945.

17. Russell, written comments to the author, February 1991.

18. The captive flight information comes from the Dow and Frost flight logs. This is supplemented by R. H. Frost to AMC Wright Field, MX-653 Bell Progress Report, September 30, 1946.

19. Office of Engineering Liaison Officer at Langley to ATSC Wright Field, TSEXL/ JAR Air Technical Service Command Progress Report, March 14, 1946.

20. Col. George Smith to Dr. G. W. Lewis NACA Washington, Air Materiel Command TSESA-4/JHV/, April 11, 1946.

21. J. W. Crowley (Acting Director of Aeronautical Research) to NACA Langley, NACA memorandum R.A. 1347, April 24, 1946. Walter Williams to Langley Chief of Research, NACA memorandum R.A. 1347, April 26, 1946. The author believes this memo is the source of later confusion over the date of the first XS-1 flight. Williams mentions the flight test program as dating from January 19. While Williams certainly understood that this date referred to the NACA arrival in Orlando, it appears that subsequent readers believed this date referred to the first flight. January 19 is the date listed in the official NACA/NASA summary (Mulac Chronology) of the program.

22. Walter Williams to Langley Chief of Research, NACA memorandum R.A. 1347, April 30, 1946.

23. H.J.E. Reid to NACA Washington, NACA memorandum R.A. 1347, April 29, 1946.

24. H.J.E. Reid to NACA Washington, NACA memorandum R.A. 1347, May 28, 1946.

25. Walter Williams to Langley Chief of Research, NACA memorandum R.A. 1347, June 7, 1946; H.J.E. Reid to NACA Washington, NACA memorandum R.A. 1347, June 24, 1946.

26. J. W. Crowley (Acting Director of Aeronautical Research) to AMC Wright Field, NACA letter R.A. 1347, June 13, 1946.

27. Williams to Langley Chief of Research, NACA memorandum R.A. 1347, May 8, 1946. Also see Williams to Langley Chief of Research, NACA memorandum R.A. 1347, July 25, 1946, which lists the original staff as Walter Williams (Aeronautical Engineer [AE]), William Aiken (AE), Cloyce Matheny (AE), Abraham Ruvin (soon Donald Youngblood) (Electrical Engineer), Joel Baker (Test Pilot), Charles Forsyth (AE), Beverly Brown (AE), John Gardner (AE), Warren Walls (Instrument Technician), George Minalga (Electrical Engineer), Howard Hinman (Airplane Crew Chief), Roxannah Yancey (Computer), and Isabel Martin (Computer). This number was increased from the six selected as needed on May 23. See H.J.E. Reid to NACA HQ, NACA memorandum R.A. 1347, May 23, 1946.

28. Maj. Donald Eastman to AMC Wright Field, AMC letter R.A. 1347, May 7, 1946; Col. George Price to AMC Liaison Officer Langley, AMC letter R.A. 1347, May 28, 1946. Also see H.J.E. Reid to AMC Wright Field, NACA letter R.A. 1347, July 10, 1946. The reply is Col. George Price to AMC Liaison Langley, AMC letter R.A. 1347, August 2, 1946. Radar information is found in Col. George Price to AMC Liaison Officer at Langley, Army teletype TSESA-4/JHV, July 25, 1946. On Muroc see Walter Williams to Langley Chief of Research, NACA memorandum R.A. 1347, July 19, 1946. Maj. Donald Eastman to the Files, Army memorandum 2256 TSEXL/DRE, September 11, 1946.

29. H.J.E. Reid to NACA HQ, NACA letter R.A. 1347, May 23, 1946; H.J.E. Reid to NACA HQ, NACA letter R.A. 1347, June 5, 1946; H.J.E. Reid to NACA HQ, NACA letter R.A. 1347, July 30, 1946; Dr. George Lewis to Commanding General AMC Wright Field, NACA letter R.A. 1347, August 1, 1946; Walter Williams to Langley Chief of Research, NACA memorandum R.A. 1347, August 16, 1946. Maj. Donald Eastman to AMC Wright Field, Army teletype TSEXL/DRE R.A. 1347, September 10, 1946; Maj. Donald Eastman to AAF personnel, Army teletype TSEXL 2253 R.A. 1347, September 11, 1946; J. F. Victory (NACA HQ Executive Secretary) to Langley Laboratory, NACA letter R.A. 1347, September 27, 1946.

30. The NEA story is by Douglas Larsen; the UP story had no author listed. Both were supplied by Mrs. Woolams.

31. See Maj. Donald Eastman to Commanding General AMC Wright Field, AMC

TSEXL 2170, June 27, 1946. The response is Col. George Patterson to AMC Engineering Liaison Officer (Eastman) Langley Field, Army TXEXS, July 16, 1946. Both documents are found in "Sara Clark" Collection RD 3753, File 000.7 labeled " XS-1 Bell Publicity Release."

32. Frederick Neely, ed., "1,000 Miles an Hour?" *Collier's* (August 17 and 24, 1946): 78.

33. Woolams, "How We Are Preparing to Reach Supersonic Speeds," 38–39.

34. Eugene Draley to Langley Chief of Research, NACA memorandum R.A. 1347, July 10, 1946 (typed July 15, 1946). The turbine pump would allow the return to the original cylindrical propellant tanks and lower nitrogen requirements. The lower weight and increased propellant capacity would allow a full 4.2-minute four-cylinder flight instead of the 2.5 minutes available with the nitrogen system.

35. See the NACA press release R.A. 1347 entitled, "First Supersonic Airplane, XS-1, Built for NACA Research," August 11, 1946.

36. See Miller, *The X-Planes*, 35–36.

37. Details on the Cleveland Air Race are from Birch Mathews, "Cobra" in *AAHS Journal* 8, no. 1 (Spring 1963). Details on the coin-toss agreement are from interviews with Goodlin and Mary Woolams and cited in "Jet Plane Falls in Flames," *Buffalo Evening News*, September 4, 1946, p. 1.

38. Details are found in an unattributed article, "Search Lake for Woolams Body," *Niagara Falls Gazette*, August 31, 1946, p. 1. Brig. Gen. Laurence Craigie to Lawrence Bell, personal letter, September 4, 1946. Lawrence Bell to Brig. Gen. Laurence Craigie, Bell Aircraft Corporation letter, September 6, 1946. Both letters found at Edwards AFFTC-HO. Norton, *Larry*, 185.

39. Biographical details are from Chalmers Goodlin. Employment dates from Bell Aircraft Corporation letter dated October 14, 1947, supplemented by unattributed article, "Jet Plane Falls in Flames."

40. Details for the September 16–17, 1946, meeting are found in Walter Williams to Langley Chief of Research, NACA memorandum R.A. 1347, September 20, 1946. The Langley rebuttal comes from H.J.E. Reid to NACA HQ, NACA memorandum R.A. 1347, September 26, 1946.

41. Air Materiel Command Correspondence Summary of Project MX-653 History, prepared by James Voyles, January 14, 1947, and transmitted to the author by Ezra Kotcher. The "Stanley" letter is actually the Bell memorandum, August 28, 1946, signed by Frost. Marchese was assistant project engineer. He had also been Stan Smith's deputy.

42. Norton, *Larry*, 197–98, 202–3, 210.

43. See H.J.E. Reid to NACA HQ, NACA memorandum R.A. 1347, September 26, 1946.

44. H. Soule to Langley Chief of Research, NACA memorandum R.A. 1347, October 1, 1946.

45. Col. George Patterson to the AMC Liaison Officer at Langley, TSEOL/TSEOP-9-866, September 26, 1946, in "Sara Clark" RD 3753, File 000.7, titled "XS-1 Bell Publicity Release." The Bell reply is found in the files of the Edwards AFFTC-HO under Lawrence Bell to Maj. Gen. L. C. Craigie, Bell Aircraft Corporation letter, October 4, 1946.

46. R. H. Frost to AMC Wright Field, MX-653 Bell Progress Report, September 30, 1946; Goodlin, telephone interview with author, July 29, 1991.

47. See H.J.E. Reid to Commanding Officer Muroc Army Air Base, NACA letter R.A. 1347, September 20, 1946. Also Walter Williams to Melvin Gough, NACA letter R.A. 1347, October 1 and 2, 1946.

48. See Walter Williams to Melvin Gough, NACA letter R.A. 1347, October 3, 1946; Williams to Gough, NACA letter R.A. 1347, October 4, 1946.

49. Accident details are found in "Jet Plane Falls in Flames," 1. Contemporary accounts credit Goodlin with heroic action in staying in the plane to allow his passenger Fay to jump.

50. Hallion, *Supersonic Flight*, 1–3.

51. The details of the Stanley agreement were provided to the author in interviews with Chalmers Goodlin (March 9, July 19 and 21, and December 12, 1991) and Edmund Burke (December 4, 1990, and January 3, 1991). This information was supplemented in a telephone interview with Charles Hall (February 5 and July 18, 1991). Also see the transcript of the C. H. Goodlin speech, March 13, 1986, before the Dallas/Fort Worth Society of Experimental Test Pilots, the Society of Flight Test Engineers, the American Helicopter Society, and the American Institute of Aeronautics and Astronautics. Goodlin told the author that Stanley "possibly" prepared the graph *prior* to the meeting and that the bonus figures were verbally added during the discussion. If correct, this would indicate that some outline of a follow-on test program by Bell Aircraft had already been developed. This would be consistent with the May 31, 1945, ATSC (AMC) letter to the NACA indicating the follow-on procedure for the supersonic flights.

Chapter 4. The Goodlin Era

1. The date and time of the trip to Muroc come from the Dow and the Frost flight logs. The author has been unable to identify the stopover base (symbol DTE). Details for the Bell and NACA arrival and October 8, 1946, preparations come from Walter Williams to Mel Gough, NACA letter R.A. 1347, October 8, 1946, and Walter Williams to Langley Chief of Research, NACA memorandum R.A. 1347, October 10, 1946. Also see Williams to Langley Chief of Research, NACA memorandum R.A. 1347, October 11, 1946.

2. Crew details from Goodlin, "Twenty Minutes of Eternity," 21. Details for the October 9, 1946, aborted flight are from Goodlin's official Bell interoffice memorandum of flight 10/9/46, Chalmers Goodlin to R. H. Frost et. al., no date listed (supplied by C. Goodlin from his personal files). Other sources include: Walter Williams to Mel Gough, NACA letter R.A. 1347, October 9, 1946; and Joel Baker to H. H. Hoover (LMAL Pilot Office), NACA letter R.A. 1347, October 9, 1946. This information is supplemented from a special XS-1 (#2) flight log (R.A. 1347) for tests at Muroc; no other heading, author, or date is listed on the fifteen-page document. The details of the log present data that clearly comes from the NACA instruments. However, additional details on B-29 and Bell actions lend support to the concept that the author had access to Bell personnel. The listing for the aborted flight erroneously gives the date as October 10, 1946.

3. Repairs and preparations for the October 11 flight come from Joel Baker to Herb Hoover, NACA memorandum R.A. 1347 of October 10, 1946 and NACA letter R.A. 1347 of October 10, 1946. Also see XS-1 (#2) flight log. The attitude remarks come from the above sources and Joel Baker to H. H. Hoover, NACA letter R.A. 1347, October 9, 1946. For the Goodlin meeting see Walter Williams to Mel Gough, NACA letter R.A. 1347, October 15, 1946.

4. The other aircraft research is found in Walter Williams to Mel Gough, NACA letter R.A. 1347, October 4, 1946; and Joel Baker to Herb Hoover, NACA letter R.A. 1347, October 10, 1946.

5. Information on the October 11, 1946, flight comes from Chalmers Goodlin's official

Pilot Report #1 to Bell Aircraft Corporation, October 18, 1946. This information is supplemented by Walter Williams to Langley Chief of Research, NACA memorandum R.A. 1347, October 11, 1946. In the NACA files is a letter to Melvin Gough also dated October 11, 1946, that contains additional details of the flight. Interestingly, Williams put the time of flight at 3:00 P.M. This may be the time of drop. Details on the B-29 setup come from the Dow and the XS-1 (#2) flight logs. The official *History of Muroc Army Air Field, October 1, 1946–December 31, 1946*, p. 17 (AFFTC-HO), is incorrect. It appears to give the time for the October 9, 1946, flight, but labels it October 11. This is the first documented instance of Goodlin attempting to land on the runway. Although unsuccessful on this occasion due to brake failure, Goodlin succeeded many other times in confining the XS-1 landing to the Muroc hard runway. The Air Force program strictly utilized the dry lakebed.

6. Information for the October 14, 1946, flight comes from Goodlin's official Pilot Report #2 to Bell Aircraft Corporation, October 16, 1946. This information is supplemented by Walter Williams to Mel Gough, NACA letter R.A. 1347, October 14, 1946; Walter Williams to Langley Chief of Research, NACA memorandum R.A. 1347, November 8, 1946, included some technical data from the flight. This is supplemented by the XS-1 (#2), Dow, and Frost flight logs, and "Pertinent Information Regarding the Landings of the Bell XS-1 Airplane," NACA document, November 19, 1946, in "Sara Clark" RD 4316 File 452.1 (hereafter Bell XS-1 Landing Information).

7. See Hartley Soule to Langley Chief of Research, NACA memorandum R.A. 1347, October 15, 1946.

8. Walter Williams to Langley Chief of Research, NACA memorandum R.A. 1347, October 18, 1946.

9. Ibid. Also see Walter Williams to Langley Chief of Research, NACA memorandum R.A. 1347, October 21, 1946.

10. Information on the October 17, 1946, flight of the XS-1 comes from Goodlin's undated official Pilot Report #3 to Bell Aircraft Corporation. (Goodlin advised the author that if only a flight date was listed, as on this report, the document was prepared shortly after the test.) Also see the Dow, Frost, and XS-1 (#2) flight logs; and "Summary of XS-1 Flights at Muroc Army Air Base," NACA report R.A. 1347, October 17, 1946, with the initials LPS (hereafter XS-1 Flights Summary). Also see Bell XS-1 Landing Information.

11. Walter Williams to Mel Gough, NACA memorandum R.A. 1347, October 17, 1946. Also see the Dow flight log and Williams to Langley Chief of Research, October 18, 1946. A. J. Marchese to AMC Wright Field, MX-653 Bell Progress Report, October 15, 1946; R. H. Frost to AMC Wright Field, MX-653 Bell Progress Report, October 21, 1946. Many people believe the wing/tail from the number 1 aircraft was put on the number 2 airplane. The author determined this was not the case as evidenced by the chronology of the construction of the two aircraft. However, A. J. Marchese to AMC Wright Field, MX-653 Bell Progress Report, November 1, 1946, definitively proves that the number 2 aircraft had the spare wing/tail pieces constructed early in the program and not the components from the number 1 airplane. Progress report from the files of R. H. Frost.

12. Details on the events during late October–early November are found in Walter Williams to Langley Chief of Research, NACA memorandum R.A. 1347, November 1, 1946; Walter Williams to Langley Chief of Research, NACA memorandum R.A. 1347, November 8, 1946; Walter Williams to Mel Gough, NACA letter R.A. 1347, November 9, 1946; Walter Williams to Langley Chief of Research, NACA memorandum R.A. 1347, November 15, 1946; Walter Williams to Mel Gough, NACA letter R.A. 1347, November 15, 1946.

13. Details for the preparations for the fourth flight come from Walter Williams to

Langley Chief of Research, NACA memorandum R.A. 1347, November 29, 1946. This information is supplemented by the Dow flight log and Walt Williams to Melvin Gough, NACA memorandum R.A. 1347, December 1, 1946. Also Frost, written comments to the author, March 14, 1991, and telephone interview, April 16, 1991 (hereafter Frost interview). Also see A. J. Marchese to Major Eastman, Bell Aircraft Corporation letter, November 12, 1946; and H.J.E. Reid to AMC Wright Field, NACA memorandum, November 27, 1946, for the lack of instruments (both are from "Sara Clark" RD 4316 File 452.1).

14. The information on the December 2, 1946, flight of XS-1 number 2 comes from Goodlin's official Pilot Report #4 to Bell Aircraft Corporation, December 4, 1946. This is supplemented by a letter to Hartley Soule (unsigned), NACA letter R.A. 1347, December 3, 1946. (The contents seem to indicate Williams as the author.) Frost interview. Also see the XS-1 (#2), Dow, and Frost flight logs, and Walter Williams to Langley Chief of Research, NACA memorandum R.A. 1347, December 7, 1946.

15. Details for the efforts to drop test on December 6 come from Walter Williams to Langley Chief of Research, NACA memorandum R.A. 1347, December 7, 1946. This is supplemented by Walter Williams to Mel Gough, NACA letter R.A. 1347, December 8, 1946; and Joel Baker to H. H. Hoover, NACA letter R.A. 1347, December 12, 1946. Flight details are from the Dow and Frost flight logs. The Captain Smith may be Martin L. Smith of Wright Field.

16. Details for the first powered flight on December 9, 1946, are from Goodlin's official Pilot Report #5 to Bell Aircraft Corporation, December 11, 1946. This is supplemented by Walter Williams to Langley Chief of Research, NACA memorandum R.A. 1347, December 11, 1946; and *History of Muroc Army Air Field*, 18. Also see Walter Williams to Langley Chief of Research, NACA memorandum R.A. 1347, December 15, 1946; and the XS-1 (#2), Dow, and Frost flight logs, and Richard Frost, the official Bell Aircraft Corporation Preliminary Report, December 11, 1946. Some details may also be found in a letter sent by Henry Aiken and reported in Henry Pearson to the Langley files, NACA memorandum R.A. 1347, December 18, 1946 (Pearson was head of the Aerodynamic Flight Loads Section at Langley).

17. See Walter Williams to Mel Gough, NACA memorandum R.A. 1347, December 11, 1946; Goodlin, taped interview with the author, March 9, 1991 (hereafter Goodlin interview), and telephone interview, February 20, 1991.

18. *New York Times*, December 11, 1946, p. 1. According to Goodlin, the correct quote is from "Army and Navy," *Time*, December 23, 1946, p. 21.

19. "Army and Navy."

20. "Supersonics," *Life*, January 6, 1947, pp. 49–56.

21. Williams to Gough, December 11, 1946. Also see Walter Williams to Hartley Soule, NACA letter R.A. 1347, December 12, 1947.

22. See the three Army teletypes (undated, January 17, and January 20, 1947) dealing with the issue. The three documents first give approval and then rescind the authorization to take photos. Also see Maj. Gen. Laurence Craigie to Commanding General AAF AC/AS-4, AMC teletype TSEOA/MPE/, January 21, 1947. The four documents are found in "XS-1 Airplane—Bell—Publicity Releases 1946–47," in "Sara Clark" RD 3753, File 000.7. Also see the preprint of the speech by Maj. Gen. Laurence Craigie entitled "Engineering and Research Planning of the Army Air Forces" delivered to the Society of Automotive Engineers annual meeting, January 6–10, 1947, AFFTC-HO. (Craigie lists the wrong date, September 9, 1946 [p. 16], for the first powered flight.)

23. See Walter Williams to Mel Gough, NACA letter R.A. 1347, January 5, 1947.

24. See ibid. and Williams to Gough, December 11, 1947. Goodlin interview. Goodlin had a book agent, not a movie agent. Also see Walter Bonney to Chalmers Goodlin, Bell

Aircraft Corporation letter, May 22, 1947, and a copy of the Bell release for the article "Twenty Minutes of Eternity" (both supplied by C. Goodlin).

25. The damage assessment is found in Frost Preliminary Report, December 11, 1946, and XS-1 (#2) flight log. Also see Walter Williams to Mel Gough, NACA letter R.A. 1347, December 16, 1946. Final analysis is found in John Gardner to Langley Chief of Research, NACA memorandum R.A. 1347, March 28, 1947. Frost interview, and written comments of March 14, 1991.

26. Williams to Gough, December 16, 1946.

27. Frederick Robinson to Chalmers Goodlin, personal letter, December 12, 1946 (provided by C. Goodlin). Williams to Gough, December 16, 1946. XS-1 (#2) flight log. Goodlin Pilot Report #5. Dow flight log.

28. Details for the December 20, 1946, flight of the XS-1 come from Goodlin's official Pilot Report #6 to Bell Aircraft Corporation, December 24, 1946. This is supplemented by the Dow, XS-1 (#2), Frost, and Baker flight logs, Williams to Gough, December 11, 1946, and Walter Williams to Langley Chief of Research, NACA memorandum R.A. 1347, December 23, 1946. The Dow flight log lists Heaney as the copilot, but the Frost flight log indicates specifically that Frost was copilot for 30 minutes of the flight. The claim that Goodlin "forgot" to turn on the NACA instruments is the first of a number of incidents involving both Goodlin and Yeager.

As Goodlin stated to the author, his checklist of predrop mandatory activities required activation of the NACA instruments. Thus, it is extremely unlikely that he or Yeager would have "forgotten" to turn on the instruments. More probably, the instruments failed to function or the power system failed to perform.

29. Williams to Gough, January 5, 1947. Also see the Frost flight log.

30. A. J. Marchese to AMC Wright Field, MX-653 Bell Progress Report, January 6, 1947.

31. Dow flight log.

32. Williams to Gough, January 5, 1947. Preparations for the January 8 flight are from Walter Williams to Langley Chief of Research, NACA memorandum R.A. 1347, January 6, 1947.

33. Details for the January 8, 1947, flight of the XS-1 come from Goodlin's official Pilot Report #7 to Bell Aircraft Corporation, January 22, 1947. This is supplemented by the Dow and XS-1 (#2) flight logs, and Charles Forsyth to Langley Chief of Research, NACA memorandum R.A. 1347, January 13, 1947. Beeler (from LMAL) had a long involvement with the XS-1 program going back to 1945. He was assigned as a project engineer in charge of the aircraft loads program.

34. Preparations are found in Walter Williams to Langley Chief of Research, NACA memorandum R.A. 1347, January 20, 1947. Also see Lawrence Clousing to the Engineer-in-Charge, NACA memorandum R.A. 1347, January 29, 1947.

Details for the January 17, 1947, flight and wrap-up come from Goodlin's official Pilot Report #8 to Bell Aircraft, January 22, 1947. This information is supplemented by Williams to Langley Chief of Research, January 20, 1947, and the XS-1 (#2) flight log. Also see Clousing to the Engineer-in-Charge, January 29, 1947. The speed mark is found in XS-1 Flights Summary.

35. See Charles Forsyth to Langley Chief of Research, NACA memorandum R.A. 1347, January 27, 1947. There is some confusion about this January 20, 1947, aborted flight. The XS-1 (#2) flight log clearly discusses a first aborted flight on January 20. No record exists in the Dow flight log. The Baker flight log lists a one-hour *local* flight at Muroc in the C-45! I am inclined to believe this is the date for a first aborted flight.

36. Details for the January 21, 1947, second aborted flight come from the XS-1 (#2) flight log. This is confirmed by the Dow and Baker flight logs.

37. Details for the January 22, 1947, flight are found in Goodlin's official Pilot Report

#9, January 23, 1947. This is supplemented by the XS-1 (#2), Dow, and Baker flight logs, and Forsyth to Langley Chief of Research, January 27, 1947.

38. Details for the January 23, 1947, flight are found in Goodlin's official Pilot Report #10, January 27, 1947. This is supplemented by the XS-1 (#2), Dow, and Baker flight logs, and Forsyth to Langley Chief of Research, January 27, 1947. There is a discrepancy on the issue of the scoops. Goodlin's Pilot Report #10 states the scoops were reinstalled prior to the flight. The XS-1 (#2) flight log indicates they were reinstalled after the flight. The author believes the latter statement is correct since the scoops were removed after the ninth flight. Thus, it would seem illogical to not make one flight without them. The Goodlin report was prepared some days after the #10 flight and I believe it is in error.

39. Details for the preparations, January 30, 1947, flight, and wrap-up are found in Goodlin's official Pilot Report #11, date illegible. No takeoff time was listed. This information is supplemented by the XS-1 (#2), Dow, and Baker flight logs, and Walter Williams to Langley Chief of Research, NACA memorandum R.A. 1347, February 10, 1947.

40. Details for the January 31, 1947, flight are found in Goodlin's official Pilot Report #12, date illegible. No takeoff time was listed. Also see the XS-1 (#2) and Dow flight logs, and Williams to Langley Chief of Research, February 10, 1947.

41. Details for the February 5, 1947, flight are found in Goodlin's official Pilot Report #13, February 7, 1947. No takeoff time was listed. This is supplemented by the XS-1 (#2), Dow, and Baker flight logs, and Williams to Langley Chief of Research, February 10, 1947. The author believes this flight is the basis for the *Air Trails* article. The official reports list 8.8 g. But as the cockpit photograph reveals, the true indicated number was 8.7 g.

42. Frost interview. Frost was quite emphatic that the noise heard that day was a sonic boom. Walt Williams, telephone interview with the author, June 17, 1992. Williams remembered the incident and told the author he assumed it was a "normal" event—that is to say, not a big deal. Russell did not hear the boom but did remember Frost instructing him to tear the plane apart to find the source of the noise (telephone interview, April 11, 1993). However, several test pilots interviewed by the author believed that it was not possible for the XS-1 to have created a boom at speeds as low as Mach 0.80. The author requested that the National Aeronautics and Space Administration (NASA) comment on this issue based upon their vast experience at measuring such phenomenon. In a NASA eleven-page letter dated March 16, 1993, Christine Darden to the author; retired Langley expert Domenic Maglieri replied for the agency. He stated that "shock waves can and do exist on bodies (airfoils and aircraft) at freestream Mach numbers in the 0.75 to 0.99 range." Maglieri reported that for planes in a diving configuration, these shock waves can sometimes be heard on the ground. The accompanying charts and mathematical computations reinforced the point.

43. Details for the February 7, 1947, flight and wrap-up are found in Goodlin's official Pilot Report #14, February 11, 1947. This is supplemented by the XS-1 (#2), Dow, and Baker flight logs, and Williams to Langley Chief of Research, February 10, 1947. There is no record of the chase pilot for this mission.

44. A. J. Marchese to AMC Wright Field, MX-653 Bell Progress Report, February 5, 1947.

45. John Gardner to Langley Chief of Research, NACA memorandum R.A. 1347, April 9, 1947.

46. Williams to Langley Chief of Research, February 10 and 18, 1947.

47. Walter Williams to Hartley Soule, NACA letter R.A. 1347, February 14, 1947. Also see XS-1 flight log.

48. See Williams to Langley Chief of Research, February 18, 1947.

49. Details for the February 19, 1947, flight and wrap-up are found in Goodlin's official Pilot Report #15, March 6, 1947. No takeoff time was listed. This is supplemented by the XS-1 (#2) and Dow flight logs, and Walter Williams to Langley Chief of Research, NACA memorandum R.A. 1347, February 25, 1947 (hereafter Williams to Langley Chief of Research, NACA, February 25, 1947).

50. Hartley Soule to Langley Chief of Research, NACA memorandum R.A. 1347, February 24, 1947.

51. Details for the February 21, 1947, flight are found in Goodlin's official Pilot Report #16, March 7, 1947. No takeoff time was listed. This information is supplemented by the XS-1 (#2) flight log (final entry), the Dow and Baker flight logs, and Williams to Langley Chief of Research, NACA, February 25, 1947.

52. Dow flight log. See Charles Forsyth to Langley Chief of Research, NACA memorandum R.A. 1347, March 3, 1947. Also see A. J. Marchese to AMC Wright Field, MX-653 Bell Progress Reports, March 10, March 17, and April 1, 1947.

Chapter 5. The Changing of the Guard

1. See "NACA Reveals New Results of Supersonic Research Program," *Aviation Week* 47, no. 1 (July 7, 1947): 18, 21. Also see Roland, *Model Research*, 1:247–49; and "Research Men Clash on Program," *Aviation Week* 47, no. 18 (November 3, 1947): 17.

2. See Hartley Soule to Langley Chief of Research, NACA memorandum RG 355 (Suitland Record Center, Suitland, Md.), December 20, 1946.

3. Information on the February 6 meeting comes from two primary sources. These are: Hartley Soule to Walter Williams, NACA memorandum R.A. 1347, February 11, 1947; and J. W. Crowley to the Commanding General Army Air Forces, NACA memorandum R.A. 1347, February 19, 1947. Two X-3s were eventually ordered although only one was built.

4. Details for the February 13 meeting at Muroc come from Walter Williams to Hartley Soule, NACA letter R.A. 1347, February 14, 1947. Putt was head of Bomber Aircraft Subsection at Wright Field.

5. Hartley Soule to Langley Chief of Research, NACA memorandum R.A. 1347, February 24, 1946.

6. See Baker to Langley Chief of Research, NACA memorandum, February 25, 1947.

7. See "Flight Test Programs on Fighter Airplanes," Air Materiel Command TSEOA-4, February 28, 1947; this document, prepared by Charles Hall, is in AFFTC-HO.

8. Cost details are listed in the "Green Book." This and the discussion by Hall are from AFFTC-HO.

9. See "New AAF Jet-Rocket Fighters Designed for Supersonic Speeds," *Aviation Week* 47, no. 3 (July 21, 1947): 10. Much of the anticipated progress outlined in the article turned out to be a case of very wishful thinking. Most of the aircraft did not go into production.

10. Information on the important March 5 meeting comes from three primary sources. First, H.J.E. Reid to Air Materiel Command Liaison Langley Field, NACA memorandum R.A. 1347, March 21, 1947. Second is Brig. Gen. S. R. Brentnall (Chief, Engineering Operations/Engineering Division) to NACA Washington HQ for J. W.

Crowley, Army letter TSEOA-4/JHV/, March 18, 1947. Finally, Robert Stanley to D. R. Shoults et al., internal Bell memorandum, March 6, 1947 (document supplied by C. Goodlin from his personal files). It is a businessman's summary to his bosses and thus may be tinged with some bias as to the strength of his remarks and the level of acceptance of his arguments. Stanley's comments regarding Baker's financial request for a raise from $5,200 per year and bonus to cover Muroc expenses and flight hazards was confirmed to the author by Baker, telephone interview, December 29, 1992.

11. In a telephone interview with the author on December 17, 1990, General Ritland disputed the point that he took Stanley's side in the debate. Rather, he stated that Stanley may have inferred that Ritland's remarks, dealing with the past practice of allowing contractors to fly new aircraft until acceptance by the government, constituted adopting Bell's position. This was not Ritland's intent but was rather simply an acknowledgment of the traditional method of doing business on aircraft projects.

12. Ritland, telephone interview with the author, November 13, 1990. Also see Young, ed., *Supersonic Symposium*, 35 and n. 43. Soule's reservations seemed to largely involve the air loads and stress to be found in testing the XS-1 below 60,000 feet. With the XS-1's limited fuel capacity, Soule did not believe it was capable of tests above that altitude.

13. See Brig. Gen. S. R. Brentnall to NACA Washington, Air Materiel Command TSEOA-4/JHV/, March 18, 1947, in "Sara Clark" RD 2306. (This is not the same letter as mentioned in reference to the March 5, 1947, meeting.)

14. A brief discussion of the Washington meeting may be found in Floyd Thompson to Files regarding telephone conversation with J. Crowley, NACA memorandum R.A. 1347, April 11, 1947. The minutes of the meeting may be found in G. W. Lewis and J. W. Crowley to the Files, NACA memorandum, March 26, 1947. This document was found in RG 355, File XS-1, folder labeled January–June 1947 at Washington National Records Center, Suitland, Maryland. General Craigie indicated to the author he did not remember this meeting, but stated it was possible he was there at the request of Larry Bell. Craigie, telephone interview with the author, June 9, 1992.

15. The Muroc meeting details are from Walter Williams to H. Soule, NACA letter R.A. 1347, April 9, 1947. No record of the Bell contract proposal has been found. However, the basic outline of the Bell test program is described in "AAF Begins Tests with Bell XS-1," *Aviation Week* 47, no. 7 (August 18, 1947): 12. In telephone conversations with the author on December 12, 1990, and March 26, 1991, Frost indicated his naiveté in assuming the AMC letter indicated that the problems for the program were resolved. He was not being informed of the larger issues at play.

16. Modifications are found in Marchese to AMC Wright Field, April 1, 1947. Details on the return to Muroc are in the Dow flight log. The NACA record is found in D. E. Beeler and Charles Forsyth to Langley Chief of Research, NACA memorandum R.A. 1347, April 14, 1947.

17. Details for the preparations for the first flight are in A. J. Marchese to AMC Wright Field, MX-653 Bell Progress Report, April 21, 1947. Also see Beeler and Forsyth to Langley Chief of Research, April 14, 1947.

18. Details for the April 10, 1947, first Muroc flight of the number 1 aircraft are found in Goodlin's official Pilot Report #11, April 10, 1947. The 11 refers to the sequence number of the number 1 aircraft flight series. This is supplemented by the Dow flight log and De E. Beeler and Charles M. Forsyth to Langley Chief of Research, NACA memorandum, April 21, 1947.

19. Details for the April 11, 1947, power flight are found in Goodlin's official Pilot Report #12, April 15, 1947. This information is supplemented by the Dow and Frost flight logs, Marchese to AMC Wright Field, April 21, 1947, and Beeler and Forsyth to

Langley Chief of Research, April 14, 1947. Also see A. J. Marchese to AMC Wright Field, MX-653 Bell Progress Report, May 1, 1947.

20. Details of the April 29, 1947, flight are found in Goodlin's official Pilot Report #13, date illegible (provided by C. Goodlin from his personal files). Further details are found in Charles Forsyth to Langley Chief of Research, NACA memorandum R.A. 1347, May 5, 1947. This is supplemented by the Dow flight log and Marchese to AMC Wright Field, May 1, 1947.

21. Details of the April 30, 1947, flight are found in Goodlin's official Pilot Report #14, date illegible (provided by C. Goodlin from his personal files). Further details are found in Forsyth to Langley Chief of Research, May 5, 1947. Also see the Dow and Frost flight logs, and Marchese to AMC Wright Field, May 1, 1947.

22. Details of the two aborted flights are found in the Dow flight log (which only lists the May 1 flight), the Frost flight log, and Forsyth to Langley Chief of Research, May 5, 1947.

23. No record has been found of Goodlin's official Pilot Report #15. Details of the flight are found in the Dow and Frost flight logs, the Baker flight log (mislabeled as May 3), and Forsyth to Langley Chief of Research, May 5, 1947. Also see Charles Forsyth to Langley Chief of Research, NACA memorandum R.A. 1347, May 12, 1947.

24. See the Dow flight log and A. J. Marchese to AMC Wright Field, MX-653 Bell Progress Report, May 16, 1947.

25. Baker interview, December 29, 1992. Marchese to AMC Wright Field, April 1, 1947. Joel Baker, letter to the author, November 18, 1990. J. P. Reeder, interview with the author, May 29, 1993. Reeder stated that he and Grey both declined the opportunity to serve as the NACA XS-1 pilot due to the expected birth of their respective children later that fall and the harsh living conditions at Muroc. Reeder was unable to pinpoint the date he declined the offer, but since Herb Hoover was identified as the prime pilot in early June, it would seem logical to accept late May as the time frame.

26. Charles Hall, interviews with the author, June 24 and July 18, 1991, and written comments, June 26, 1991. The interesting question is how the graph made its way to Wright Field. Hall says he doesn't remember seeing the Bell proposal, only the graph. A far more serious question is why Bell Aircraft attempted to insert a personal services contract for Slick Goodlin into a contract proposal to AMC. One logical answer is that Stanley was attempting to capitalize on the NACA/AMC bonus-payment conversations of October 14, 1946, at Wright Field. The contract negotiations for additional aircraft became formal in November and were concluded in the spring of 1948. The four new planes were the X-1A, X-1B, and X-1D. The C model was later canceled.

27. The proposed Air Force flight program is summarized in an unsigned document entitled *XS-1*, April 20, 1947. It is part of the Edwards AFFTC-HO collection on the X-1 airplane. Walter Williams to Mel Gough, NACA memorandum R.A. 1347, May 29, 1947 (actually written June 3, 1947).

28. For the Bell denial of the contract see Col. George Smith to the Commanding General Army Air Forces, AAF TSEOP-5-652, May 1, 1947. Also see Marchese to AMC Wright Field, May 1, 1947.

29. Details for the AMC decision come from the Air Materiel Command HQ letter TSEOA-4/CLH/ (copy unsigned but initials indicate Charles Hall) to Dr. Lewis at NACA HQ Washington, May 6, 1947.

30. Colonel George Smith to Hartley Soule, Air Materiel Command letter TSEOA-4/JHV/, May 8, 1947.

31. Dow flight log. Frost, telephone interview with the author, March 20, 1991. Frost flight log.

32. Details for the May 15, 1947, flight are found in Goodlin's official Pilot Report #16, May 16, 1947. This is supplemented by the Dow and Frost flight logs, and Charles Forsyth to Langley Chief of Research, NACA memorandum R.A. 1347, May 16, 1947 (actually typed May 20, 1947).

33. See Forsyth to Langley Chief of Research, May 16, 1947; Frost and Dow flight logs.

34. Details on the May 19, 1947, flight and wrap-up are found in Goodlin's official Pilot Report #17 to Bell Aircraft Corporation, date illegible. This information is supplemented by the Dow and Frost flight logs, and Forsyth to Langley Chief of Research, May 16, 1947.

35. Details on the May 21, 1947, flight are found in Goodlin's official Pilot Report #18 to Bell Aircraft Corporation, date illegible. This is supplemented by the Dow flight log.

36. Details for the May 22, 1947, flight are found in Johnston's official Pilot Report #17 to Bell Aircraft Corporation, date illegible. This information is supplemented by the Dow and Frost flight logs. The details in Johnston's biography *Jet Age Test Pilot* (Washington, D.C.: Smithsonian Press, 1991), 116, imply that he discovered this problem. While he did devote attention to it in his pilot report, the issue was already well recognized. For example, the issue of stabilizer control may be found in J. S. Limage (Bell Chief Structural Engineer) to AMC Wright Field, Bell memorandum, February 5, 1947. AMC concerns are found in AMC internal memorandum TSEOA4/TSEAC12B, signed Col. C. K. Moore (Chief Aircraft Laboratory Engineering Division), May 15, 1947. Both memos can be found in "Sara Clark" RD 2306. For Frost's complaints see R. H. Frost to AMC Wright Field, Bell Aircraft Corporation memorandum, March 28, 1946, in "Sara Clark" RD 4026. Certainly Goodlin had noted the problem in his October 14 Pilot Report and Baker identified the issue in his February 25 memo. NACA demanded a fix for the problem at the March 5 meeting at Wright Field. It appears cost and time may have been Bell's constraints in fixing the stabilizer problem during the acceptance flights. Once the follow-on contract was worked out, it can be assumed that the Bell engineering team solution would have been adopted for their flights as did finally occur for the Air Force flights.

37. See the unsigned memorandum apparently prepared by Flight Test Division and dated May 16. However, details in the memo indicate the date of preparation was probably around May 1. The memo is a detailed history and status report on the program and its future. Also see Capt. Robert Fackler to Commanding General AAF Washington, AC/AS-4, AMC Teletype TSFCO-5-4, May 7, 1947. Both memos are found in the records of the AFFTC-HO.

38. See R. H. Frost to AMC Wright Field, Bell Progress Report, June 2, 1947. Williams to Gough, May 29, 1947. (Captain Walker was in charge of the powerplant laboratory at Muroc.) Also see Joseph Vensel to Langley Chief of Research, NACA memorandum R.A. 1347, June 2, 1947.

39. Details for the May 29, 1947, flight are found in Goodlin's official Pilot Report #18 to Bell Aircraft Corporation, May 29, 1947. This is supplemented by the official NACA Flight Request Record; the NACA Flight Report; Williams to Gough, May 29, 1947; Vensel to Langley Chief of Research, June 2, 1947; and the Dow and Frost flight logs. Also see D. E. Beeler and Charles Forsyth to Langley Chief of Research, NACA memorandum R.A. 1347, June 3, 1947.

40. See Williams to Gough, May 29, 1947. Also see Vensel to Langley Chief of Research, June 2, 1947, and Beeler and Forsyth to Langley Chief of Research, June 3, 1947. Further information is in Frost to AMC Wright Field, June 2, 1947. It appears

Herb Hoover accepted the duties as prime NACA pilot after Reeder and Grey withdrew. In an interview with the author on May 29, 1993, Jack Reeder stated Hoover accepted out of responsibility to NACA since he was the chief of the test pilots. Gough had indicated he wanted a pilot immediately and Hoover was now the choice. But Hoover made it clear to LMAL that he did not intend to live at Muroc. This decision opened the door for a second NACA pilot.

41. Details for the June 5, 1947, events are found in Goodlin's official Pilot Report #19 to Bell Aircraft Corporation, date illegible. Frost, telephone interview with the author, March 20, 1991. Lundgren, *Across the High Frontier*, 53–54. Further details are in John Gardner to Langley Chief of Research, NACA memorandum R.A. 1347, June 11, 1947; Charles Forsyth to Langley Chief of Research, NACA memorandum R.A. 1347, June 9, 1947; and Walter Williams to Mel Gough, NACA letter R.A. 1347, June 11, 1947. Joseph Vensel from AERL arrived in April 1947 as NACA's chief of flight operations; "The X-1 Story," *X-PRESS* (the newspaper of the NACA High Speed Flight Station). The term "reported" is used to notify any researcher to take the notated times on the pilot reports with a great deal of caution. In a letter to the author on March 7, 1991, Frost indicated the remarkable number of 16-minute flights, regardless of maneuvers, drop altitudes, and difficulties. Simple calculation of flights 15 and 16 reveal only 15 minutes total time!

42. Edmund Burke, interview with the author, December 4, 1990, and January 3, 1991, and Goodlin interview.

43. See Joseph Conners to Robert Stanley, Bell Aircraft Corporation memorandum, June 4, 1947; and accompanying Bell Aircraft Corporation letter, Robert Stanley to Chalmers Goodlin, June 6, 1947. This information was supplemented by the author through interviews with Goodlin and Burke. Bell test pilot Jay Demming was also offered the job to fly the XS-1 after Goodlin quit the program. Demming stated he was offered $165,000 by Stanley to fly the accelerated program, but was told in late June that the Air Force had taken over the program (Demming, interview with the author, July 31, 1991). This incident gives an important behind-the-scenes look at Bell Aircraft decisionmaking. It is possible to speculate that on June 10 Bell Aircraft believed that they didn't have a XS-1 program so there was no need to pay Goodlin. But something might have happened between the end of the Goodlin conversation and the initiation of talks with Demming. Perhaps Bell Aircraft believed that Air Force HQ in Washington was going to reverse the Wright Field decision. This had been the Bell hope earlier and it is possible that additional conversations with Washington had renewed that hope.

44. Dow flight log. Also see Williams to Gough, June 11, 1947. Chalmers Goodlin to Robert Stanley, Bell Aircraft Corporation memorandum, June 10, 1947 (resignation memo from Goodlin's personal files). Goodlin interview.

45. Charles Hall, telephone interview with the author, June 24, 1991. The agent was Frank Nichols of the William Morris Agency. He was hired to handle book and article contracts. Goodlin interview.

46. Ritland, telephone interview with author, December 17, 1990 (hereafter Ritland interview); Smith, telephone interview with author, December 4, 1990.

47. H.J.E. Reid to the NACA HQ, NACA memorandum R.A. 1347, June 13, 1947.

48. Walter Williams to Langley Chief of Research, NACA memorandum R.A. 1347, July 1, 1947. Also see R. H. Frost to AMC Wright Field, MX-653 Bell Progress Report (supplemental), June 16, 1947, found in Box 310b, Container 111325, File "X-1 #1 Research File Data" at the USAF History Museum, Wright-Patterson AFB. Stefan Cavallo, interview with the author, May 26, 1993. Also Cavallo letters to the author, February 23 and March 13, 1993. Cavallo was an experienced high-speed (Mach 0.83) test pilot for NACA. On June 9–11, 1947, Cavallo traveled to Wright Field for indoc-

trination work on seat ejectors. After visiting Bell, he traveled to Kitanniny Arsenal and RMI for rocket-firing demonstration work.

Chapter 6. The Air Force Takes Over the Program

1. Boyd biographic information supplied by AFFTC-HO. Ritland (Ritland interview) discusses the desire for the program. Also Chuck Yeager and Leo Janos, *Yeager* (New York: Bantam Books, 1985). The Yeager/Janos book must be taken with some caution as several of the flight details are incorrect. Also see Lundgren, *Across the High Frontier*, 29–35, and Hallion, *Supersonic Flight*, 100–102.

2. Lundgren, *Across the High Frontier*, 41

3. The Ridley and Hoover biographic details are from the files of the AFFTC-HO. Also see Lundgren, *Across the High Frontier*, 124; Hallion, *Supersonic Flight*, 102; and Young, ed., *Supersonic Symposium*, 96, 104.

4. "Notes on the XS-1 Project" undated memo in the files of the AFFTC-HO. The Bell trip is from Lundgren, *Across the High Frontier*, 46–58. At that time, Hoover was a lieutenant, not a captain as Lundgren reports.

5. Brig. Gen. Alden Crawford to Commanding General AMC, HQ AAF memorandum, June 24, 1947, AFFTC-HO.

6. Details for the Flight Test Division internal meeting are found in the handwritten (initialed L.H.S.) notes of participant Louis H. Sibilsky, dated June 25, 1947 (Sibilsky was the acting chief of the Test Engineering subdivision at Wright Field). These are supplemented by the formal minutes contained in the Army report of Col. P. B. Klein to Col. G. F. Smith, "Conference, Accelerated Transonic Program XS-1 Airplane," June 25, 1947. The notes were prepared by Voyles.

7. Col. George F. Smith to Bell Aircraft Corporation, Army letter TSEOA-4, June 27, 1947.

8. Details for the Wright Field conference are found in the records of the AFFTC-HO. Their files include a copy of an annotated agenda for the meeting dated June 30, 1947, and an unsigned copy of the minutes of the meeting apparently prepared by Flight Test Division and dated July 1, 1947. These primary sources are supplemented by Walter Williams to Langley Chief of Research, NACA memorandum R.A. 1347, July 1, 1947. Also see Col. Fred Dent to AMC Liaison Langley, Aircraft Projects Section, Wright Field TSEOA-4/JHV/, July 15, 1947. Further information is in Hartley Soule to Langley Director, NACA memorandum R.A. 1347, July 8, 1947. On July 21, 1947, Soule wrote a somewhat longer and different memorandum for Langley Chief of Research. In the memo, Soule again expressed his reservations about the AAF program. In an interview with the author on December 17, 1990, Colonel Ritland cited the continuing concerns of NACA as a major reason that AMC never seriously considered letting them run the follow-on program. Also see Young, ed., *Supersonic Symposium*, 80, nn. 43, 44.

9. The Air Force wanted seventy groups in fiscal year 1948; they received authorization for fifty-five. The allusion to Buck Rogers is from "More Buck Rogers Planes" *Aviation Week* 47, no. 2 (July 14, 1947): 7. The author has not found the origin of the phrase "No bucks, no Buck Rogers" popularized by *The Right Stuff*. This might be it. The Symington quote is from Wolk, *Planning and Organizing*, 182, and 214 for Arnold/Spaatz views on publicity.

10. See Walter Williams to Mel Gough, NACA letter R.A. 1347, July 17, 1947. Also

see the Klein speech of June 16–19 from "Research Aircraft," 819. Hallion, *Supersonic Flight*, 129. The Soviet threat is from "Independent Observer," *Aviation Week* 47, no. 2 (July 14, 1947): 22.

11. See R. H. Frost to AMC Wright Field, Bell Aircraft XS-1 Progress Report, July 1, 1947. Also see A. J. Marchese to AMC Wright Field, Bell Aircraft XS-1 Progress Report, July 22, 1947. Additional information supplied in Jack Russell in a telephone interview with the author, December 17, 1990. Hamilton was hired by the NACA on August 12, 1947.

12. See Capt. Robert Fackler (actually Thomas Reynolds) to Flight Test Division, Army TSFCO EO#43-260, July 10, 1947, AFFTC-HO. See "Notes on the XS-1 Project" for the Cardenas assignment. Cardenas biographic sheet from AFFTC-HO.

13. Yeager and Janos, *Yeager*, 98–99; Lundgren, *Across the High Frontier*, 58–62.

14. See Lundgren, *Across the High Frontier*, 127. (The Lundgren book erroneously lists July 4 as the date of the flight.) Details for the initial arrival and preparations for the glide flights come from a Army teletype Muroc Field to AMC Wright Field, Colonel Gilkey to Commanding General AMC, July 29, 1947. Also see the Progress Report XS-1 Project prepared by Maj. R. L. Cardenas and sent to Flight Test Division (hereafter TSFLT) Wright Field. These reports are dated July 28, 29, 30, 31 (two reports), and August 1, 1947, AFFTC-HO.

15. Details for the aborted flight efforts of August 4–5, 1947, may be found in the XS-1 status report, unsigned but dated July 10, 1947. Details for the efforts to ready the XS-1 may be found in Maj. R. L. Cardenas to AMC Wright Field TSFLT, Progress Report XS-1 Project, August 4 and 5, 1947, AFFTC-HO.

16. See "Industry Observer," *Aviation Week* 47, no. 5 (August 4, 1947): 15.

17. See Hallion, *Supersonic Flight*, 129–30.

18. Details for the first glide flight come from the official flight plan for the No. 1 Glide Flight signed by Maj. R. L. Cardenas. No record exists of a pilot report. The official Progress Report XS-1 Project, Maj. R. L. Cardenas to AMC Wright Field TSFLT, August 6, 1947, indicates the glide flight followed the above referenced plan. Further details in Charles Forsyth to Langley Chief of Research, NACA memorandum R.A. 1347, August 18, 1947. Also see Lundgren, *Across the High Frontier*, 145–50.

19. Yeager and Janos, *Yeager*, 113.

20. For the second glide flight see Maj. R. L. Cardenas to AMC Wright Field TSFLT, Progress Report XS-1 Project, August 7, 1947, AFFTC-HO. This is supplemented by Yeager and Janos, *Yeager*, 113.

21. For the third glide test see telex, Col. S. A. Gilkey to AMC Wright Field TSFLT, August 8, 1947, AFFTC-HO. Also see Yeager and Janos, *Yeager*, 113, and Herbert Hoover to Langley Chief of Research, NACA memorandum R.A. 1347, August 22, 1947.

22. See Edward Swindell to AMC Wright Field TSFLT, Progress Report XS-1 Project, August 15, 1947, AFFTC-HO. Also see Charles Forsyth to Langley Chief of Research, NACA memorandum R.A. 1347, August 18, 1947. Cavallo, letter to the author, February 23, 1993. Lilly had replaced Cavallo between August 7, 1947, when the latter started his P-80 checkout work, and August 15. Apparently, during that period Cavallo was called to a private meeting with Hartley Soule. Soule asked why NACA pilots would not fly the plane. Cavallo mentioned the XS-1 access/egress problem and suggested the installation of a seat ejector. He also indicated the need for a raise from his $5,000 per year salary. Soule declined both requests and Gough was furious when he found out about the meeting. Cavallo believed his position at NACA was fatally compromised and resigned in November.

23. Details for the return to Muroc appear in AMC teletype, AMC Wright Field to CO Muroc AAF, August 19, 1947. Also see Maj. R. L. Cardenas to AMC Wright Field TSFLT, Progress Report XS-1 Project, August 22, 1947, AFFTC-HO.

24. Walter Williams to Hartley A. Soule, NACA memorandum R.A. 1347, August 15, 1947.

25. See AMC teletype, AMC Wright Field to CO Muroc AAF, August 20, 1947, AFFTC-HO. Also see Walter Williams to H. A. Soule, NACA memorandum R.A. 1347 (no date is listed but from the information available the NACA document was prepared on either August 20 or 21, 1947).

26. Williams to Soule, August 15, 1947.

27. The details for the delays come from Maj. R. L. Cardenas to AMC Wright Field TSFLT, Progress Report XS-1 Project, August 25 and 26, 1947, AFFTC-HO.

28. Unlike the Bell Aircraft Pilot Reports, the Air Force reports only concern the actual flight of the XS-1. Details on drop altitude, takeoff times, and other pertinent information are lacking. However, Capt. Jack Ridley prepared a flight plan for each of the first four flights. Unless otherwise noted, it is assumed by the author that the flight plan was followed as written. Details for the August 29, 1947, flight come from XS-1 official Pilot Report #1 Powered Flight (undated) prepared by Charles Yeager and the Flight Plan #1 (undated) by Jack Ridley. This information is supplemented by van Lonkhuzyen, "Problems Faced in Designing Famed X-1," 24; and Lundgren, *Across the High Frontier*, 163–69, 170–72. Also see Army teletype, Col. S. A. Gilkey to AMC Wright Field TSFLT, August 29, 1947; Yeager and Janos, *Yeager*, 122; and Maj. R. L. Cardenas to AMC Wright Field TSFLT, Progress Report XS-1 Project, September 2, 1947. The Yeager quotation is from his letter dated October 6, 1947, printed in Clarence L. "Bud" Anderson and Joseph Hamlin, *To Fly and Fight* (New York: St. Martin's Press, 1990), 167. For the perceived NACA condescension, see Yeager and Janos, *Yeager*, 108, 180–81, and 86, 106.

29. Details for the September 4, 1947, flight are from XS-1 official Pilot Report #2 Powered Flight (undated) prepared by Charles Yeager and the Flight Plan #2 (undated) by Jack Ridley. Also see Colonel Gilkey to AMC Wright Field TSFLT, Army teletype, September 4, 1947; Maj. R. L. Cardenas to AMC Wright Field TSFLT, Post-flight information from Progress Report XS-1 Program, September 5, 1947. The correction is found in an AAF teletype, Colonel Gilkey to AMC Wright Field TSFLT (undated), AFFTC-HO.

30. See Milton Ames to the Washington NACA Associate Director of Aeronautical Research, NACA memorandum R.A. 1347, September 5, 1947.

31. Yeager, telephone interview with author, July 24, 1992 (hereafter Yeager interview); Young, ed., *Supersonic Symposium*, 107–9. Yeager's Form 5 (individual flight record), p. 23, for October 1947, lists P-84 flights on October 1 and 16. The latter is out of sequence in the chronological listing and appears to really be the sixth. Copy from AFFTC-HO.

32. Details for the September 8, 1947, flight are from the official Pilot Report #3 (undated) prepared by Charles Yeager and Flight Plan #3 (undated) by Jack Ridley. Also see Maj. R. L. Cardenas to AMC Wright Field TSFLT, Progress Report XS-1 Program, September 9, 1947, AFFTC-HO. Further details are in Walter Williams to H. A. Soule, NACA letter R.A. 1347, September 10, 1947.

33. See Maj. R. L. Cardenas to AMC Wright Field TSFLT, Progress Report XS-1 Program, September 9, 1947. Also see Frank Danis to Maj. Robert Cardenas, RMI memorandum R.A. 1347, September 11, 1947.

34. This is my estimate of the Ridley-Williams meeting based upon the telephone interview with Walt Williams of June 17, 1992, and the information in the September 12, 1947, flight. Also see Walter Williams to the LMAL Chief of Research, NACA memorandum R.A. 1347, September 15, 1947, for the elevator remarks.

35. Details for the September 10, 1947, flight are from the official Pilot Report #4 Powered Flight (undated) prepared by Charles Yeager and the Flight Plan #4 (un-

dated) by Jack Ridley. Further details are found in Joseph Vensel to AERL Chief of Research, NACA memorandum R.A. 1347, September 15, 1947; and Walter Williams to H. A. Soule, NACA letter R.A. 1347, September 10, 1947. Also see Colonel Gilkey to AMC Wright Field TSFLT, Army teletype, September 10, 1947; Maj. R. L. Cardenas to AMC Wright Field TSFLT, Progress Report XS-1 Program, September 11, 1947; and Colonel Gilkey to AMC Wright Field TSFLT, Army teletype, September 11, 1947.

36. No pilot report seems to exist for the September 12, 1947, flight. The reason none may ever have existed in the Edwards records for the flight may be found in Maj. R. L. Cardenas to AMC Wright Field TSFLT, Progress Report XS-1 Program, September 15, 1947, AFFTC-HO. The document indicates Yeager will report *personally* to AMC TSFLT on his imminent return to Wright Field for a pressure-suit check. Other details may be found in Walter Williams to Langley Chief of Research, NACA memorandum R. A. 1347, September 22, 1947. Also see Joseph Vensel to AERL Chief of Research, NACA memorandum R.A. 1347, September 15, 1947.

37. R. H. Frost to AMC Wright Field, XS-1 Bell Aircraft Progress Report R.A. 1347, August 14, 1947.

38. Carlton Kemper to NACA HQ, NACA letter R.A. 1347, July 30, 1947. Also see Addison Rothrock to the AERL, NACA letter R.A. 1347, August 12, 1947.

39. Henry Pearson to Langley Chief of Research, NACA memorandum R.A. 1347, September 4, 1947; Charles Forsyth to Langley Chief of Research, NACA memorandum R.A. 1347, August 18, 1947. Also see "The X-1 Story," on the tenth anniversary of the supersonic flight. Dated October 14, 1957, it provides details of the NACA early days at Edwards. For conditions at Muroc, see Herbert Hoover to Langley Chief of Research, NACA memorandum R.A. 1347, August 22, 1947.

40. Lt. Col. William K. Pfingst to AMC Wright Field TSFLT, Progress Report XS-1 Project, September 24, 1947 (Pfingst was Projects Officer Flight Test Division), AFFTC-HO. Walter Williams to H. Soule, NACA letter R.A. 1347, September 22, 1947.

41. Details for the September 25, 1947, flight of the XS-1 (the Yeager/NACA number 1) may be found in Walter Williams to Langley Chief of Research, NACA memorandum R.A. 1347, October 6, 1947. No Yeager pilot report has been found in either the Langley archives or at AFFTC-HO.

42. Details for the preparations for the next AAF flight come from the official Progress Reports of the XS-1 Project, dated September 23, 24, 25, 26, 29, and October 1. Williams to Langley Chief of Research, October 6, 1947, discusses the use of the two degree per second actuator. Frost, interview with the author. This is another case of how much more informal most decisionmaking was in the early days of high-speed flight testing.

43. Williams to Langley Chief of Research, October 6, 1947. The October 3, 1947, flight details are from the official XS-1 Pilot Report #6 Powered Flight, October 5, 1947, by Charles Yeager (AFFTC-HO). The date is incorrect and is the day Yeager prepared the document. His Form 5 at AFFTC-HO has the correct date on it. Yeager also omitted the NACA acceptance flight from his numbering sequence as this flight constituted his seventh flight. It was his sixth flight in XS-1 number 1 and the forty-seventh of the program (AFFTC-HO). Time estimate to break the sound barrier from Anderson and Hamlin, *To Fly and Fight*, 167.

44. Details for the Wright Field visit are found in Lundgren, *Across the High Frontier*, 181–84. This is the author's estimate of the date of the visit since the details of the conversation imply that this is the correct time frame.

45. The burning dream is from Lundgren, *Across the High Frontier*, 177–78. The flying tail dream is from Young, ed., *Supersonic Symposium*, 112–13.

46. Williams to Langley Chief of Research, October 6, 1947, and October 20, 1947.

47. Details for the October 8, 1947, flight of the Air Force program are from the official XS-1 Pilot Report #7 Powered Flight of October 8, 1947, by Charles Yeager. This information is supplemented by Williams to Langley Chief of Research, October 6, 1947, and October 20, 1947; and Joseph Vensel to the AERL Chief of Research, NACA memorandum R.A. 1347, October 13, 1947. Also see Col. S. A. Gilkey to AMC Wright Field TSFLT, Army teletype, October 9, 1947 (the flight discussed is misdated October 15).

48. Boyd visit from *Yeager*, p. 124. Also see Vensel to the AERL Chief of Research, October 13, 1947, and Williams to Langley Chief of Research, October 20, 1947.

49. Details for the October 10, 1947, flight are from the official Pilot Report #8 Powered Flight of October 10, 1947, by Charles Yeager. Also Yeager interview. Yeager's anger at the lack of press coverage is from his letter found in Anderson and Hamlin, *To Fly and Fight*, 167, 172. This letter indicates that Yeager believed that the lack of coverage was due to Goodlin's boasting about the initial flights. Actually, the press decision was certainly influenced by the sheer number of groups interested in reporting on anything involving the program. Press coverage did continue after the Air Force took over the program. Further details are in Lundgren, *Across the High Frontier*, 184–86; "The X-1 Story," 3.

50. Details are from Yeager and Janos, *Yeager*, 127–29. Tom Wolfe, *The Right Stuff* (New York: Bantam Books, 1980), 43–44. Also see Lundgren, *Across the High Frontier*, 186–90. Lundgren records the day of the accident as Friday rather than the correct Sunday.

51. This is the author's estimate of the date of the meeting. Also Yeager interview, and Williams, interview with the author, June 17, 1992. It is possible that the meeting occurred on October 9, and that the elevator ineffectiveness mentioned in the report to Wright Field was simply a statement of fact rather than a revelation. Yeager's pilot reports #7 and #8 are confusing since the elevator control problems reported on the eighth flight were already suspected as far back as flight #4 and certainly known after flight #6. Details are found in Young, ed., *Supersonic Symposium*, 104–10; Yeager and Janos, *Yeager*, 124–27. Also see Walter Williams, "Instrumentation, Airspeed Calibration Tests, Results, and Conclusions" in *AFSRA XS-1*, 23–25. Note that Goodlin (on May 29) and Johnston (on May 22) attempted to use the existing stabilizer control system to fly the plane. The slack in the old system prevented them from doing what Yeager would achieve with the new Bell and Dick Frost–improved system.

52. Details for the October 14, 1947, flight appear in the official Pilot Report #9 Powered Flight prepared October 28, 1947 (crossed out on the original and redated October 14) by Charles Yeager. The application of the nickname "Glamorous Glennis" was in keeping with pilots' tradition of naming their planes. Yeager had used this logo as his good luck talisman on his P-51 in Europe. Anderson and Hamlin, *To Fly and Fight*, 167, indicates it was applied shortly before October 10. Lundgren implies incorrectly that it was much earlier. Details of the drop conversation are from the official "Transmission Transcript, First Supersonic Flight" currently in the XS-1 archives at the Smithsonian Institution National Air and Space Museum. This information is supplemented by the Yeager interview, and Frost written comments of March 14, 1991; and by the participant interviews in Young, ed., *Supersonic Symposium;* Lundgren, *Across the High Frontier*, 190–202; Wolfe, *The Right Stuff*, 44–47; Charles Yeager, "The Operation of the XS-1 Airplane" in *AFSRA XS-1*, 19; Colonel Gilkey to AMC Wright Field TSFLT, Army teletype, October 14, 1947; and Williams to Langley Chief of Research, October 20, 1947. Also see Milton McLaughlin to Langley Chief of Research, NACA memorandum R.A. 1347, December 9, 1947; Floyd Thompson to Files, NACA memorandum R.A. 1347, November 13, 1947.

53. Hubert Drake (stability and control engineer) to Langley Chief of Research,

NACA memorandum R.A. 1347, December 22, 1947 (actually January 8, 1948). Drake was working on the number 2 airplane and expressed surprise after *reading* of the supersonic flight. The NACA number 2 airplane team apparently had not even heard the rumors. "Douglas Skystreak Hits 680 MPH," in *Aviation Week* 47, no. 18 (November 3, 1947): 17. Also Robert McLarren, "Douglas Unveils Its New Research Plane," *Aviation Week* 47, no. 20 (November 17, 1947): 12. Col. S. A. Gilkey to AMC Wright Field, Air Force teletype Mucpe-11, November 11, 1947, and see Lt. Col. William Pfingst to AMC Wright Field, AAF letter XS-1 Project, November 4, 1947, AFFTC-HO.

54. Herbert Hoover to Langley Chief of Research, NACA memorandum R.A. 1347, October 22, 1947 (typed October 31, 1947); Hubert Drake and Harold Goodman to Langley Chief of Research, NACA memorandum R.A. 1347, December 9, 1947. Miller, *The X-Planes*, 21. Walter Williams to Langley Chief of Research, NACA memorandum R.A. 1347, November 20, 1947.

55. Robert McLarren, "Bell XS-1 Makes Supersonic Flight," *Aviation Week* 47, no. 25 (December 22, 1947): 9–10. See Drake to Langley Chief of Research, December 22, 1947. Jay Waltz, "New U.S. Plane Said to Fly Faster than Speed of Sound," *New York Times*, December 22, 1947, pp. 1, 26. Marvin Miles, "U.S. Mystery Plane Tops Speed of Sound," *Los Angeles Times*, December 22, 1947, pp. 1–2. The *Times* story also repeated the use of the term "Buck Rogers" in reference to the event.

56. *AFSRA XS-1*. The report misstates the date of the first flight by day (p. 5) and by month (p. 15) as well as the drop altitude and flight time.

57. Robert McLarren, "XS-1: Design and Development," *Aviation Week* 49, no. 4 (July 26, 1948): 22–27. McLarren repeats the earlier story that NACA "shortly" after the Yeager flight also achieved supersonic speeds. Actually, Herb Hoover made the first supersonic NACA flight on March 10, 1948.

58. "Navy Blocks Supersonic Stories," *Aviation Week* 48, no. 24 (June 14, 1948): 54. Also see Robert Wood, "Supersonic Flight News—A Dilemma," *Aviation Week* 48, no. 23 (June 7, 1948): 50.

59. Young, ed., *Supersonic Symposium*, 144.

Epilogue

1. Becker, *High-Speed Frontier*, 96.
2. Hallion, *Supersonic Flight*, 120.

Index

Aberdeen Ordnance Proving Grounds, 22
Accademia d'Italia, 7
AeroMedical Laboratory, 100, 123
Aiken, W. S., 105, 123, 161
Aircraft Engine Research Lab (AERL), 16, 161, 241, 262, 263, 292
Air Engineering Development Center, 99
Air Force Aircraft and Weapons Board, 242
Air Force Development Command, 14, 15
Air Materiel Command (AMC), 41–43, 48, 49, 51, 69, 93, 94, 98, 99, 102, 103, 105–11, 113, 114, 116, 117, 121, 124, 126, 128, 136–38, 145, 154, 182–97, 207, 208, 210, 211, 216, 218, 219, 225, 227–32, 236, 237, 239–41, 245, 246, 250, 286
Airplanes
—B-17, 55, 58, 60, 63, 66, 71, 82
—B-29, 34, 35, 38, 41, 44, 46–49, 51, 52, 55, 58, 59, 61–66, 69, 71, 72, 74, 77, 78, 81–84, 86, 87, 89, 95, 98–101, 104, 123, 127, 129, 130, 133, 135, 137, 138, 140–44, 146, 151, 154, 157, 159, 161, 163–67, 169, 174, 176–78, 183, 185, 196–99, 202–7, 212, 214–17, 220, 222, 224–26, 231, 238, 239, 241, 244, 245, 248–50, 252, 254, 256, 258, 260, 261, 264, 269, 272, 276, 278, 280, 281
—B-35, 132
—B-43, 132
—B-48, 98
—B-50, 286
—C-45, 42, 55, 56, 69, 70, 81, 88, 111, 128, 177
—C-47, 89, 101, 111, 141, 173, 177, 178, 198, 227, 243
—D-558, 26, 122, 181, 184, 188, 191, 242, 243, 247, 263, 280, 281, 285
—De-108, 124, 125, 138
—F4U, 115
—F5U, 132
—F6U, 132
—F8F, 115
—FJ1, 132
—Heinkel 162, 67
—L-39, 114, 115
—Me-163, 27, 237
—Me-262, 242
—Mosquito, 55, 125
—P-38, 2, 6, 115
—P-39, 9, 25, 115, 141
—P-42, 2
—P-47, 6, 9, 71, 73, 74, 78, 79, 81–83, 87
—P-51, 6, 51, 56, 58, 63, 66, 68, 101, 102, 104–6, 115, 127, 133, 135, 143, 146, 149, 157, 166, 167, 172, 204, 212, 231
—P-59, 10, 25, 95, 96, 244
—P-63, 30, 84, 114, 141
—P-77, 67
—P-79, 11
—P-80, 10, 16, 20, 29, 31, 132, 146, 163–65, 169, 172, 176, 190, 191, 200, 204–7, 214, 217–20, 224, 231, 234, 239, 246–48, 251, 253, 254, 262
—P-83, 124, 253
—P-84, 132, 191, 256
—SB2C, 9
—Spitfire, 55, 117
—XP-85, 187
—XP-86, 187
—XP-87, 187
—XP-88, 187
—XP-89, 187

Airplanes (*cont.*)
—XP-90, 188
—XP-91, 188
—XP-92, 188
Air Policy Board, 180, 181, 271
Air Technical Services Command, 18–27,
    30–34, 36–38, 54, 56, 95, 192
Allen, H. J., 10
Ames, E. J., 122
Ames, Milton, 182, 188
Ames Laboratory, 161, 162
Angebilt Hotel, 57
Arnold, Maj. Gen. Henry H. "Hap," 9,
    14, 15, 262
Ascani, Lt. Col. Frederick J., 238, 245
Aviation Writers Association, 222–34

Baker, Joel Robert, 55, 57, 59–61, 69–71,
    73–75, 77, 78, 81–84, 88, 111, 131,
    142, 157, 163–66, 176–78, 186, 189,
    191, 207, 217
Bailey, Clyde, 161
Bailey, Capt. G. W., 13, 18, 19
Barnes, Florence "Pancho," 228, 274, 279
Becker, John, 23, 24
Beeler, De E., 161, 263
Bell Aircraft Corporation, 9, 15–32, 34–
    38, 41–44, 47–49, 51, 52, 54, 55, 57,
    58, 63, 68, 69, 70, 72, 73, 77, 85–88,
    94–97, 99–105, 107–20, 123, 124, 126,
    128, 129, 131, 133, 136–40, 145, 149,
    150–53, 155, 156, 158, 163, 172–78,
    181–83, 185–98, 202, 204, 207–12,
    215, 219, 222, 223, 225, 226, 228–32,
    237–41, 243–47, 254, 262, 263, 279–
    81, 286
Bell, Lawrence "Larry," 15, 22, 68, 115,
    116, 120, 122, 142, 143, 145, 156,
    190–93, 195–97, 208, 237, 255, 279,
    285
Bloom, Trig, 136
Boeing Aircraft, 38, 55, 98, 295
Bonney, Walter, 149
Boyd, Col. Albert, 123, 209–11, 219, 229,
    230, 234–38, 241, 245–47, 255, 262,
    267–69, 271, 275, 279, 286
Buckley, E. C., 161
Bureau of Aeronautics, 13, 21, 26, 184
Bureau of the Budget, 92, 111, 137
Burke, Edmund, 226
Busemann, Adolf, 7, 32

Bush, Dr. Vannevar, 99, 180
Buttman, Maj., 238

Caldwell, Comm. Turner, 247
Cannon, Joseph, 44, 54, 63, 65, 71, 73,
    74, 78, 81, 83, 87, 95, 101, 104, 286
Cardenas, Maj. Roberto "Bob," 244–50,
    252, 256, 259–61, 269, 272, 276, 277
Carl, Maj. Marion, 247, 280
Carlson, Milton, 124
Carroll, Brig. Gen. Franklin O., 9
Cavallo, Stefan R., 111, 117, 118, 231
Chidlaw, Maj. Gen. Benjamin, 88, 268
Clayton, Harry, 244
Cleveland National Air Races, 115
Clousing, Lawrence, 162
Cobra I, 115
Cobra II, 115
Colchagoff, Capt. George, 77, 78, 110
Committee on Aerodynamics, 16
Connors, Joseph E., 226
Consolidated Vultee Aircraft, 98, 236
Convair Aircraft, 188
Craigie, Brig. Gen. Laurence, 33, 69, 93,
    99, 116, 122, 145, 152, 192, 195–97,
    237, 238, 292
Crawford, Brig. Gen. Alden, 237, 238
Crowley, John, 10, 12, 16, 29, 31, 32, 50,
    122, 182, 195
Curtiss-Wright Aircraft, 55, 98, 187

Davidson, Milton, 18, 21, 55, 102, 122
Davis, Frank, 259
de Havilland, Geoffrey, 124, 125
Douglas Aircraft, 21, 22, 26, 27, 55, 92,
    132, 242, 280, 285
Dow, Harold, 44, 51, 54, 59, 63, 65, 71,
    73, 74, 77, 78, 81, 83, 87, 88, 95, 99,
    101, 104, 129, 130, 140, 141, 143,
    144, 147, 156, 157, 159–61, 163–66,
    170, 171, 177, 178, 198–200, 202,
    203, 207, 212–14, 217, 224–26
Draley, Eugene, 18
Dryden, Dr. Hugh, 181
Dryden Flight Research Center, 286

Eastman, Maj. Donald, 49, 51, 55–58,
    69, 93, 110, 112, 122
Emmons, Paul, 15, 20, 35, 67, 237

Fay, Charles, 124
Forrestal, James, 283
Forsyth, Charles, 142
Flight Test Division, 209–11, 219, 220,
    229, 230, 234–40, 244, 251
Frost, Richard "Dick," 86, 102–4, 115,
    117–20, 127, 128, 131, 132, 135, 140,
    142–44, 146, 147, 149, 151, 154, 159,
    161, 163, 166, 177, 184, 185, 188,
    192, 196, 198, 199, 201, 202, 204,
    206, 207, 212, 214, 217–20, 222–25,
    227, 228, 237–41, 244, 247–49, 252–
    54, 263, 265, 272, 273, 275, 279

Gardner, John, 117, 118
Garman, Col. R. S., 136, 138
Garrison, Capt. K. M., 238
General Electric Corporation, 87, 93,
    101, 159, 210
George, Ellie, 244
George, James, 244
Gilkey, Col. Signa A., 97, 141, 150, 247,
    250, 251, 254, 271, 279, 282
Gilruth, Robert, 18, 32
Glenn Martin Aircraft, 85, 98
Goodlin, Chalmers "Slick," 115–17, 123–
    25, 127, 129, 130, 131, 133–36, 138–
    44, 147–71, 173–77, 186, 191, 198–
    201, 203–6, 207, 208, 212–16, 218,
    220–29, 232, 248, 253, 275, 280,
    289–90
Gough, Melvin, 38, 39, 42, 55, 61, 69, 78,
    101, 105, 106, 108, 109, 111, 123,
    141, 160–62, 185, 186, 188, 190, 191,
    207, 217, 225
Gracey, William, 34, 49
Gray, William E., 111, 189, 207
Greene, Col. Carl F., 18

Hall, Charles, 128, 182, 186–88, 208, 238,
    250, 251, 282
Hamilton, Charles MacLean "Mac," 45,
    64, 65, 129, 130, 133, 144, 244, 263
Hamlin, Benson, 20, 67
Harper, Paul, 117–19
Hartman, Edward, 145
Hauptmann, Ivan, 45, 54, 104
Hayes, O. N. "Norm," 49, 51, 55, 57, 69,
    70, 73, 74, 77, 88, 117–19
Heaney, Mark, 104, 129, 143, 144, 156,

159, 161, 163, 167, 170, 177, 198,
    203, 224
High-Speed Panel, 10, 16, 17, 20
Hilton, W. F., 6
Hinman, H. N., 77, 88, 142
Holtonor, Col. J. S., 182
Hoover, Herbert, 111, 207, 222, 231,
    239, 249, 258, 262–64, 281
Hoover, Lt. Robert A. "Bob," 236–39,
    241, 248–50, 252, 253, 258, 262, 272–
    74, 279
Horn, Lieutenant, 248
Horning, Major, 56
Householder, John, 70
Hyatt, 1st. Lt. Abraham, 21

Iwanowsky, Frank, 154

Jacobs, Eastman, 9, 10
Johnston, Alvin "Tex," 31, 87, 88, 114–
    16, 198, 200, 201, 212, 215–20, 232,
    290
Joint Research and Development Board,
    180
Jones, Robert T., 19

Kilner-Lindbergh Board, 8, 9, 291
Klein, Col. P. B., 238, 239, 243
Kotcher, Capt. Maj. Ezra, 8, 9, 15–20,
    25, 33, 49, 291

Langley Memorial Aeronautical Lab
    (LMAL), 8, 10, 11, 13, 16, 18, 20,
    26, 28–35, 37, 42, 43, 49, 50, 55, 56,
    61, 69–71, 77, 88, 89, 135, 136, 162,
    175, 177, 181, 182, 184–87, 189, 196–
    98, 207, 208, 211, 225, 230, 231, 251,
    263, 265
Lekas, John, 49, 53, 56, 57, 69, 70, 74,
    77
Lemay, Maj. Gen. Curtis, 99
Lewis, Dr. George, 10, 14, 30, 37, 43, 105,
    107, 110, 122, 181, 192, 194–96, 211
Lilly, Howard C. "Tick," 249, 263, 264,
    285
Linde Company, 231
Littell, R. E., 10
Lockheed Aircraft, 6, 132, 188

McDonnell Aircraft, 18, 19, 35, 187
Mach, Ernst, 5
"Mach 0, 999 study," 11
Marchese, A. Joseph "Joe," 120
Materiel Command, 11, 13, 16–18
Matheny, C. E., 123
May, Eugene F. "Gene," 242, 280
Mead, Sen. James M., 39
Means, William, 45, 54
Miller, Elton, 69
Miller, William "Bill," 45, 54, 129, 212
Minalga, George, 123, 142
Morkovin, Vladimir, 20
Muroc Dry Lake: 12, 20, 23, 29–31, 35,
     36, 41, 95–97, 100, 102, 104, 106,
     110, 112, 118, 119, 123, 126, 127,
     129–34, 138–41, 144–46, 149, 150,
     152, 154, 160–62, 165, 172, 174–78,
     184, 185, 189, 197–99, 204, 206, 212,
     214, 215, 218, 222, 224, 226–28, 232,
     234, 237, 244, 245, 247–52, 255, 257,
     258, 262, 264, 268, 270, 271, 273,
     275, 276, 279–81, 300
Muroc Flight Test Unit, 263

National Advisory Committee for Aero-
     nautics: research funding, 179–81,
     299; research policy, 181, 182, 299;
     XS-1 pilot evaluation, 186; XS-1 re-
     quested changes, 182–83, 188–95,
     222, 231; XS-1 team, 53, 55, 56, 70,
     301; XS-1 test program, 62, 181–82
National Air Museum, 286
National Defense Research Council, 2
National Physical Laboratory, 6
National Supersonic Research Center, 122
Nicholas, Frank, 104, 129
North American Aviation, 6, 17, 98, 132,
     187
Northrup, 98, 132, 187

Orlando Army Air Base, 42, 49, 50, 53,
     56, 57, 60, 69, 70, 71, 74
Orzanio, Capt. F. D., 13

Patterson, Col. George, 122
Patterson Field, 173
Pearsall, Capt. David, 23, 37, 49–51, 56,
     58, 60

Pearson, Henry, 34, 105, 263
Phillips, W. H., 106
Pierce, Capt. R. B., 18
Pinecastle Field, 38, 41, 43, 47, 49–51,
     53, 54, 56, 57, 59–63, 70, 71, 74, 79,
     81, 85, 87, 88, 90, 95, 98, 102, 112,
     121, 128, 129, 133, 134, 143, 198,
     202, 212, 228, 263, 295
Price, Col. George, 43
Proctor, LeRoy, 161
Prodanovich, Maj. George, 138
Project designations: B-7, 15; MCD-520,
     19; MCD-524, 19, 26; Model 44, 26;
     MX-324, 11; MX-653, 23, 26, 27, 29–
     37, 39, 41, 43, 90, 104, 136; XS-1,
     26, 43
Putt, Col. Donald, 184

Raymond, Arthur E., 92
Raymond panel, 92, 99, 179, 180
Reaction Motors, Inc. (RMI), 23, 24,
     27–29, 38, 44, 48, 87, 97, 100, 118,
     123, 150, 154, 155, 177, 232, 259, 262
Reeder, John P. "Jack," 189, 207
Reid, Dr. Henry J. E., 10, 33, 107, 108,
     110, 120–22, 230, 231
Republic Aviation, 17, 124, 188
Research and Development Board, 179,
     180
Rhode, R. W., 105
Ridley, Capt. Jack, 236–39, 241, 248,
     252, 255, 256, 258, 260, 261, 267,
     268, 271, 272, 274–79, 286
Ritland, Col. Osmond, 188, 193, 194,
     225, 230, 238
Robert J. Collier Trophy, 285
Robbins Air Force Base, 84
Robinson, Frederick, 145, 156
Robinson, Russell, 10, 40, 41, 182
Roche, Jean, 18
Rodert, Lewis, 161
Ruegg, Lt. Col. R. G., 238
Rumsey, Douglas, 54, 63, 71, 80
Russell, Jack, 104, 129, 244, 263, 275
Ruvin, Abraham "Ace," 77, 78, 88, 89

Sandstrom, Roy, 27, 231, 282
Schneider, Herman, 45, 54
Scientific Advisory Board, 179
Scientific Advisory Group (SAG), 15

Shoop, Maj. Clarence, 127, 138, 184

Shoults, D. R., 61, 145, 190

Sibert, Lt. Col. H., 23

Sibilsky, L. H., 238

Smith, Col. George F., 30, 37, 105, 136, 174, 182, 184–88, 192–96, 208–10, 219, 228, 230, 235, 239, 251

Smith, Capt. R. D., 146, 250

Smith, Stan, 30, 34–36, 41, 44, 48, 51, 58, 77, 86, 114

Soule, Hartley, 105, 122, 136, 151, 174, 181, 184, 185, 188, 190, 211, 239, 240, 250, 251, 258

Spaatz, Gen. Carl, 193, 194, 210, 219, 222

Stack, John, 8–12, 18, 19, 21, 23, 29, 31, 32, 50, 55, 101, 116, 181, 265, 285

Stanley, Robert "Bob," 22, 23, 25, 28, 31, 36, 37, 39, 42, 44, 51, 54, 57–60, 63, 66, 68, 71, 76, 77, 80, 83, 85, 106, 118, 119, 125–29, 131–33, 136–38, 140–43, 145, 149, 150, 153, 156, 183, 188, 190–95, 197, 207, 208, 225–27, 237, 282

Sullivan, J. L., 271

Swindell, Lt. Edward, 245

Symington, Stuart, 242

Taylor, Charles, 53, 56, 58, 69

Thompson, Floyd, 32, 37

Truman, Harry, 271, 283, 285

Truszynski, Gerald M., 53, 56, 57, 63, 65, 69–71, 74, 77, 80, 83, 85, 263

Turner, Harold, 55

Twining, Lt. Gen. Nathan, 111

Vandenburg, Gen. Hoyt, 283, 286

Vensel, Joseph, 225, 241, 262

Volta Congress on High-Speeds in Aviation, 7, 32

von Kármán, Theodore, 7–9, 15, 19, 40, 278

Vought Aircraft, 132, 207

Voyles, James, 110, 112, 128, 136, 138, 145, 150, 153, 182, 188, 220, 222, 238, 239, 247

Walker, Capt., 220

Wallis, William, 117–19

Where We Stand, 15, 40

Williams, Walter, 55, 57, 58, 61, 63, 69, 70, 74, 77, 78, 105–10, 117–20, 123, 127, 128, 131, 132, 136, 138, 141, 143, 150–54, 156, 158, 161, 162, 174, 178, 181, 184, 185, 188, 196, 197, 202, 222, 225, 231, 239, 241, 250, 251, 258, 260, 261, 263, 264, 271, 276, 282

Willow Springs, 96

Wolf, Robert E., 9

Wood, Clotaire, 122, 136, 182, 239

Woods, Robert, 17, 18, 20–22, 27, 209

Woolams, Jack, 25, 27, 35, 44, 46–48, 51, 54, 57–61, 63–69, 71–85, 87, 88, 95–97, 99–101, 111–16, 124, 125, 134, 148, 153, 189, 215, 218, 228, 232, 248, 289, 293

Wright Field, 8, 11, 13, 14, 16–18, 20, 22, 27, 29, 30, 33, 38, 42, 43, 49–51, 58, 60–62, 77, 93, 94, 98, 100, 102, 108, 111, 112, 123, 128, 129, 133, 136–38, 145, 146, 150, 174, 181–84, 186, 188, 193, 196, 197, 208–11, 219, 222, 225, 228, 229, 234–39, 241, 244, 247, 250, 251, 254, 255, 257, 259, 261, 262, 267, 268, 271, 279, 280, 282

XS-1

—Air Force flight program, 209, 210, 237, 239, 240, 241, 313

—Bell flight program, 102, 103, 196, 197

—Construction, 27, 28, 32–34, 41, 45–49, 51, 86, 140, 158–59

—Contract, 26

—Design, 11, 16–17, 19–24, 27, 28, 44, 67, 295, 296, 304, 311, 316

—Disposition: number 1 airplane, 286; number 2 airplane, 286; number 3 airplane, 286

—Ejection seat, 98

—Fire report, 154–56

—Flight costs, 172–73

—Flights (Bell)
Pinecastle (number 1 airplane): no. 1, 63–69; no. 2, 71–73; no. 3, 73; no. 4, 74–76; no. 5, 78–80; no. 6, 81–82; no. 7, 82–83; no. 8, 83; no. 9, 84–85; no. 10, 87–88
Muroc (number 2 airplane): aborted flights, 129–30, 143–44, 146, 157,

XS-1 (*cont.*)
163; no. 1, 133–34; no. 2, 135; no. 3, 138–39; no. 4, 144; no. 5, 147–49; no. 6, 157–58; no. 7, 159–60; no. 8, 161–62; no. 9, 163–64; no. 10, 164–65; no. 11, 166–67; no. 12, 167–68; no. 13, 169–70; no. 14, 170–71; no. 15, 174–75; no. 16, 176–77; no. 17, 217–19; no. 18, 220–21
Muroc (number 1 airplane): aborted flights, 206; no. 11, 198–99; no. 12, 200–201; no. 13, 203–4; no. 14, 204–5; no. 15, 207; no. 16, 212–13; no. 17, 214–15; no. 18, 215–16; no. 19, 224
—Flights (Air Force)
Muroc (number 1 and 2 airplanes): no. 1, 247–48; no. 2, 248–49; no. 3, 249; no. 4, 252–54; no. 5, 256–57; no. 6, 258–59; no. 7, 260–61; no. 8, 261–62; no. 9, 264; no. 10, 265–66; no. 11, 269–70; no. 12; 271–73; no. 13, 275–79
—Pilot selection, 25, 125, 126, 137, 208, 227–29, 303, 310–12, 314
—Preliminary Date Report #1, 79
—Press coverage, 98, 99, 111–14, 150–54, 222, 223, 317, 282
—Program costs, 26, 186, 293, 294
—Program transfer, 208–11, 220, 225, 232, 233, 238, 239
—Proposed test sites: Cherry Point, 31, 39; Daytona Beach, 35, 38; Langley, 31, 35, 37, 43, 162; Marietta Field, 35; Niagara Falls, 25; Salina Field, 35; Wendover Field, 29–31, 35
—Sonic boom, 170, 307
—Test program, 36, 38, 50, 102, 103, 108, 195, 240, 308, 309
—Travel to Muroc: number 1 airplane, 198; number 2 airplane, 127, 212, 245
—Travel to Pinecastle: number 1 airplane, 54, 77
XS-2, 19, 33, 50, 86, 114, 181, 232, 243
XS-3, 181
XS-4, 125, 281

Yeager, Capt. Charles E. "Chuck," 116, 235, 237–39, 241, 245, 246, 248–50, 252–62, 264–80, 282, 285, 290
Yeager, Glennis, 274, 275, 279
Youngblood, Donald, 124